Missouri Breeding Bird Atlas

1986–1992

Missouri Breeding Bird Atlas

1986–1992

Brad Jacobs
James D. Wilson

Natural History Series, No. 6
Missouri Department of Conservation, Jefferson City

Project Coordination Assistance by Sharon Hoerner
Illustrations by George Miksch Sutton and authors
Cover Art by David Plank

Editing by Margaret Castrey
Layout and design by Bruce Helm

ISBN 1-887247-13-0

Missouri Department of Conservation
P.O. Box 180
Jefferson City, MO 65102-0180

To all the birders who made this book possible

Contents

Illustrations

Tables

Preface

In 1986, the Missouri Department of Conservation commenced an ambitious statewide survey to determine the distribution of Missouri's breeding bird species—the Missouri Breeding Bird Atlas Project. The department's ornithologist, James D. Wilson, initiated the project and about 50 dedicated birders participated in the first year's effort. As word spread among the birding community, more people volunteered to be Atlasers, enough to survey 1207 breeding bird atlas blocks.

During the second year, the Department of Conservation hired Brad Jacobs, a 1986 block surveyor, and Jim and Brad worked together on the many tasks associated with data management, volunteer workshops, special survey weekends, newsletters and block surveys. The Atlas Project took much of the authors' time, including seven years of field work and four years of data analysis and writing. Sharon Hoerner skillfully handled volunteer communications and coordination, and the day-to-day tasks that kept this project on target.

This report would not have been possible without the help of more than 400 dedicated volunteer birders who contributed more than 21,000 hours to advance our understanding of Missouri birds. Their efforts have given us a benchmark set of data from which to compare future bird studies and to monitor changes in bird distribution. Other bird species information that was collected during the Atlas Project, including relative abundance, breeding season phenology and brood parasitism heightened our understanding of Missouri's birds.

The Department of Conservation is pleased to make the Missouri Breeding Bird Atlas available for scientists, land managers, students, birders and interested citizens.

Richard H. Thom, Chief
Natural History Division
Missouri Department of Conservation
December 1996

Acknowledgements

The Missouri Breeding Bird Atlas Project was sponsored by the Missouri Department of Conservation (MDC). Without the support of the Audubon Society of Missouri, the 14 chapters of the National Audubon Society and their members, and many additional birders from within and outside of Missouri, the Missouri Breeding Bird Atlas Project would not have been possible. The individuals listed at the end of this section contributed an average of 49.3 hours of field work to complete surveys in 1,207 blocks. A special thank you goes to the regional coordinators for their effort in assigning blocks to volunteers in their regions. While we have made every attempt to include all participants, observers, recorders, drivers and beginning bird-watching enthusiasts who participated, we realize many others, unknown to us, participated in field surveys.

This project and publication would not have been possible without the support of several people in the Missouri Department of Conservation. Special thanks to three Natural History Division Chiefs whose administrations this project spanned: John Wylie, James H. Wilson and Richard H. Thom. Data entry and data management support was provided by Sharon Hoerner, Nancy Priddy and Shawna Taylor. Special recognition goes to Sharon Hoerner for her dedication and assistance with all aspects of volunteer communications, and construction of graphics and tables. Special thanks also to Debra Hardin for proofreading, editing and document management and to Paul McKenzie for his excellent advice on content. Clerical staff who offered continual support to the project included Rhonda Allen, Freida Fisher, Sandy Fischer, Kelly Gentges, Debra Hardin, Verita Hayden, Amy Linsenbardt, Diana Munsterman, Barbara Singleton, Jean Ruggiero and Debbi Welschmeyer.

We are obligated to the many Atlasers who contributed articles to the Atlas Project newsletter and to Martha Daniels for her help and expertise in newsletter layout, design and production. We thank the United States Fish and Wildlife Service, especially Bruce Peterjohn and Sam Droege, for their help with the Breeding Bird Survey data. Also, thanks to Steve Lewis, United States Fish and Wildlife Service Region 3 Non-game Bird Coordinator, for funding the Atlas Project in the early years.

Thanks to the United States Forest Service, Mark Twain National Forest, for funding graduate student and summer employee Debbie East to survey 11 blocks on the Mark Twain National Forest near Eminence. Thanks also to MDC temporary summer employees Jerry Sowers, Jack Hilsabeck, Daniel Hatch, Chris Wilson and Tim Barksdale who conducted surveys.

Computer programming and mapping support was supplied by Rick Wolters, Jeff Kunz, Kevin Borisenko, Mike Klein, Terry Showers, Kirk Keller and Todd Larivee of the Missouri Department of Conservation.

Grateful acknowledgment goes to members of the Conservation Department's Public Affairs Division including Division Chief Kathy Love, Publications Editor Supervisor Bernadette Dryden, Art Department Supervisor Dickson Stauffer and Phototypesetter Libby Block.

Of the 167 species illustrated in pen and ink, 120 drawings were created by the late George Miksch Sutton for his 1928 *An introduction to the birds of Pennsylvania*. Illustrations sketched by James D. Wilson are found on pages 32-38, 44-46, 50, 66, 78, 80, 92, 96, 100, 110, 118, 120, 128, 142, 156, 166, 168, 176, 180, 198, 200. Illustrations by Brad Jacobs are found on pages 94, 204, 242-244, 260, 272, 276, 280, 284, 290, 298, 302, 304, 308, 316, 332-336, 344, 346, 352, 354.

And finally, our thanks to David Plank of Salem for donating his beautiful Wood Thrush painting for use on the Atlas cover.

REGIONAL COORDINATORS

REGION 1 NORTHWEST
Mike McKenzie, 1986-1988
Ann Webb, 1989-1992

REGION 2 NORTH CENTRAL
Pete Goldman, 1986-1992

REGION 3 NORTHEAST
Art Suchland, 1986-1992

REGION 4-WEST CENTRAL
JoAnn Garrett, 1986-1992

REGION 5-CENTRAL
Tim Barksdale, 1986-1988
James D. Wilson, 1986-1992

REGION 6-EAST CENTRAL
Carmen Patterson, 1986-1992

REGION 7-SOUTHWEST
Mark Goodman, 1986-1988
David Blevins, 1989-1992

REGION 8-SOUTH CENTRAL
Brad Jacobs, 1986-1992

REGION 9 SOUTHEAST
William Reeves, 1986-1992

SURVEYORS

Karen Adams
Nelson G. Allen
Liz Anderson
Fred Anesi
Don Arney
Darla Banman
Timothy Barksdale
Felicia Bart
Paul Bauer
Brad Beard
Mark Belwood
Bob Bever
Jean Blackwood
Carol Boehringer
Kathy Brady
Lynne Breakstone
Robert Brundage
Donna Burris
Bonita Camp
Pat Cason
Marie Chouteau
Mary Conrad
Phil Covington
Barbara Crouser
Mary Daniel
Jerri Davis
John Degenhardt
John Doggett
Lorna Domke
Susan Dornfeld
Marilyn Drieling
Debbie East
Greg Eddy
Tony Elliot
Scott Ellis
Leonard Fair
Bob Farr
Mike Ferro
Dennis E. Figg
Robert G. Fisher
Shirley Flood
Martha Gaddy
Leo Galloway
Rich George
Tammy Gilmore
Cathy Glueck
Deborah Good
Mary Frances Goodloe
Nicholas Goodman

Tim Adkerson
Cornelius Alwood
Mitzi Anderson
Joanna Anesi
Hazel Ayers
Jamie Barger
Jeanne Barr
Craig Barwick
Oda Beall
Jeff Belshe
Jerome F. Besser
Danny Billings
Barbara Blevins
Richard Boehringer
Patrick Brady
Cindy Bridges
Chana Burney
Louise Bushnell
Ann Sue Campbell
Ruth Chenhall
Pierre Chouteau
Jackie Cook
Frances Cramer
Frankie Cuculich
Martha Daniels
Randall P. Davis
Steve Dilks
Mary Doggett
Rod Doolen
Bruce Dorries
Donald P. Duncan
David A. Easterla
Todd Eichholz
Kathy Elliott
Vern Elsberry
Otto F. Fajen
Hal Ferris
Floyd Ficken
Nancy Findley
Doris Fitchett
Terry Dean Fockler
Eleanor Gaines
JoAnn Garrett
Bill Gilges
Villa Ann Glenn
F. Darryl Goade
Bill Goodge
John Goodman
April Gordon

Ken Allen
Sybill Amelon
Richard Anderson
Valerie Arndt
George Banfield
Silvey Barker
Vernon Barr
Carla Bascom
Roger Beall
John Belshe
Emily Bever
Shelby Birch
David Blevins
Dennis S. Bozzay
Tim Bray
Tom B. Brown
Mike Burney
Paul Calvert
Margaret Cason
Linda J. Childers
Charles Clodfelter
Susan Cook
Helen Creal
Brian Culpepper
Charlie Davidson
Samuel Michael Davis
Jerry Dobbs
Ida Domazlicky
Wanda Doolen
Anne Downing
Annette Dupree
Jody Eberle
Jo Ann Eldridge
Carol Ellis
Billie Fair
Larry Farley
Merrlyn Ferro
Virginia Ficken
Chris Fisher
Jane Fitzgerald
Bob Gaddy
Ron Gaines
Bob Gentle
Len Gilmore
Virginia Glover
Peter Goldman
Eleanor Goodge
Mark Goodman
Troy Gordon

Cheryl Gowan
Jean Graebner
Josephine Green
Betty Gundersen
Violet Hallet
Susan Haney
Azan Hayes, J.P.
Susan Hazelwood
Bjorn Held
Lawrence Herbert
Jack Hilsabeck
Peter Hoell
David Holman
Greg Hoss
Todd Houf
Dawn Huckins
Hal Huff
Kenneth R. Jackson
Brad Jacobs
Joan Jefferson
Charles A. Johnson
Barry Jones
Mrs. Howard Jones
William N. Kelley
Mary Kenner
Karl Kleen
David D. Kneir
Jerry Koch
Renee Krummrich
Eugenia Larson
Jane Leo
John Loomis
Ernestine Magner
Christine Mansfield
Rebecca Matthews
J. Knox McCrory
Mark McKenzie
Virginia McKenzie
Joseph McMahan
Marvin McNeely
Tim Menard
Mike Meyer
Nancy Miller
David Moore
Herbert Morris
Michael A. Muller
Tom Nagel
Jan Neale
Tom Nichols
Ken Ornburn
Max Parker

Howard Gowan
Ken Grannemann
Melody Green
Jenny Gunn
Rodney Hallgren
Mary L. Hartnett
Jim Hazelman
Jim Helbig
Mike Held
John Hess
Steve Hilty
Sharon Hoerner
Frank Holmes
Garry Houf
Charles R. Howard
Steve Hudlemeyer
Blanche Hutchison
Lyle Jackson
Nathan Jacobs
Vada Jenkins
Evelyn Johnson
David Jones
Jane Kasten
Larry Kennard
Colleen E. Kimmel
Laurie Kleen
Brian Knowles
Randy Korotev
Michael Laird
Lloyd Lashley
Robert D. Lewis
Cathy Lower
Marshall Magner
Neal Mansfield
Bob McCann
Maureen McHale
Mike McKenzie
Brad McKinney
Miriam McMurray
Terry McNeely
Anne Meyer
Berdy Miller
Terry Miller
Lloyd D. Moore
Abe Moshkovski
Esther Myers
Joseph C. Neal
Gene Neff
Bonnie Noble
Babs Padelford
Tom Parmeter

Jim Grace
Harrison Green
Greg Gremaud
Dorothy Hagewood
Burton Handy
Daniel Hatch
Don Hazelwood
Belinka Held
Bob Hely
Emily Hickey
Patrick Hnilicka
Kevin Hogan
Sylvia Hosler
Larry J. Houf
Martha Howard
Stanton Hudson
James P. Jackson
Marcella Jackson
Juanita James
Harold F. John
Marie Johnson
Mr. Howard Jones
Elvan Keightley
Leona Kennard
Pat Klaas
Valerie Kling
Donna Koch
Philip Krummrich
Susie Lansdown
Melda Lashley
Vivian Liddell
Brenda Madison
Patrick Mahnkey
David D. Martin
Loretta McClure
Mick McHugh
Paul McKenzie
Connie McKinney
Rob McMurray
David Mead
Arlon Meyer
Charles Miller
Jeannie Moe
Barbara Moran
Jan Moshkovski
Richard Myers
Charles Neale
Cathern Nelson
Charles Noble
Helen Parker
Carmen Patterson

Sebastian Patti
Eileen K. Powell
Deb Priest
Margaret Ptacek
William Reeves
Sandra Reiter
Jean Retzinger
Lester Richesin
Merle Rogers
Simon Rositsky
Mike Rues
Catherine I. Sandell
Nancy Schanda
Bruce Schuette
Napier Shelton
Amy Simmerman
Mary Jane Smith
Estelle Snow
Albert L. Solomon
Paul Spence
Catherine Stickann
Arthur Suchland
Dan Swofford
Sheryl Tatom
Norman Thompson
Dorothy Thurman
Jack Toll
Connie Tyndall
Claire van der Linde
Jude Vickery
Jewel Wagner
Jane Walker
Paul Watson
John Webber
Carmela Werner
Phil Weston
Dennis Wheeler
Terry Wilcox
Paul Williams
Sally Williamson
Chris Wilson
George Wisdom
John Witherspoon
Kent Woodruff
Agnes Wylie
Fred Young

Roger Phillips
Louis Prawitz
Sandy Priman
Lyle Pursell
Carol F. Reigle
Mary Rellergert-Taylor
Lynda L. Richards
Mark Robbins
Robert Rogers
Cynthia Roth
John Rushin
Tim Schallberg
Judy Schmidt
Jack Scrivner
Allison Shock
Diana Simon
Sarah Smith
Geneice Snow
Marguerite J. Solomon
Millie Stephens
Richard Stiehl
Eugene Sumter
Cherry Taber
Irene Taylor
Linda D. Thomsen
Russell Titus
Marty Toll
Blaine Ulmer
Mia Van Horn
Jim Voltz
Mary Ann Waisanen
Melvin Walls
Marilyn Wayman
Susan Wedenoja
Judy Wertz
Ann Wethington
Mary Wiese
Dayle Wilcoxen
Beth Williamson
Tim Williamson
Christopher J. Wilson
Naomi Wisdom
David Witten
Robert Wykes
John Wylie
Michael F. Zeloski

Galen Pittman
Dave Priest
Jim Ptacek
Ron Redden
Norm Reigle
Rochelle B. Renken
Fran Richesin
Mark A. Rogers
Dave Rogles
Cal Royall
Skip Russell
Hammons Schanda
Dorothy Schoech
Mrs. Jo Seaman
Steve Shupe
Carolyn Smith
Phoebe Snetsinger
John N. Soderberg
Jerry Sowers
Mariel Stephenson
Cack Strickler
Mike Sweet
John Tate
Richard H. Thom
Carolyn Thurman
Bob Todd
Harry Trickey
Eliza Valk
Sarah Vasse
Peggy Voltz
Ann Wakeman
Harold R. Ward
Anne Webb
Harriett Weger
Thomas A. Westhoff
Bob Wethington
Michelle Wiggins
Ann Wilder
Randy Williamson
Barbara Wilson
James D. Wilson
David Wissehr
Tracy Wohl
Rosalyn Wykes
William H. Yoder
Bob Ziehmer

Introduction

The Missouri Breeding Bird Atlas Project, conducted from 1986 through 1992, sought to document the status and distribution of the bird species that breed in Missouri. The primary goal was to develop a distributional map for each species that depicts as accurately as possible its true breeding range in the state. The resultant information was intended to: 1) provide baseline data against which future changes in the status and distribution of Missouri's breeding birds could be measured, 2) determine the location of rare species, 3) identify significant habitats and 4) develop a factual database to assist environmental planners in making wise decisions about resource use in Missouri. During the process of collecting the distributional and status information, data were also obtained on species' abundance, breeding phenology and Brown-headed Cowbird brood parasitism.

The Atlas Project relied on the cooperation and participation of 438 field surveyors (primarily volunteers), who recorded 21,577.7 hours collecting data during the seven-year period. Additional benefits of the project were the heightened awareness of Missouri's summer bird life and the cooperative spirit that developed among birders and scientists statewide. This project highlighted the value of volunteer-based bird population monitoring.

Methodology

Sampling Procedure

Data collection was conducted using a sampling process, established by the North American Ornithological Atlas Committee, wherein one survey area was designated within each of the 1,210 United States Geological Survey 7.5-minute quadrangles that encompass Missouri. Survey areas, termed blocks, were established by dividing each 7.5-minute quadrangle into sixths and then randomly selecting one of these sixths (fig. 1). Each block contains approximately 25 square kilometers. The random selection process ensured no habitat bias. Blocks bisected by the state's border were surveyed only if more than 50 percent of the block fell within Missouri. Of the 1,210 established blocks, 1,207 were actually surveyed. The 7.5-minute quadrangles and blocks are listed in Appendix A.

Block Assignments

Atlas surveyors, termed Atlasers, typically received their block assignments by contacting the Missouri Department of Conservation and designating a region that was convenient or that they believed had interesting bird potential. They were provided with a packet containing field cards (fig. 2a, 2b) that listed the assigned block's name and number, a topographic map and a county map detailing the block, a Missouri Breeding Bird Atlas Project Handbook and calling cards and a vehicle sign to inform landowners and passersby about the project. While most block assignments were given before the breeding season each year, some assignments were made as late as midsummer with the understanding that the survey of the block could continue in the following year. Blocks usually were assigned to one surveyor at a time, with reassignments made when an Atlaser was unable to complete a block.

Breeding Criteria and Codes

The North American Ornithological Atlas Committee's Standardized Breeding Code Criteria were used to classify species (fig. 3). Codes were arranged on the field card from less to greater certainty of breeding within three main categories: possible, probable and confirmed breeding. Possible and probable categories required observations that occurred within "safe dates," a period when migrants of that species are expected to be absent from Missouri. Within safe dates, occurrence in suitable habitat alone provided elementary evidence that the species was potentially breeding in the block. Safe dates established for each species were listed following the species' name on the field card. Confirmed observations were recorded on the field card regardless of safe dates.

Based on breeding evidence observed in the field, volunteers inserted the appropriate code in the possible, probable or confirmed column on a field card. The date associated with the highest level of breeding evidence was entered in

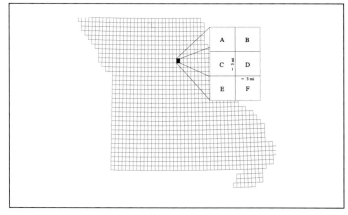

*Figure 1. **U. S. Geological Survey 7.5 Minute Quadrangles with Expanded Survey Blocks A–F***

ADDITIONAL SPECIES

*Species	Date of Highest Evidence	PO	PR	CO

'Verification Required on all Additional Species

NOTES

Field Card
Missouri
Breeding Bird
Atlas

REGION	QUADRANGLE	
	Quad Name	Quad No. & Block
Latitude and Longitude		Year
Observer's Name:		
Address:		
Phone:		

BLOCK VISITS

Date	Location	Start	End	Hours

Total Species _____

Total Hours _____

VERIFICATION

For all species asterisked on the Field Card additional details are required.
Special Verification Forms are available or the following items can simply be
listed on a sheet of paper and accompany the Field Card when it is returned.

1. Species.
2. Describe plummage, song, size, shape, behavior, etc., especially as it
   would eliminate similar species.
3. Date and time of sighting.
4. Habitat (general and specific).
5. Breeding behavior (details which justified code).
6. Distance and description of optical aids.
7. Length of observation, lighting and weather conditions.
8. Previous experience with species.
9. Books, records and advice consulted.
10. Your name, address, phone and Quad Name and Number.

If your sighting indicates breeding of a species not listed on the Field
Card (in "Additional Species" above), try to support it with a photograph,
sketch or names of other observers. Send these along when returning the
Field Card. Call or write your Regional Coordinator or State Coordinator
immediately.

Return card by September 15 to:
James D. Wilson, Ornithologist
Missouri Department of Conservation
P.O. Box 180
Jefferson City, MO 65102

314/751-4115   10/86

Breeding Criteria & Codes

PO   Possible
  O   Species *observed* or heard, but not in breeding habitat.
  X   Species heard or seen in breeding habitat

PR   Probable
  S   Seven *singing* males detected in one visit.
  P   *Pair* seen (male & female together—not a flock)
  T   Bird holding *territory* (exclude colonial nesters)
  C   *Courtship* or *copulation*
  N   Visiting probable *nest* site
  A   *Agitated* behavior or anxiety call (not intentionally provoked)
  B   Nest *building* by wrens or woodpeckers
(All of the above codes must be used within a species'
safe dates and all but "O" must be within preferred
breeding habitat—see TABLE 1).

CO   Confirmed
  NB   Nest building (by all except wrens and woodpeckers.)
  PE   *Physiological evidence* of breeding based on bird in the hand.
  DD   *Distraction display* or injury feigning.
  UN   *Used nest* or eggshells found.
  FL   Recently *fledged* young, include precocial young out of nest.
  ON   Adults entering or leaving nest site in circumstances indicating nesting.
  FS   Adult carrying *fecal sac*.
  FY   Adult seen carrying *food* for *young.*
  NE   *Nest* with *egg*(s).
  NY   Nest with *young* seen or heard.

Higher Evidence

Figure 2a. **Atlas Project Field Data Card**

the column following the code. Survey dates, times of day
and number of hours spent during each visit were recorded.
Comments on Brown-headed Cowbird brood parasitism or
other observations were placed in the notes section.

Species recorded only by the "O" code under possible
breeding were not included on the Atlas Project distribution
maps. Because these species were not in breeding habitat, it
was considered unlikely they were breeding in that block.

Asterisked Species

Unusual or difficult-to-distinguish species were asterisked on the field card, indicating that a completed verification form was required to document the sighting. Species
recorded that were not printed on the field card also
required a completed verification form.

Coverage Goals

The Missouri Breeding Bird Atlas Project sought to
record evidence of breeding for the majority of species present within each block. A minimum standard of coverage
was necessary in every block in order for Atlas Project maps
to show meaningful patterns of distribution. Atlasers were
instructed to search their block's area and habitat thoroughly
enough to be reasonably certain that they had recorded most
of the species expected to be nesting. Field tests and other
atlas projects had indicated that 15 hours of field work was
the optimal amount of time needed to record 75 percent of
the species present in the block. Depending on habitat diversity, this level of coverage would typically record 50–70
species for the block. Atlasers sought to place 25 percent of
the species in the possible breeding category, 50 percent in
the probable category and at least 25 percent in the confirmed category. Once these approximate levels were
reached, volunteers were encouraged to start a new block in
the interest of spreading coverage evenly among the blocks.

Species	Safe Dates	Highest Evidence	PO	PR	CO
Great Blue Heron (194)	5/1-7/1				
Great Egret (196)	5/20-7/1				
Green-backed Heron (201)	5/15-7/15				
*Black-cr. Night-Heron (202)	5/10-7/1				
*Yellow-cr. Night-Heron (203)	5/10-7/1				
Canada Goose (172)	5/1-8/3				
Wood Duck (144)	5/1-8/15				
Mallard (132)	5/1-8/30				
Turkey Vulture (325)	5/1-8/1				
*Northern Harrier (331)	5/15-7/30				
*Sharp-shinned Hawk (332)	6/1-8/15				
*Cooper's Hawk (333)	5/20-8/15				
*Red-shouldered Hawk (339)	5/1-8/30				
*Broad-winged Hawk (343)	5/15-8/15				
Red-tailed Hawk (337)	5/1-8/30				
American Kestrel (360)	4/30-7/31				
Ring-necked Pheasant (309)	4/15-9/30				
*Ruffed Grouse (300)	4/1-7/15				
*Grtr. Prairie-Chicken (305)	3/30-8/30				
Wild Turkey (310)	4/30-9/30				
Northern Bobwhite (289)	4/30-8/30				
*King Rail (208)	5/30-7/31				
*Common Moorhen (219)	5/30-8/31				
*American Coot (221)	6/10-8/31				
Killdeer (273)	4/20-7/5				
Spotted Sandpiper (263)	6/5-6/25				
Upland Sandpiper (261)	5/20-6/25				
American Woodcock (228)	4/15-9/20				
Rock Dove (313)	all year				
Mourning Dove (316)	5/1-7/20				
*Black-billed Cuckoo (388)	6/15-7/20				
*Yellow-billed Cuckoo (387)	6/15-7/31				
Eastern Screech-Owl (373)	4/30-8/15				
Great Horned Owl (375)	3/15-8/31				
Barred Owl (368)	3/15-8/31				
Common Nighthawk (420)	6/5-7/15				
Chuck-will's-widow (416)	5/25-8/10				
Whip-poor-will (417)	5/25-8/10				
Chimney Swift (423)	5/10-8/15				
Ruby-thr. Hummingbird (428)	5/25-7/31				
Belted Kingfisher (390)	5/10-7/20				
Red-headed Woodpecker (406)	5/25-8/20				
Red-bellied Woodpecker (409)	3/31-8/31				
Downy Woodpecker (394)	3/15-8/31				
Hairy Woodpecker (393)	4/1-8/31				
Northern Flicker (412)	5/10-8/25				

Species	Safe Dates	Highest Evidence	PO	PR	CO
Pileated Woodpecker (405)	4/1-8/31				
Eastern Wood-Pewee (461)	5/25-8/1				
*Acadian Flycatcher (465)	5/25-8/5				
*Willow Flycatcher (466)	6/10-7/25				
Eastern Phoebe (456)	5/15-8/31				
Grt. Crested Flycatcher (452)	5/25-8/1				
*Western Kingbird (447)	5/20-7/25				
Eastern Kingbird (444)	5/20-7/25				
Sc.-tailed Flycatcher (443)	5/1-6/20				
Horned Lark (474)	4/10-9/1				
Purple Martin (611)	5/25-6/25				
Tree Swallow (614)	5/25-6/25				
No. Rough-winged Sw. (617)	5/25-6/25				
Bank Swallow (616)	5/25-6/25				
Cliff Swallow (612)	6/5-6/21				
Barn Swallow (613)	5/25-7/25				
Blue Jay (477)	5/1-8/31				
American Crow (488)	5/1-8/31				
*Black-capped Chickadee (735)	5/1-9/20				
*Carolina Chickadee (736)	4/15-8/31				
Tufted Titmouse (731)	4/15-8/31				
Wh.-breasted Nuthatch (727)	5/1-8/15				
Carolina Wren (718)	4/1-9/30				
Bewick's Wren (719)	5/10-8/31				
House Wren (721)	5/15-8/15				
*Sedge Wren (724)	6/10-9/10				
*Marsh Wren (725)	5/25-8/25				
Blue-gray Gnatcatcher (751)	5/15-8/31				
Eastern Bluebird (766)	5/15-8/31				
Wood Thrush (755)	5/30-8/20				
American Robin (761)	5/1-8/31				
Gray Catbird (704)	5/20-8/30				
Northern Mockingbird (703)	4/15-9/1				
Brown Thrasher (705)	5/1-7/31				
Cedar Waxwing (619)	6/15-7/31				
Loggerhead Shrike (622)	4/20-7/20				
European Starling (493)	4/10-9/5				
White-eyed Vireo (631)	5/20-8/15				
*Bell's Vireo (633)	5/25-8/15				
Yellow-throated Vireo (628)	5/25-8/15				
Warbling Vireo (627)	6/1-8/10				
Red-eyed Vireo (624)	6/1-7/31				
Blue-winged Warbler (641)	5/20-7/20				
Northern Parula (648)	5/15-8/15				
Yellow Warbler (652)	6/1-7/10				
Yellow-thr. Warbler (663)	5/1-7/15				

Species	Safe Dates	Highest Evidence	PO	PR	CO
*Pine Warbler (671)	4/20-8/15				
Prairie Warbler (673)	5/25-7/20				
Cerulean Warbler (658)	5/25-8/15				
Black-and-white Warb. (636)	5/20-7/31				
American Redstart (687)	6/10-7/20				
Prothonotary Warbler (637)	5/20-7/20				
Worm-eating Warbler (639)	5/20-7/20				
Ovenbird (674)	5/25-8/5				
Louisiana Waterthrush (676)	5/15-7/10				
Kentucky Warbler (677)	5/25-7/15				
Common Yellowthroat (681)	5/20-8/10				
*Hooded Warbler (684)	5/25-7/25				
Yellow-breasted Chat (683)	5/25-8/5				
Summer Tanager (610)	6/1-8/10				
Scarlet Tanager (608)	6/1-8/10				
Northern Cardinal (593)	3/15-9/15				
Rose-breasted Grosbeak (595)	5/25-8/10				
Blue Grosbeak (597)	5/25-8/10				
Indigo Bunting (598)	5/25-8/10				
Dickcissel (604)	5/25-8/15				
Rufous-sided Towhee (587)	5/20-8/31				
*Bachman's Sparrow (575)	6/1-8/15				
Chipping Sparrow (560)	5/10-8/15				
Field Sparrow (563)	5/1-8/31				
Lark Sparrow (552)	6/1-7/31				
Savannah Sparrow (542)	6/1-8/31				
Grasshopper Sparrow (546)	5/15-8/31				
*Henslow's Sparrow (547)	5/15-8/31				
Song Sparrow (581)	5/15-9/10				
Bobolink (494)	6/1-7/20				
Red-winged Blackbird (498)	5/1-8/1				
*Eastern Meadowlark (501)	5/1-9/10				
*Western Meadowlark (502)	5/1-9/10				
*Great-tailed Grackle (512)	5/10-8/1				
Common Grackle (511)	4/15-7/10				
(A)Brown-headed Cowbird (495)	4/20-7/10				
Orchard Oriole (506)	6/1-7/5				
Northern Oriole (507)	6/1-7/25				
*House Finch (519)	5/15-8/31				
American Goldfinch (529)	6/1-9/1				
House Sparrow (688)	2/1-9/30				

*Verification Form Required
(A) If known, list host in notes.

Total ___ ___ ___

Figure 2b. **Atlas Project Field Data Card**

The Missouri Breeding Bird Atlas Handbook instructed Atlasers to visit their assigned blocks in various months because bird species vary seasonally in ease of detection. Atlasers were encouraged to use June as the primary month for building a species list, and July and August as optimal months to record birds in the probable and confirmed categories. Figure 4 depicts the frequency of various breeding codes from April through August. The handbook recommended early morning and evening as the most productive periods but also asked Atlasers to visit their blocks at night to record nocturnal species. Hours spent per block are compiled in Appendix A and graphically illustrated in figure 5.

Block-Busting

Within the first two years of the Atlas Project, it became apparent that blocks in the vicinity of Kansas City, St. Joseph and St. Louis were being assigned more rapidly and receiving more complete coverage than blocks elsewhere in the state. To promote coverage in under-surveyed regions, special Atlasing events, termed "Block-Busting" weekends, were held in 1990, 1991 and 1992 during the height of the breeding season. Atlasers were invited to work together in these regions to accomplish at least cursory coverage of these blocks. Blocks covered by Block-Busting are noted in Appendix A.

Abundance Surveys

Species relative abundance maps were generated using data obtained during the years of the Atlas Project from 37 Breeding Bird Survey (BBS) routes established throughout Missouri. The BBS was organized by the United States Fish and Wildlife Service and has been conducted annually in Missouri since 1967. These surveys are conducted by road once each June. Each route consists of 50, three-minute observation stops, one-half mile apart.

Atlas Codes and Criteria for Breeding Evidence

POSSIBLE CODES | BREEDING EVIDENCE

POSSIBLE CODES   <u>BREEDING EVIDENCE</u>

0   Species *observed* in block, but not in breeding habitat. (These are not mapped. Please see page 2 under Breeding Criteria and Codes.)

X   Species heard or seen in breeding habitat within safe dates.

PROBABLE CODES   <u>BREEDING EVIDENCE</u>

S   Seven *singing* males detected in one visit.

P   *Pair* observed in suitable breeding habitat within safe dates.

T   *Territorial* behavior or singing male present at same location, on at least two different days, a week or more apart. Territoriality can be presumed from defensive encounters between individuals of the same species, or by observing a male singing from a variety of perches within a small area.

C   *Courtship or copulation* observed. This includes displays, courtship feeding and birds mating.

N   Visiting probable *nest site*. Primarily applies to hole-nesters. This code applies when a bird is observed visiting the site repeatedly, but no further evidence is seen.

A   *Agitated* behavior or anxiety calls from adult. Parent birds respond to threats with distress calls or by attacking intruders.

B   Nest *building* by wrens or excavation by woodpeckers. Both groups build dummy or roosting nests.

CONFIRMED CODES   <u>BREEDING EVIDENCE</u>

NB   *Nest building* (except wrens and woodpeckers) or adult carrying nesting material.

PE   *Physiological evidence* of breeding based on bird in the hand.

DD   *Distraction display*; including injury-feigning.

UN   *Used nest* found.

FL   Recently *fledged* young or downy young. This includes dependent young only. Young cowbirds begging for food confirm both the cowbird and the host species.

FS   Adult bird seen carrying *fecal sac*.

FY   Adult carrying *food for young*.

ON   *Occupied nest* presumed by activity of parents: entering nest holes and staying, parents exchanging incubation responsibility, etc.

NE   *Nest* with *eggs* or eggshells on ground. Cowbird eggs in nests confirm both the cowbird and the host species.

NY   *Nest* with *young* seen or heard. A cowbird chick in a nest confirms both the cowbird and the host species.

Figure 3. ***Atlas Codes and Criteria for Breeding Evidence***

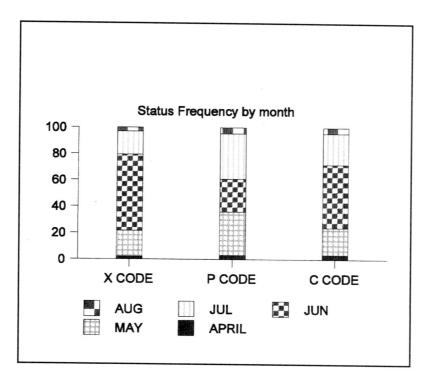

Figure 4. **Frequency of Breeding Codes by Month**

During stops, surveyors identify and record every individual bird detected. Survey of a BBS route begins one-half hour before sunrise and is completed in about four hours. Volunteer BBS surveyors included many Breeding Bird Atlas Project volunteers.

To obtain additional relative abundance data and provide a more thorough distribution of data collection locations, 60 Miniroute surveys were established specifically for the Atlas Project. Methodology for Miniroute surveys was similar to the BBS except that they 1) were run twice during the breeding season, 2) included only data collected after sunrise and 3) consisted of 15 rather than 50 observation stops. Miniroutes were necessarily shorter because they fit entirely within a Breeding Bird Atlas block. This allowed sightings along the route to be recorded on the Atlas field card. However, because Miniroutes were run twice, a total of 30, three-minute observations provided data for each Miniroute. The two runs were about a week apart, with stops sampled in reverse on the second run. The one-week interval enabled surveyors to consider whether individuals sighted on the second run could be coded as "territorial" on the Atlas field card.

Miniroute and BBS data from 1986–92 were converted to birds per stop, then combined and multiplied by 100. This created the unit measure used for the relative abundance maps included in the species accounts. Where inadequate abundance data were obtained, the abundance map has been omitted. Both BBS and Miniroutes are identified in Appendix B.

Data Processing and Verification

Atlasers were asked to submit field cards to state coordinators at the completion of the field season each year, regardless of whether the survey of the block had been completed. The majority of field cards were submitted as requested by September 15, the date by which the majority of breeding activity was concluded. State Coordinators and staff checked cards to verify all necessary information was included, reviewed data and occasionally contacted Atlasers for clarification.

At the start of the Atlas Project, eight data fields were digitized: topographic quadrangle name, assigned survey block number, bird species common name, bird species' AOU (American Ornithologists' Union) number, breeding code observed, date of observation, latitude and longitude. By the end of the project, several additional databases had been created to track volunteers, hours of effort, Brown-headed Cowbird host species and abundance information.

Over the years of the Atlas Project, various computers were employed, including IBM PCs, XTs and Pentium 90s using Microsoft DOS and IBM OS2 platforms. Dbase IIIC+, Word Perfect and ArcView software were used to store, analyze and compile the data. Additional software including Adobe Illustrator, ArcPlot, and Quark XPress were involved in publication of this report.

The codes and dates recorded for each species, together with the latitude/longitude coordinates for that block, were entered into a computer. Accuracy was checked by

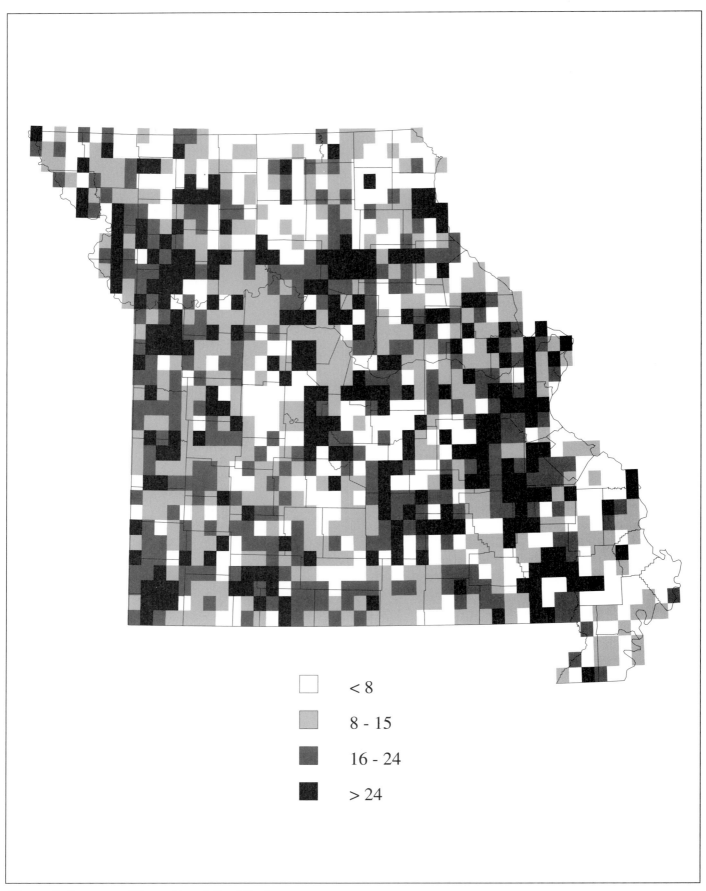

Figure 5. **Surveyor Hours Per Atlas Block**

Legend:
- < 8
- 8 - 15
- 16 - 24
- > 24

comparing print-outs with field cards. A determination was made as to whether survey of the block was complete. If additional coverage was desired, the Atlaser was sent a new field card for the following year together with print-outs of data recorded to date, suggestions of habitats to cover, expected species and other recommendations to improve coverage. Occasionally, another Atlaser was assigned to finish an incomplete block.

Limitations and Biases

The Missouri Breeding Bird Atlas Project was designed to sample one-sixth of the state's area via randomly selected survey blocks, each representing one-sixth of a 7.5-minute quadrangle. Because of this sampling procedure, and because the survey of a block was considered complete when approximately 75 percent of the block's species had been recorded, small, patchily-distributed habitats and rare, local breeding populations were likely missed entirely or under-surveyed. Rare, colonial species were especially misrepresented by the results depending on whether their few sites fell within blocks and were found.

Even within the surveyed blocks, results almost certainly do not include every species that was actually present during the Atlas Project. Although the project was conducted over a seven-year period, individual blocks were typically surveyed during only one or two years, and species that only intermittently inhabited these sites were often missed. Species requiring special efforts to locate, particularly nocturnal birds or those occupying difficult-to-survey habitats, were also under-represented. As a result, the Atlas Project counted only a small fraction of the total populations within such groups as bitterns, rails, accipiters and owls.

Blocks surveyed during a single intensive effort, such as during Block-Busting weekends, were subject to the greatest amount of bias. Because Block-Busting occurred at the average height of the avian breeding season, those species that may have been more easily detected earlier or later were likely under-represented in the results. Additionally, Block-Busters spent an average of 9.1 hours per block compared to 17.9 hours for a normally-assigned block. Block-Busting and other cursory surveys were used primarily in northeastern and southeastern Missouri and in the Ozarks. Statewide differences in hours spent per block are shown in figure 5.

There were also vast differences among Atlasers in effort and identification skills. This appears to have occasionally biased distributional results when an Atlaser was assigned several neighboring blocks. A bias towards lower species' counts near state borders resulted from the decision to include data from fractions of blocks bisected by the state line.

Relative abundance data generated from BBS and Miniroute surveys were most significant for more populous species. The true abundance of rarer species, or of those that are clumped in distribution, was poorly evaluated by the surveys. For this reason, abundance maps were not included in species accounts for rarer, unevenly distributed species. Even where abundance maps and narratives on abundance were included, the results may have been biased by the number of survey routes in each natural division. The Big Rivers natural Division contained only one route and so was subject to the greatest sampling error. Therefore, caution should be used when comparing species abundance among the natural divisions.

Results and Discussion

Of Missouri's 1,210 blocks, a total of 1,207 were surveyed during the Atlas Project. Atlasers documented 71,969 records of which 34.2 percent were possible, 32.8

Table 1. **Species Counts by Natural Division**

Natural Division	# of Blocks	# of Species	Average # of Species/Block	High Count	Low Count
Glaciated Plains	397	146	57.2	94	17
Big Rivers	59	139	57.4	86	40
Ozark Border	153	136	62.7	85	26
Osage Plains	96	136	63.5	86	39
Ozark	438	148	62.0	96	21
Mississippi Lowlands	64	131	44.8	80	21
Statewide Totals	**1207**	**167**	**59.6**	**96**	**17**

Table 2. *Species Recorded and Not Confirmed, but Known to Breed in Missouri*

Recorded as possible breeders	Recorded as probable breeders
Double-crested Cormorant	Little Blue Heron
Snowy Egret	Northern Shoveler
Cattle Egret	Swainson's Hawk
Black Vulture	Least Tern
Gray Partridge	Greater Roadrunner
Sora	Fish Crow
Common Moorhen	Chestnut-sided Warbler
	Hooded Warbler

percent were probable and 33.0 percent were confirmed breeders.

In 71.8 percent of the blocks, 50–70 species were identified. In only four blocks were 90 or more species recorded. The statewide average number of species per block was 59.6, with a high of 96 species and a low of 17 (unfinished block). The highest average number of species per block was 63.5 in the Osage Plains and 62.7 in the Ozark Border natural divisions. The Mississippi Lowlands had the lowest average at 44.8 species per block. See table 1 for species counts by natural division. Figure 6 shows the number of species per block statewide.

A total of 181 species of birds were reported during the Atlas Project. Fourteen of these were neither expected to breed in the state nor confirmed to breed by the Atlas Project (see Appendix D). Of the remaining 167 species,

all expected to breed in Missouri, seven were recorded as possible breeders, eight as probable breeders and 149 confirmed as breeders. Table 2 lists species not confirmed to breed by the Atlas Project, but which are known from other sources to have bred in Missouri (Robbins and Easterla 1992).

Only the Black-necked Stilt was for the first time confirmed to breed in Missouri as a result of the Atlas Project. In addition, as a result of a restoration effort, Peregrine Falcons bred in the state during the later Atlas Project years. Although they were not located within Atlas Blocks, their species account has been included. Two records of breeding Gadwall and one record of breeding Northern Pintail occurred outside Atlas Blocks in 1990 during the Atlas Project (Robbins and Easterla 1992).

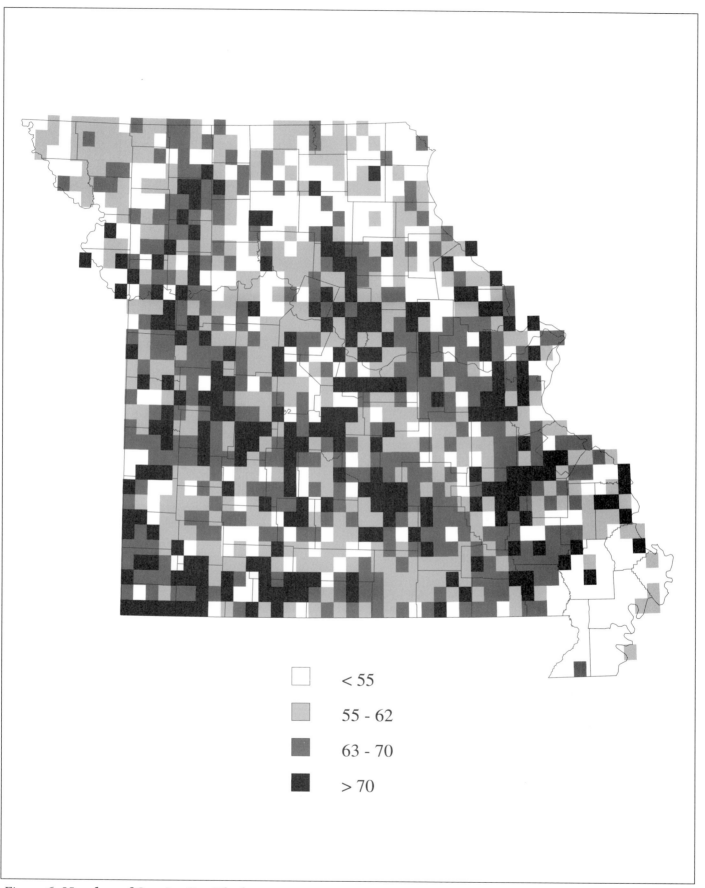

Figure 6. **Number of Species Per Block**

☐	< 55
▨	55 - 62
▨	63 - 70
■	> 70

The Natural Divisions of Missouri

The Natural Divisions of Missouri (fig. 7) divide the state into six major regions about which generalizations regarding bird distribution and relative abundance can be made: Glaciated Plains, Ozark Border, Ozark, Osage Plains, Big Rivers and Mississippi Lowlands (Thom and Wilson 1980). Geologic history, soils, bedrock geology, topography, plant and animal distribution, pre-settlement vegetation and other natural factors are all used to define division boundaries. Each division represents part of a larger physiographic region which extends beyond Missouri's borders. Because each division contains a variety of landscapes and natural communities which add considerable diversity, each is divided into natural sections.

Glaciated Plains Natural Division

Roughly encompassing the northern one-third of the state, the Glaciated Plains Natural Division is characterized by soils and topography that resulted from the influence of Pleistocene glaciation about 400,000 years ago. Soils are formed from loess and glacial till or from alluvium. The topography is younger than that of the unglaciated portions of the state, although much of this division is moderately dissected. Upland and bottomland deciduous forest and prairie were the main pre-settlement vegetation, with prairie comprising about 45 percent of this division.

The Western and Central natural sections of this division are characterized by loess-dominated topography and soils. The Western Natural Section is more highly dissected and has the driest climate in the state. The Central Natural Section contains soils derived from glacial till and is more moderately dissected. Today, agriculture dominates the landscape with large fields of corn, soybeans and pasture. Woodlands suitable for forest interior breeding birds are rare except at a few of the region's state parks and conservation areas.

The Eastern Natural Section includes a region of flat, claypan soils typified by the Audrain plains, where corn and soybean production predominates today. This natural section also includes rugged river breaks such as those along Salt River. Mark Twain Lake, an impoundment of the Salt River, is a major feature of this section. Farther to the east, the Lincoln Hills Natural Section extends along the Mississippi River in Lincoln, Pike, Ralls and Marion counties. The Lincoln Hills, which may have partially escaped glaciation, are distinguished from the rest of the division by bedrock geology, caves, and steep, heavily-forested areas.

Ozark Border Natural Division

The Ozark Border Natural Division comprises about 13 percent of the state. It extends along both sides of the lower Missouri River and the lower Mississippi River to the Mississippi Lowlands Natural Division. It includes rugged river hills with deep, relatively productive soils and a few isolated rolling plains. Although upland deciduous forest was the main pre-settlement vegetation, glade, marsh, prairie and bottomland forest communities were also present. Prairie accounted for less than 10 percent of the pre-settlement vegetation. Many of the soils are derived from loess and are relatively productive. Today, rolling pastures, shrubby, fragmented forests and forested bottomlands characterize this division.

Ozark Natural Division

The Ozark Natural Division is a large, unglaciated region of greater relief and elevation than the surrounding areas. This division comprises almost 40 percent of the state. It is characterized by thin, often stony, residual soils. Topography is often very steep. Caves, springs, bluffs, and high-gradient, clear-flowing streams with entrenched meanders are characteristic features. Deciduous, pine-oak and pine forests formed the predominant vegetation in pre-settlement times. Glades, some of them extensive, commonly occur where bedrock surfaces. Bottomland deciduous forests are common along many of the streams. The great age and physiographic diversity of the Ozarks make it the region of greatest species diversity in Missouri. This division is divided into six sections.

The Salem Plateau Natural Section is a broad, gentle plain that was originally forested to a great extent but is now characterized by open pastureland and scattered trees. Bottomland deciduous forests remain along many streams.

The Lower Ozarks Natural Section is richly forested and characterized by springs, caves, sinkholes, calcareous wet meadows, glades, clear, high-gradient streams and steep-sided hills with narrow, chert-covered ridges. Shortleaf pine (*Pinus echinata*) is also characteristic. Streams flow generally southward and include the St. Francis, Black, Current and Eleven Point rivers.

The St. Francois Mountains Natural Section is the area of greatest relief and elevation in Missouri. There are few springs, caves or other karst features in this natural section. The pre-settlement vegetation of pine and mixed pine-oak still predominates.

The White River and Elk River natural sections are

both characterized by steep terrain and deciduous forest mixed with some pine. Glades are common, especially in the White River drainage. Construction of reservoirs (Table Rock, Taneycomo and Bull Shoals lakes), has created habitat for a few wetland-associated breeding birds while inundating habitat for bottomland forest species.

The Springfield Plateau Natural Section is physiographically the most distinct in the Ozark Natural Division. Historically, its landform was not highly dissected and its topography, soils, and pre-settlement vegetation were characterized by a mosaic of Osage Plains prairies grading into Ozarks forests. Prairie once occupied about 29 percent of this section, which today is less than 1 percent prairie. Today, this natural section is characterized by fragmented forests, pasture and early successional shrub-scrub habitats.

Osage Plains Natural Division

The Osage Plains Natural Division, which occupies about 8 percent of the state, is an unglaciated region in central western Missouri with an open, grassland aspect and gently rolling topography. Compared to the prairies of the Glaciated Plains, the upland prairie of this natural division has a greater proportion of southwestern plants and animals, fewer northern species, and a greater diversity in stream-side woody vegetation. More than 70 percent of the Osage Plains was prairie in pre-settlement times. Savanna, upland and bottomland deciduous forest and marsh also occurred. Streams commonly had shallow valleys and broad floodplains with many sloughs and marshes.

Today, much of the Osage Plains Natural Division is used for agriculture. Much of the area is pastureland with row cropland dominating in only a few areas. With the absence of frequent prairie fires, many savannas have succeeded to scrub-land and second-growth forests. The upper reaches of Truman Reservoir provide some wetland habitat for breeding birds at the Schell-Osage, Four Rivers and Montrose conservation areas. The highest quality prairie habitat remaining in the Osage Plains is on public prairies, including those on the Taberville and Paint Brush Prairie conservation areas and at Prairie State Park.

Big Rivers Natural Division

The Big Rivers Natural Division, comprising about 5 percent of the state, includes the floodplains and terraces of the largest rivers, primarily the Missouri and Mississippi, but also the lower Grand and the lower Des Moines rivers. Soils are mostly alluvial, deep and productive. Pre-settlement natural features included mesic to wet prairie, bottomland and upland forests, marshes, sloughs, islands, sandbars, mud flats, oxbow ponds and rivers. In pre-settlement times, and until extensive channel modification began in the early 1900s, the Missouri River was a braided stream with many chutes, sloughs, islands and seasonal wetlands.

Between 1879 and 1972, about 50 percent of the original surface area of the Missouri River was lost, backwater habitat was eliminated and the main channel was deepened and narrowed. Most of the original natural communities are no longer present. The Mississippi River, although less changed than the Missouri, has been modified and the addition of locks and dams has converted portions of the Mississippi into a series of large pools.

Today, this is the second-most intensively farmed natural division of the state, with corn, soybeans and wheat the most common crops. Wildlife habitat is preserved within a number of managed wetland areas, including Squaw Creek, Swan Lake and Mark Twain national wildlife refuges, and Bob Brown, Ted Shanks and Fountain Grove conservation areas.

Mississippi Lowlands Natural Division

The Mississippi Lowlands Natural Division comprises about 5 percent of the state. The Lowlands Natural Section, one of two natural sections in this natural division, is primarily composed of flat, alluvial plain and low terraces at the head of the Mississippi Embayment. This division has the highest average precipitation and temperature in Missouri. Relief is slight, with much of the division less than 100 meters above sea level. Historically, much of this division was bald cypress (*Taxodium distichum*) and tupelo gum swamp (*Nyssa aquatica*) forest, mixed deciduous bottomland forest, and low upland deciduous forest. Clearing and draining began in the early 1900s. Conversion to agriculture has been almost total and today only small remnants of natural forest and swamp remain.

The deep, alluvial soils, and water available for irrigation have made this natural division the most heavily farmed in the state. Today, much of the natural division is composed of large fields of soybeans, wheat, corn, rice and cotton. Wildlife habitat is provided by several public wetlands including Mingo National Wildlife Refuge and Duck Creek, Otter Slough, Coon Island and Ten Mile Pond conservation areas. A few large tracts of forest remain along the Mississippi River, including Ben Cash and Donaldson Point conservation areas and Big Oak Tree State Park. Drainage ditches and isolated forest fragments provide limited habitat for wetland and woodland species. Certain agricultural practices, including the flooding of rice fields, simulate natural habitat for some wetland-associated breeding birds.

Crowley's Ridge Natural Section is this division's most prominent topographic feature, rising above the surrounding lowlands in a disjunct series of forested, low hills. This section of the Mississippi Lowlands resembles the forest- and pasture-dominated Ozarks and supports many of the bird species characteristic of the Ozark and Ozark Border natural divisions.

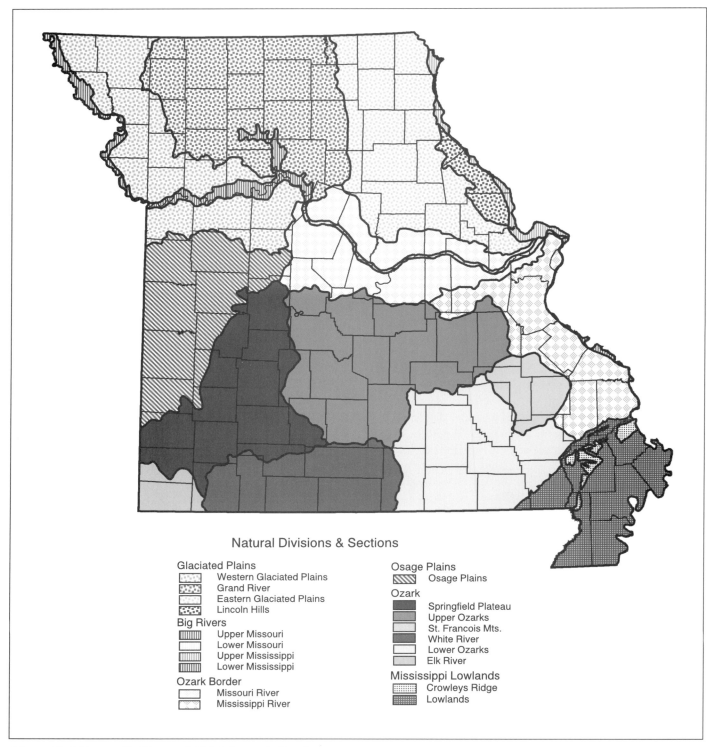

Natural Divisions & Sections

Glaciated Plains
- Western Glaciated Plains
- Grand River
- Eastern Glaciated Plains
- Lincoln Hills

Big Rivers
- Upper Missouri
- Lower Missouri
- Upper Mississippi
- Lower Mississippi

Ozark Border
- Missouri River
- Mississippi River

Osage Plains
- Osage Plains

Ozark
- Springfield Plateau
- Upper Ozarks
- St. Francois Mts.
- White River
- Lower Ozarks
- Elk River

Mississippi Lowlands
- Crowleys Ridge
- Lowlands

Figure 7. **Missouri's Natural Divisions & Sections**

Figure 8. **Missouri's Counties**

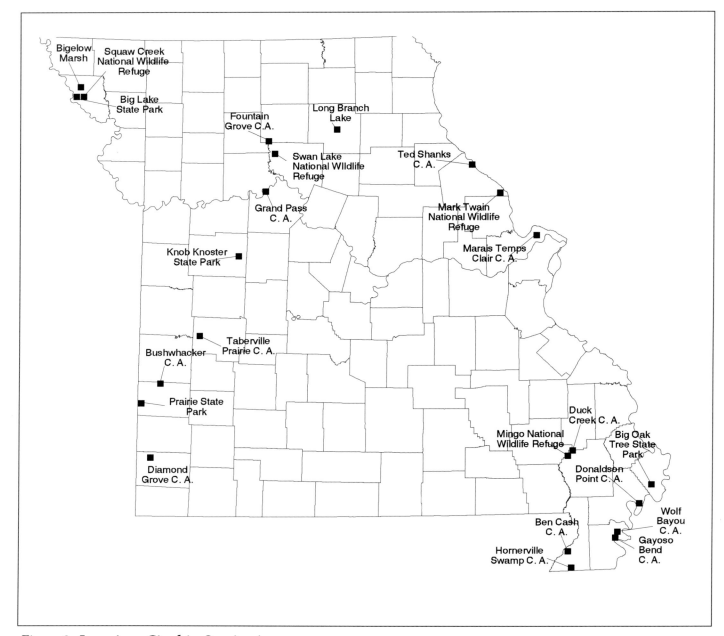

Bigelow Marsh
Squaw Creek National Wildlife Refuge
Big Lake State Park
Fountain Grove C.A.
Long Branch Lake
Swan Lake National Wildlife Refuge
Ted Shanks C. A.
Grand Pass C. A.
Mark Twain National Wildlife Refuge
Marais Temps Clair C. A.
Knob Knoster State Park
Taberville Prairie C. A.
Bushwhacker C. A.
Prairie State Park
Duck Creek C. A.
Big Oak Tree State Park
Mingo National Wildlife Refuge
Donaldson Point C. A.
Diamond Grove C. A.
Wolf Bayou C. A.
Ben Cash C. A.
Gayoso Bend C. A.
Hornerville Swamp C. A.

Figure 9. **Locations Cited in Species Accounts**

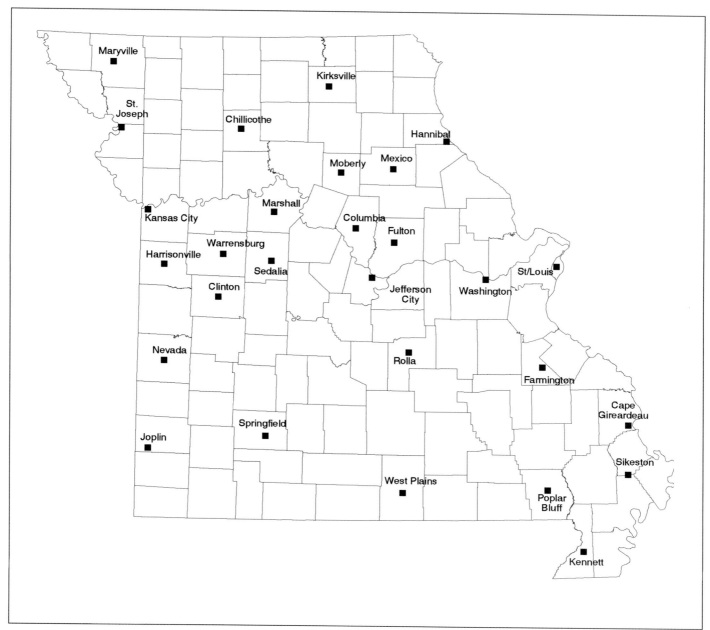

Figure 10. **Missouri's Cities and Towns cited in text**

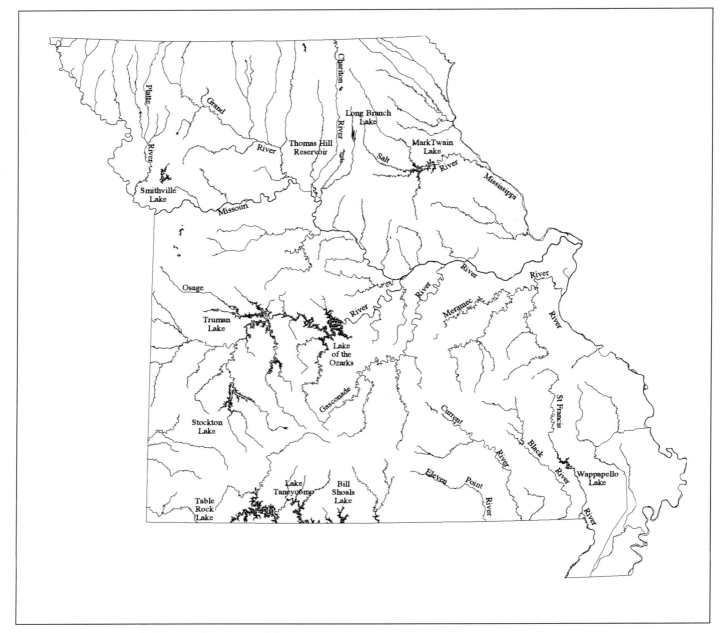

Figure 11. **Missouri's Major Rivers and Streams**

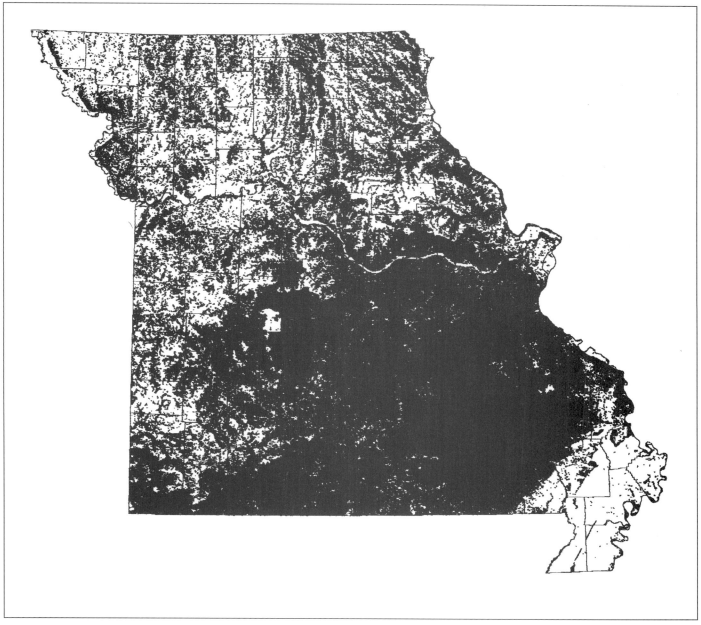

Figure 12. **Missouri's Forests (black areas)**

The map of Missouri's Forsts (Figure 12) is based on 1984TM data. Resolution is 30 meters. Minimum mapping units varied across the state with 2.5 acres in northern Missouri and 5 acres in southern Missouri.

Guide To Species Accounts

Drawing of the species.

American Goldfinch
Carduelis tristis

Rangewide Distribution: South central & southeast-ern Canada, northern, central & southeastern United States
Abundance: Common & widespread
Breeding Habitat: Edges of deciduous forest & open, weedy & cultivated areas
Nest: Tightly woven forbs & vegetation, lined with plant down, in trees
Eggs: 4–6 pale blue or bluish-white, unmarked
Incubation: 10–12 days
Fledging: 11–17 days

Discusses identification, habitat associations, range limits, detectability, historical status and behaviors that affect patterns of distribution.

During the breeding season, these widespread finches occupy rural areas, especially fallow, weedy fields inter-spersed with brushy thickets. They are prevalent in weedy margins of wetlands and brushy borders of farmsteads and forests. Small parcels such as fence rows, roadside ditches and even small forest openings are suitable if brushy cover is present. Goldfinches construct a thick-walled cup nest in an upright fork of a shrub or small sapling at a height of 1–3 meters above the ground (Peck and James 1987).

American Goldfinches typically initiate breeding in July or August, the latest of any North American passerine (Peterjohn and Rice 1991). The comparatively late nesting of this species appears to be correlated with the matura-

tion of thistles, which provide seeds for food and down for nesting material (Harrison 1975).

Code Frequency

American Goldfinches are easy to detect and, therefore, where not located in a block, they presumably occurred in extremely low numbers or were not present. Their distinc-tive twittering usually attracts attention first, and they are often then seen passing overhead in characteristic undulat-ing flight. Because flying birds are often observed in pairs, Atlasers were able to easily elevate them to the probable breeding level.

Breeding confirmations, however, were difficult to observe. Apparently Atlasers were rarely successful in their attempts to follow the movements of goldfinches suspect-ed of nesting. Perhaps the bulk of the surveying effort was conducted too early in the season to confirm this unusual-ly late-nesting species. Considering these limitations, American Goldfinches presumably bred in most blocks in which they were found.

Distribution

American Goldfinches were recorded statewide. Their only region of scarcity was in the Mississippi Lowlands where extensively tilled lands and lack of brushy, weedy fields offer few potential nest sites. Atlasers obtained only possible breeding evidence in many adjacent blocks in the lower Ozarks, perhaps because American Goldfinches rarely breed in its extensive forests which have little brush and edge habitat. Another region of fewer probable and confirmed breeding records was centered on Shelby County in northeastern Missouri, which has much favor-able habitat. Reasons for the lower detection rate are unknown. Perhaps the cursory coverage that region received via Block-Busting and other single-weekend sur-veys, combined with late-season nesting of the species, resulted in reduced detections.

Abundance

American Goldfinches were recorded as most abun-

Abundance by Natural Division
Average Number of Birds / 100 Stops

0.0   11.0

10.7   10.1

9.6   1.8

Abundance as detected on Miniroutes and United States Fish and Wildlife Breeding Bird Survey routes. When no valid data were obtained, this map is absent. (See Abundance text.)

Breeding Phenology

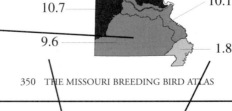

EVIDENCE (# of Records)	MAR	APR	MAY	JUN	JUL	AUG	SEP
NB (30)			5/27				8/26
NE (2)					7/10		8/29
NY (8)				6/08			8/31
FY (8)					6/28		9/03

From top, numbers on left indicate abundance in Big Rivers, Osage Plains and Ozark natural divisions: numbers on right indicate Glaciated Plains, Ozark Border and Mississippi Lowlands natural divisions.

Span of dates of four confirmed breeding behaviors: nest building, nest with eggs, nest with young and food being carried to young. (See Phenology text.)

BREEDING EVIDENCE
Reported in 1,103 (91.4%) of 1,207 blocks

Confirmed	100	9.1%
Probable	681	61.7%
Possible	322	29.2%

Distribution

Numbers and percentages of blocks in which possible, probable and confirmed breeding codes were recorded.
(See Code Frequency text.)

Shading of the 7.5 minute quadrangles indicates possible, probable or confirmed breeding within the block in that quadrangle. Unshaded blocks indicate that the species was not detected within appropriate dates and in appropriate breeding habitat. (See Distribution text.)

dant in the Glaciated Plains and Osage Plains natural divisions. This concurs with Robbins and Easterla (1992) who reported on BBS data. Goldfinches also appeared nearly as abundant in the Ozark Border and Ozark natural divisions. As expected from the distributional map, they were least abundant in the Mississippi Lowlands.

Phenology

In Ohio, pair formation continues into June but most goldfinches do not initiate nesting until after July 15 (Nice 1939). Atlasers observed confirmed breeding evidence unexpectedly early in comparison. A nest building record on May 27 was three days earlier than the earliest record reported in Ohio. A nest containing hatched young, recorded on June 8, was even more remarkable. Middleton (1993) reported a few early nests with eggs in May and

early June; however, nesting peaks for this species in the second half of July and continues into September. Paul McKenzie, United States Fish and Wildlife Service, suggested in editorial review that earlier nesting may be a response to earlier-maturing, exotic thistles that are now more abundant and may provide an early source of nesting material. The Atlas Project's latest nest with eggs was observed on August 29. Nesting events likely continued but were not reported as Atlasers were asked to conclude their field survey by September 15.

Notes

Although American Goldfinches are common Brown-headed Cowbird hosts (Ehrlich et al. 1988), Atlasers did not document any parasitism during the project.

AMERICAN GOLDFINCH   351

Other observations by Atlasers, such as Brown-headed Cowbird brood parasitism and other related information.

The phenology of nesting-related events is discussed and related to known sequences in Missouri and adjacent states.
(See Breeding Phenology chart opposite.)

Species Accounts

Pied-billed Grebe
Podilymbus podiceps

> **Rangewide Distribution:** South central Canada, all
> North America to South America
> **Abundance:** Common & not gregarious
> **Breeding Habitat:** Vegetated lakes, ponds, sluggish
> streams & marshes
> **Nest:** Shallow platform of decaying vegetation
> anchored in open water among reeds & rushes
> **Eggs:** 5–7 blue-white, chalky, nest-stained buff or
> brown
> **Incubation:** 23 days
> **Fledging:** Unknown number of days

Pied-billed Grebes, local summer residents in Missouri, have a secretive nature. Once sighted foraging in open water, they may slowly submerge and move to the vegetated edge of the pond, where they are difficult to detect. They are perhaps best detected at night, when they are most vocal.

The Pied-billed Grebe was apparently once more common and widespread. Widmann (1907) wrote, this species was "Formerly a common breeder in all reedy lakes throughout the state, but with drainage and persecution it is becoming rare every year."

Code Frequency

Pied-billed Grebes were reported with only five breeding behavior codes. Breeding was confirmed in only three blocks. Possible and probable sightings potentially represent actual breeding sites. Unless flightless young are observed later in the summer, an extensive wade in the marsh is usually required to detect a nest.

Distribution

This species was found statewide. Due to the patchy distribution of suitable wetland habitat, many areas were outside blocks and went unsurveyed. Special nest searches conducted by the Missouri Department of Conservation in wetland areas have confirmed breeding sites outside of blocks.

Abundance

In some areas, where conditions are excellent, nesting density can be high. In 1990 at Squaw Creek National Wildlife Refuge in Holt County, refuge manager Ron Bell reported 65 nesting pairs. In years when less suitable habitat was available, only a few pairs nested in the same area.

Phenology

Some individuals arrive when the ice melts from ponds and lakes with peak numbers of migrants passing through in mid-April (Robbins and Easterla 1992).

Notes

During the first 3–4 years of the Atlas Project, much of the Glaciated Plains was under severe drought conditions and little wet marsh habitat was available. With restoration of wetlands a priority in the state, this species should be found in more locations in the future.

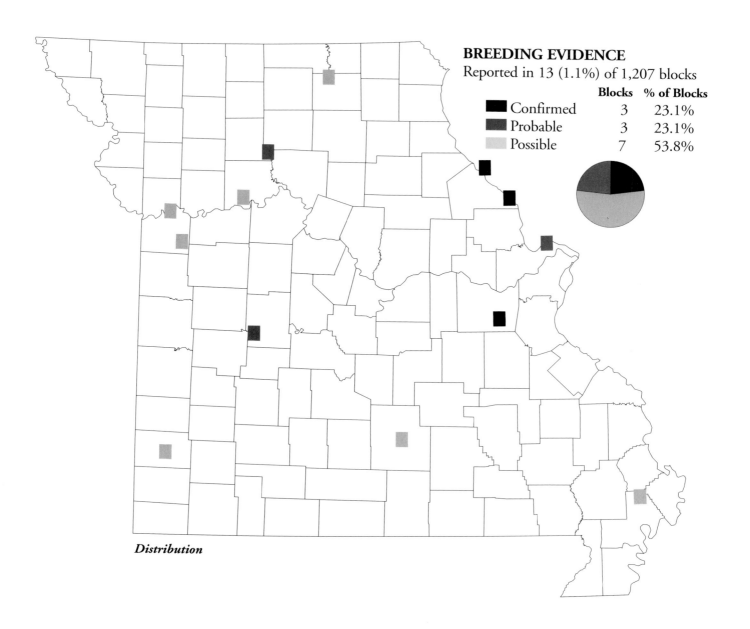

BREEDING EVIDENCE
Reported in 13 (1.1%) of 1,207 blocks

	Blocks	% of Blocks
Confirmed	3	23.1%
Probable	3	23.1%
Possible	7	53.8%

Distribution

Double-crested Cormorant
Phalacrocorax auritus

Rangewide Distribution: South central eastern coast of Canada; northern central & coastal United States
Abundance: Common & widespread
Breeding Habitat: Freshwater with snag or cliffs near water
Nest: Platform of sticks, seaweed or drift material on tree or ground by water
Eggs: 3–4 light blue or blue-white & usually nest-stained
Incubation: 25–29 days
Fledging: 35–42 days

Code Frequency

Double-crested Cormorants were recorded as possible breeders in five blocks. Because breeding confirmations are easily obtained for these large birds that nest in open situations, the absence of higher evidence suggests breeding did not occur in blocks during the Atlas Project.

Double-crested Cormorants are locally common transients and rare summer visitors in Missouri (Robbins and Easterla 1992). The increase in their numbers during migration and summer in recent years parallels an increase in breeding populations in the Great Lakes region (Cadman et al. 1987). No breeding was confirmed during the Atlas Project although breeding in the state has occurred in the past. According to Widmann (1907) the Double-crested Cormorant bred in considerable numbers in the extensive Mississippi Lowlands swamps during the 1800s. The most recent documented nesting was at Mingo National Wildlife Refuge where eight nests were photographed on June 23, 1956 (Robbins and Easterla 1992).

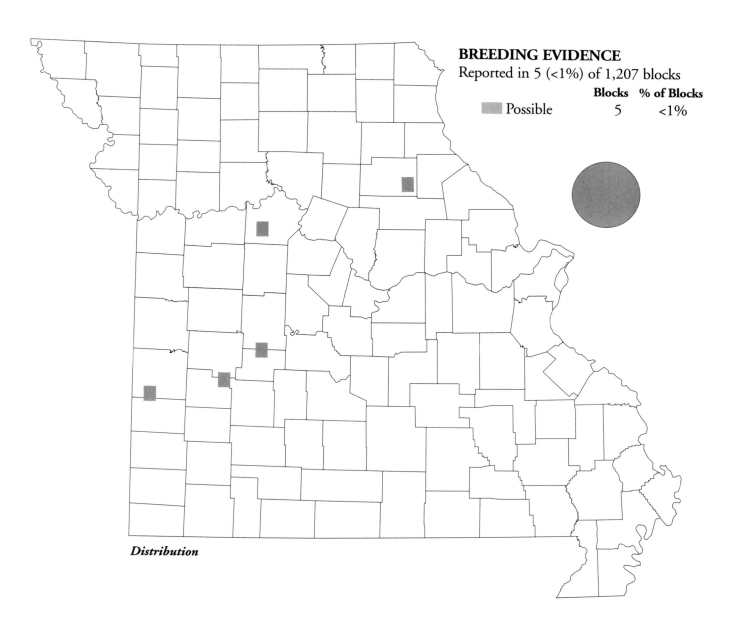

	Blocks	% of Blocks
Possible	5	<1%

Distribution

American Bittern
Botaurus lentiginosus

Rangewide Distribution: Southern & south central
    Canada; northern to north central & eastern
    United States
Abundance: Fairly common but declining in north-
    eastern United States & New England
Breeding Habitat: Marshes & wet meadows with
    emergent vegetation
Nest: Sticks, grass & sedge on dry ground or over
    water; separate exit & entrance
Eggs: 4–5 buff-brown or olive-buff
Incubation: 28–29 days
Fledging: Unknown number of days

American Bitterns once nested fairly commonly in all marshes of the state (Widmann 1907). Currently Missouri lists them as an Endangered species. They are generally restricted to undisturbed wetlands of greater than eight hectares supporting dense stands of cattails and other tall, emergent vegetation interspersed with patches of open water (Gibbs et al. 1992). They are extremely secretive except when establishing breeding territories. Territorial vocalizations include distinct pumping sounds that carry great distances over the marsh. Nests are usually platforms of dead plant material placed over shallow water and well-hidden among dense vegetation (Harrison 1975).

Code Frequency

The reported rarity of American Bitterns as a summer resident (Robbins and Easterla 1992) was supported by the findings of the Atlas Project. In only six blocks was the species recorded in habitat appropriate for breeding. Because of the difficulty in surveying marshes and detecting breeding evidence, it may be they bred in all six blocks. In one, at the Ted Shanks Conservation Area, breeding was confirmed in 1986 by the observation of a nest with eggs.

Distribution

Because of the extreme rarity of this species as a breeder, locations discovered during the Atlas Project provide little information on distribution or abundance. Gibbs et al. (1992) suggested that only the northern third of Missouri constitutes potential breeding range for this species. Early in the century, the American Bittern bred commonly through-out the state (Widmann 1907).

Phenology

Most evidence of breeding was obtained in June, however, a territorial individual was recorded on July 27. Terres (1987) reported egg dates that ranged from April through July and, according to Ehrlich et al. (1988), American Bitterns can rear two broods per season.

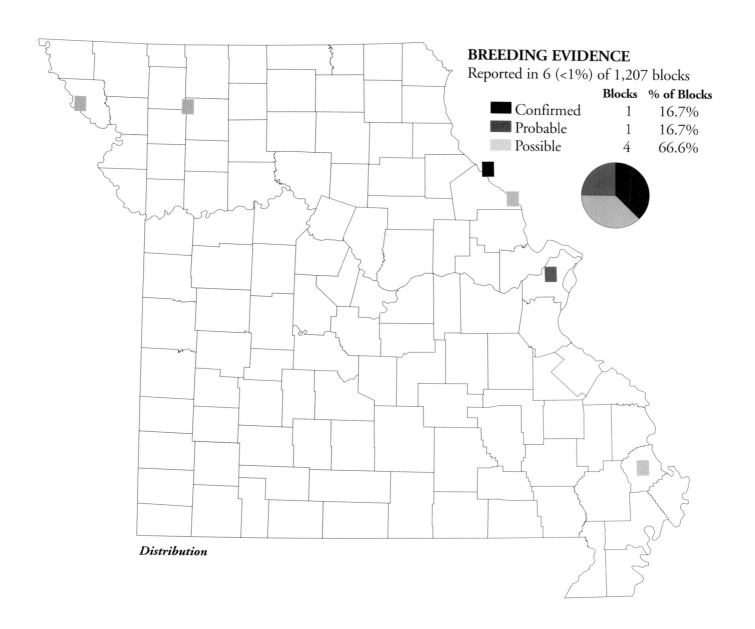

BREEDING EVIDENCE
Reported in 6 (<1%) of 1,207 blocks

	Blocks	% of Blocks
Confirmed	1	16.7%
Probable	1	16.7%
Possible	4	66.6%

Distribution

Least Bittern
Ixobrychus exilis

> **Rangewide Distribution:** Eastern United States, California & South America
> **Abundance:** Fairly common east, less common west
> **Breeding Habitat:** Dense vegetation, shallow water & cattails
> **Nest:** Aquatic vegetation & sticks near or over water, on ground or low shrub
> **Eggs:** 4–5 bluish-white to greenish-white
> **Incubation:** 19–20 days
> **Fledging:** 25 days

Extensive cattail (*Typha* spp.) marsh with open water and water depths up to 50 centimeters provide appropriate breeding habitat for this secretive marsh dweller (Gibbs et al. 1992). For about an hour at dawn and dusk, the Least Bittern's soft, ventriloquistic "coo-coo-COO-COO-Coo-coo-coo" betrays its presence. The smallest member of the heron family in Missouri, it is difficult to see. In early morning or late evening the patient observer might catch a glimpse of an individual flying low over the cattails. Wading or canoeing in a marsh during the day is the best way to census and document nest site activity.

Code Frequency

Most Atlasers likely did not adequately survey for this secretive species. The three nest and egg sightings represent efforts of surveyors who searched the right habitat at the right time and worked hard to locate nests. Special searches by the authors outside of blocks have located breeding pairs at most of the larger marshes in the state. As with most wetland species, training and concentrated effort is needed to document Least Bittern activity.

Distribution

Least Bitterns may be one of the most widespread marsh birds. This wide distribution, however, was not supported by Atlas Project findings. From sources other than the Atlas Project, it is known that many pairs nest at Squaw Creek National Wildlife Refuge in Holt County (Wilson 1992) and that potential breeding sites exist at smaller marshes. Currently, the Big Rivers and Mississippi Lowlands natural divisions still harbor the most productive marshes for Least Bitterns. In addition to Squaw Creek, Swan Lake, Mingo and the Mark Twain national wildlife refuges, Ted Shanks and Marais Temps Clair conservation areas harbor the largest known breeding populations in the state.

Abundance

While little can be gleaned from Atlas Project records, special searches revealed well over 100 nests at Squaw Creek National Wildlife Refuge. Lesser numbers have been found at the other refuges and conservation areas mentioned above, with an occasional nest found in cattails along lake margins in areas such as Bushwhacker Conservation Area. Widmann (1907) described Least Bitterns as locally common in large permanent marshes in most parts of the state in the early 1900s. As with most wetland species, the restoration of high quality, large, emergent marshes would greatly enhance breeding habitat that could support greater numbers of individuals.

Breeding Phenology

EVIDENCE (# of Records)	MAR	APR	MAY	JUN	JUL	AUG	SEP
NE (3)			6/5		6/22		

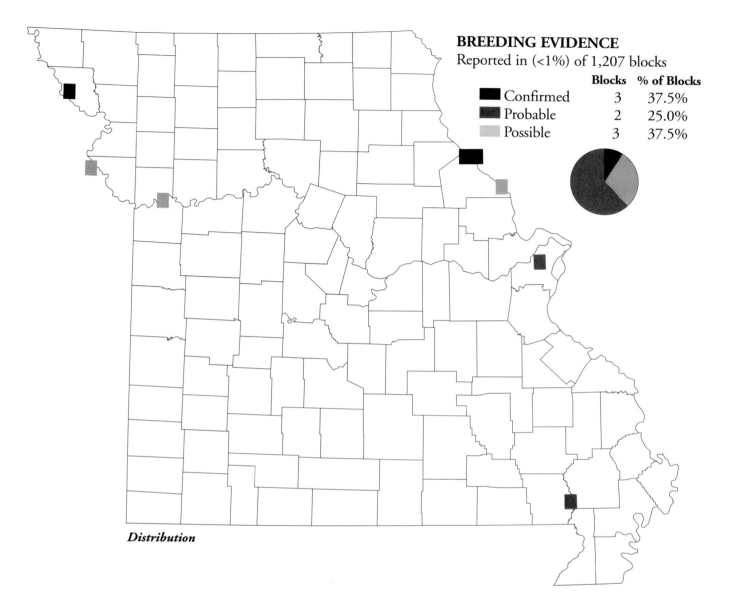

BREEDING EVIDENCE
Reported in (<1%) of 1,207 blocks

		Blocks	% of Blocks
■	Confirmed	3	37.5%
■	Probable	2	25.0%
▨	Possible	3	37.5%

Distribution

Phenology

Least Bitterns normally arrive in Missouri in late April and early May with many areas apparently unoccupied until later in May (Robbins and Easterla 1992). Although considered a potentially double-brooded species (Gibbs et al. 1992), no evidence of a second nesting was found during the Atlas Project.

Great Blue Heron
Ardea herodias

Rangewide Distribution: Western coast of Canada, entire United States to northern South America
Abundance: Widespread & common
Breeding Habitat: Bottomland forests & swamp areas
Nest: Flat, of woven sticks, lined with twigs & leaves in tree or shrub
Eggs: 3–5 light bluish-green
Incubation: 28 days
Fledging: 56–60 days

These large wading birds are commonly seen foraging for fish and other aquatic animals in shallow waters during the spring, summer and fall. They breed colonially, typically in floodplain forests. Colony sites are used annually, some for over 35 years (Butler 1992). In Missouri their stick nests are placed in the crowns of large trees, especially sycamores (*Platanus occidentalis*).

Great Blue Heron numbers declined during the '50s, '60s and '70s throughout much of their range, including Illinois (Graber et al. 1978) and the Upper Mississippi River (Thompson 1978). Beginning in the early 1980s, growth of the breeding population was documented in many areas

including Illinois (Kleen 1987) and Michigan (Butler 1992). The Missouri Department of Conservation 1978 survey of Great Blue Heron colonies recorded only 123 colonies averaging 29.3 active nests per colony. In 1992, 192 colonies were located, averaging 29.2 active nests.

Code Frequency

While Great Blue Herons are easily sighted, their presence does not necessarily indicate a breeding colony. Some of the individuals found in Missouri during the breeding season may have already concluded breeding at colonies farther south (Heitmeyer 1986). Also, although Great Blue Herons most often forage within 2.3 to 6.5 kilometers of breeding colonies, they may range up to 30 kilometers (Butler 1992).

Considering these factors, the possible and probable breeding blocks on the accompanying map provide a misleading view of breeding distribution. Fortunately, their breeding colonies are relatively conspicuous, so confirmed breeding locations suggest the true distribution of breeding Great Blue Herons in Missouri.

Distribution

Great Blue Herons bred most densely in southwestern Missouri and were more scattered elsewhere in the state. Although breeding colonies were found on a variety of streams and lakes, none were located on the Missouri River nor on the Missouri side of the Mississippi River during the Atlas Project.

Abundance

Great Blue Herons appeared to be most abundant in the Osage Plains and Ozark natural divisions and least abundant in the Ozark Border, Mississippi Lowlands and Big Rivers natural divisions.

Abundance by Natural Division
Average Number of Birds / 100 Stops

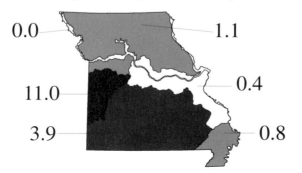

0.0   1.1
11.0   0.4
3.9   0.8

Breeding Phenology

EVIDENCE (# of Records)	MAR	APR	MAY	JUN	JUL	AUG	SEP
NB (3)		4/30			6/24		
NY (20)	4/06				7/09		
FY (4)			6/11			7/26	

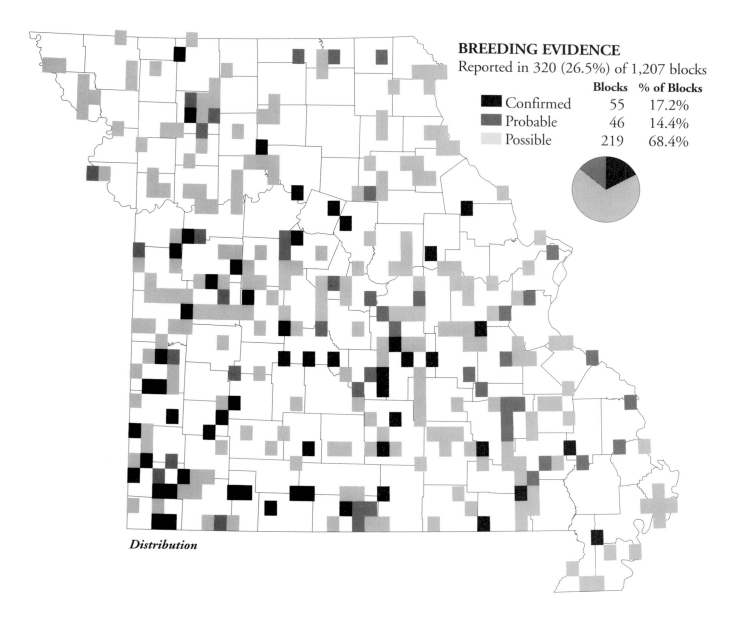

BREEDING EVIDENCE
Reported in 320 (26.5%) of 1,207 blocks

	Blocks	% of Blocks
Confirmed	55	17.2%
Probable	46	14.4%
Possible	219	68.4%

Distribution

Phenology

Great Blue Herons return to Missouri in late February and early March (Robbins and Easterla 1992) and apparently initiate breeding activities immediately. Late dates for nesting activities may have resulted from re-nesting subsequent to nest failure or been due to late nest initiation. It is unclear how many late broods are first or second nesting attempts (Butler 1992).

Great Egret

Ardea alba

> **Rangewide Distribution:** Worldwide
> **Abundance:** Widespread & common
> **Breeding Habitat:** Fresh & brackish-water marshes &
> swamps
> **Nest:** Frail construction of sticks & twigs, lined or
> unlined, in tree or shrub
> **Eggs:** 3 light blue or light bluish-green
> **Incubation:** 23–26 days
> **Fledging:** 42–49 days

These stunning white birds breed colonially, constructing stick nests in the crowns of large trees in areas well buffered from disturbance along rivers and in swamps. In Missouri, on rare occasions they share breeding colonies with Great Blue Herons and Black-crowned Night-Herons. Colony sites and nests are re-used year after year and the destruction of a site can cause a local reduction in the population (Parnell et al. 1988).

Code Frequency

As with Great Blue Herons, Great Egrets often disperse some distance after breeding. Therefore, the possible breeding locations shown on the map should be disregarded. These large, conspicuous nesters likely bred only where confirmed to breed.

Distribution

Great Egrets are rare breeders in Missouri as indicated by the map. Nests were actually observed in only one block, 12 kilometers north of Nevada in Vernon County. The three confirmed breeding sites along the Mississippi River were in Pike, Lincoln and Pemiscot counties. The latter site was a mixed colony that included Black-crowned Night-Herons, Little Blue Herons, Cattle Egrets and Snowy Egrets. In the remaining colonies, Great Egrets co-nested with Great Blue Herons.

Phenology

Great Egrets begin appearing in Missouri at the end of March and numbers peak in mid-May (Robbins and Easterla 1992). Nesting activities commence immediately.

Breeding Phenology

EVIDENCE (# of Records)	MAR	APR	MAY	JUN	JUL	AUG	SEP
NY (1)				6/19	6/19		
FY (3)			5/17		6/05		

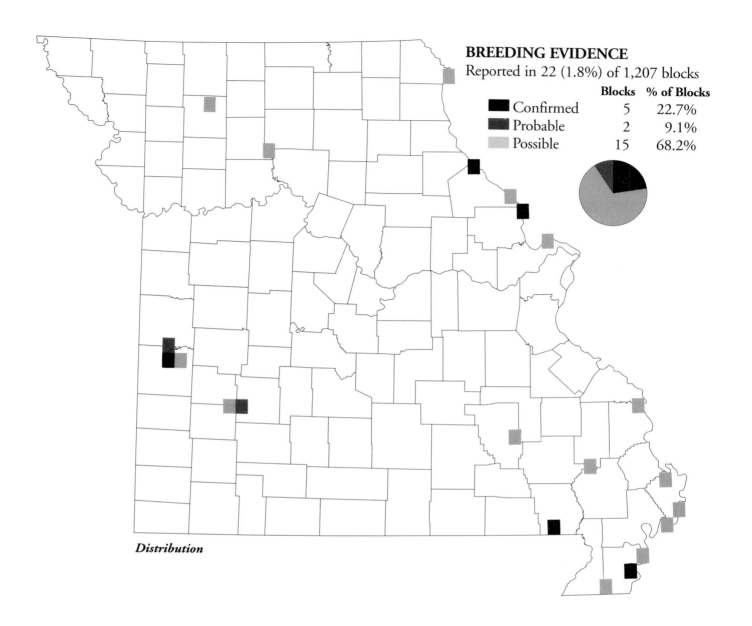

BREEDING EVIDENCE
Reported in 22 (1.8%) of 1,207 blocks

		Blocks	% of Blocks
■	Confirmed	5	22.7%
■	Probable	2	9.1%
■	Possible	15	68.2%

Distribution

Snowy Egret
Egretta thula

Rangewide Distribution: Sections of southern
United States & Gulf Coast states
Abundance: Common
Breeding Habitat: Swamps, lowland forests & marshes
with emergent vegetation
Nest: Flat, flimsy, of sticks lined with fine sticks &
rushes, on shrubs or occasionally on ground
Eggs: 3–5 light bluish-green
Incubation: 20–24 days
Fledging: 30 days

Snowy Egrets are colonial nesters, usually nesting in association with other egret and heron species. This is one of the most beautiful species of the heron family when in full display, with plumes spreading from the body off the head, neck and back. Plume hunters sought this species more than other herons for its exquisite, soft, white nuptial feathers. During early 20th century raids on heron nesting colonies, Snowy Egrets suffered greater losses than other herons, because they were less secretive and more numerous (Terres 1987). Snowy Egrets are currently listed as Endangered in Missouri.

Code Frequency

Only possible records were obtained for this species of which only one was in appropriate habitat. A sight record from McDonald County probably represents an individual from a nearby colony in Kansas.

Distribution

Except for the block in McDonald County, all locations were near known colony sites which were not in Atlas Blocks. Observations independent of the Atlas Project indicated Snowy Egrets bred at four different colonies during the term of the Atlas Project. All colonies were in the Mississippi Lowlands, with two near Sikeston in Scott County and the others near Charleston in Mississippi County and Caruthersville in Pemiscot County.

Abundance

Based on colony counts independent of the Atlas Project, 6–8 active nests were present at the Pemiscot County site and 10–30 active nests were present at the Mississippi County site in 1992. Fewer nests were at the Scott County sites, which was abandoned by 1987.

Phenology

Breeding in southeastern Missouri commences in May and continues into June (Robbins and Easterla 1992). Post-breeding wanderers begin to arrive in the state in July and continue to increase until late August and early September (Robbins and Easterla 1992).

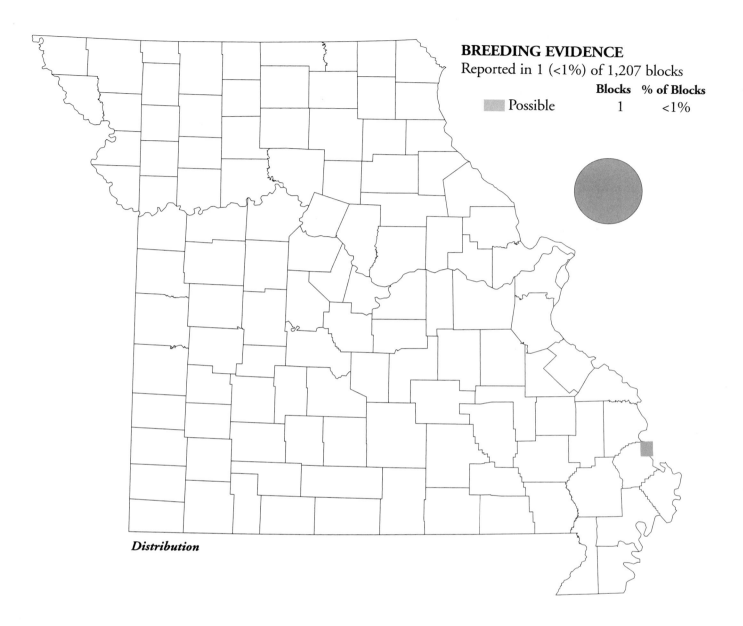

Reported in 1 (<1%) of 1,207 blocks

	Blocks	% of Blocks
Possible	1	<1%

Distribution

Little Blue Heron
Egretta caerulea

Rangewide Distribution: Southeastern United States, East Coast & coastal Middle America

Abundance: Common

Breeding Habitat: Stream banks, marshes, ponds & reservoirs

Nest: Platform of sticks & twigs in low trees or shrubs, often above water

Eggs: 2–5 light bluish-green

Incubation: 20–23 days

Fledging: 42–49 days

A colony-nesting species that disperses statewide after the breeding season, Little Blue Herons are commonly seen near water. Most sightings represent birds which have dispersed from the colony area after breeding is completed. Late summer white-plumaged birds are immatures, which develop blotched slate blue and white plumage in their first spring as they molt to adult plumage.

Code Frequency

No nests were located in Atlas Project blocks, so no confirmations resulted. Many observations were concentrated in two areas of the Mississippi Lowlands near known colonies. A colony in Oklahoma was likely the source for the records in the southwestern Ozark and Osage Plains natural divisions. Large numbers of individuals observed in April, May and June usually indicate a colony nearby. Foraging could occur over a 15–30 kilometer radius from the colony but usually much less. In late June and early July, many individuals disperse from colonies, some likely from other states.

Distribution

Independent of the Atlas Project, three breeding colonies were identified in the Mississippi Lowlands near Charleston in Mississippi County, near Caruthersville in Pemiscot County and Sikeston in Scott County. The latter colony moved north of Sikeston in 1987 before disbanding entirely. As of 1994, two years after Atlas Project data collection ended, the Mississippi County colony had disappeared, presumed to have moved to an adjacent site in the Sikeston city limits. The Pemiscot County colony moved about 1.5 kilometers east although some herons remained. Movement of colonies is common for this and other heron species because excess birds leave to form new colonies, nest trees die and fall due to guano build up, foraging areas are lost, and sometimes human activities interfere. Previous nesting colonies in the Big Rivers and other natural divisions have disappeared and relocated frequently.

Abundance

Observers independent of the Atlas Project counted 50–100 individuals in the Sikeston sites; 500–1,000 in the Caruthersville site; and 50–1,420 in the Charleston site during 1986 to 1992.

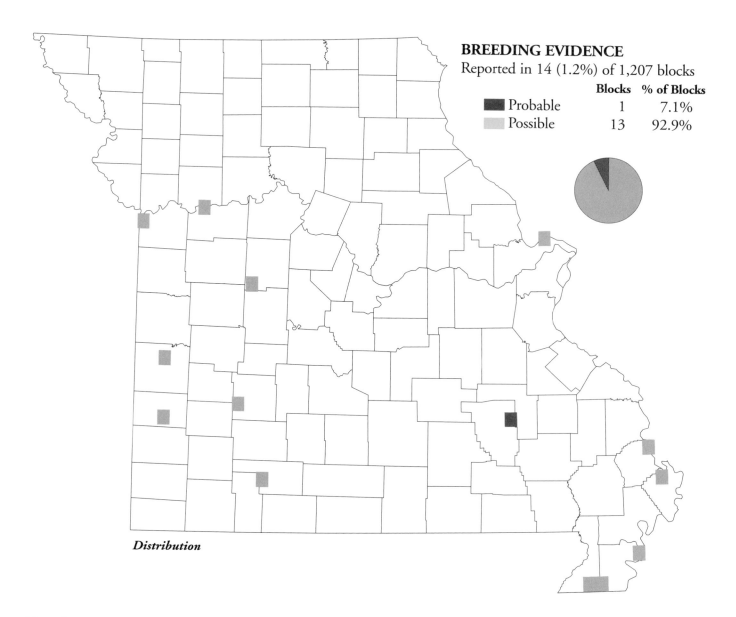

	Blocks	% of Blocks
Probable	1	7.1%
Possible	13	92.9%

Distribution

Phenology

Robbins and Easterla (1992) documented the arrival of this species from late March to mid-April. After a 62–72 day incubation and fledging period (Ehrlich et al. 1988), they subsequently disperse by mid-July and are found statewide.

Cattle Egret
Bubulcus ibis

Rangewide Distribution: Southeastern Canada; central & eastern United States to South America

Abundance: Common, with population increasing & spreading through North America

Breeding Habitat: Wet pastures, marshes & watercourses

Nest: Reeds, sticks, twigs & vines in trees or shrubs

Eggs: 3–4 light blue-green or bluish-white

Incubation: 22–26 days

Fledging: 30 days

These gregarious mid-20th century immigrants to the United States are becoming an increasingly common sight in Missouri, especially during late spring and early fall when they migrate through the state. Adults are distinguished from similar small, white ardeids by their yellow bills and legs and their tendency to forage in association with grazing animals or behind farm machinery. Adult males and females share a high breeding plumage of orange plumes on the crown, back and neck. Nesting colonies can reach densities of several hundred nests per hectare. Telfair (1983) reported a maximum nest density of one nest per 0.77 square meters in Texas.

Code Frequency

Many of the Cattle Egrets seen in Missouri are not associated with nearby breeding colonies but are likely migrating to breeding colonies in the upper Midwest. Because this species also disperses after breeding (Telfair 1994), some seen in summer likely nested elsewhere earlier in the year. Also, Cattle Egrets will travel up to 32 kilometers from breeding colonies to secure food (Bateman 1970). Considering all of the above, the sighting of Cattle Egrets in a block, with no further evidence of breeding, contributed little information about the species' breeding distribution.

Distribution

The Atlas Project did not confirm breeding despite known breeding in the state outside of Atlas Blocks. Only possible breeding codes were recorded for this species. Most were in the Mississippi Lowlands, and likely were associated with known breeding colonies in and near Sikeston in Scott County, near Charleston in Mississippi County and near Caruthersville in Pemiscot County.

Abundance

Within the Charleston colony, the number of breeding Cattle Egrets ranged from 250–1,000 during the Atlas Project while the number ranged from 70–1,200 at the Caruthersville colony. A maximum of about 100 nested at the Sikeston sites.

Phenology

Observations independent of the Atlas Project at the Charleston and Caruthersville colonies revealed Cattle Egrets initiate breeding upon arrival at the colony in late April. Egg laying occurs from early May to early June and

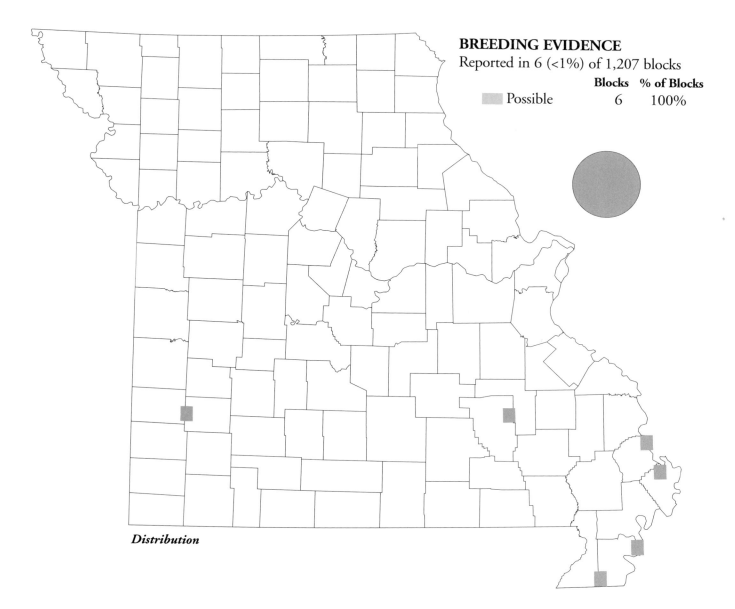

		Blocks	% of Blocks
	Possible	6	100%

Distribution

most fledging occurs from early June through August. Re-nesting is apparently commonplace, probably because of disturbance resulting from the density of nests. As a result, some clutches were still being incubated following the fledging of birds from other nests. Cattle Egrets will produce a second brood in the southern United States (Telfair 1994) but it is unclear whether Missouri's late broods were second broods or cases of re-nesting.

Green Heron
Butorides virescens

Rangewide Distribution: Southern Canada, southwestern & eastern United States to Middle America
Abundance: Widespread & common
Breeding Habitat: Wooded margins of ponds, swamps & other water
Nest: Woven sticks & twigs, occasionally lined, on trees or emergent vegetation
Eggs: 2–4 light greenish or bluish-green
Incubation: 21–25 days
Fledging: 34–35 days

Unlike the other heron species that breed in Missouri, Green Herons normally nest solitarily, preferring dense, brushy thickets near water for nest sites (Andrle and Carroll 1988). Meyerriecks (1960) considered them semi-social, intermediate between the totally solitary American Bitterns and the highly social Night-Herons. In Missouri, nesting usually occurs in the forested margins of ponds, rivers, lakes, marshes and swamps (Kaiser 1982). They normally place their nests in dense cover within 5–7 meters of the ground, but when dense thickets are unavailable, they will nest 15–25 meters high in tall trees (Williams 1950). They sometimes re-build and re-use nests from a previous season (Davis and Kushlan 1994).

Code Frequency

Green Herons are reasonably easy to detect and identify as they fly in straight-line fashion across the sky or along streams. They are also easily located by their sharp calls. As a result, Green Herons were likely absent or in low numbers in those blocks where not recorded. Several factors contributed to the lack of confirmed breeding records for this species. Green Herons are secretive around their nests and therefore nest sites are not likely to be detected until young herons fledge (Peterjohn and Rice 1991). Two-thirds of the confirmed breeding records were obtained by observing fledglings. In only 17 blocks were active nests recorded.

Distribution

Green Herons bred statewide but were found in fewer blocks in the northern one-third of the state. Their sparse presence in the Mississippi Lowlands may be due to the lack of both stream quality and appropriate stream-side vegetation in this agriculturally developed region.

Abundance

Green Herons were most abundant in the vicinity of the remnant swamplands of Mingo National Wildlife Refuge and Duck Creek Conservation Area. They were next most abundant in the Ozark Border Natural Division. They were least abundant in the heart of the Osage Plains and Glaciated Plains natural divisions. Breeding Bird Survey data in Missouri reveal a significant average annual decline of 2.1 percent in Green Heron numbers through the period 1967 to 1989. The removal of riparian forests appears to be largely responsible for reduced breeding populations.

Phenology

The initial migrants appear in mid-April, but few individuals are encountered until the very end of April or early May (Robbins and Easterla 1992). Nest construction may

Abundance by Natural Division
Average Number of Birds / 100 Stops

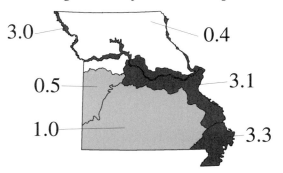

3.0   0.4

0.5   3.1

1.0   3.3

Breeding Phenology

EVIDENCE (# of Records)	MAR	APR	MAY	JUN	JUL	AUG	SEP
NB (2)			5/17	6/15			
NE (4)			5/25	6/08			
NY (7)			6/04		7/14		
FY (1)				6/27	6/27		

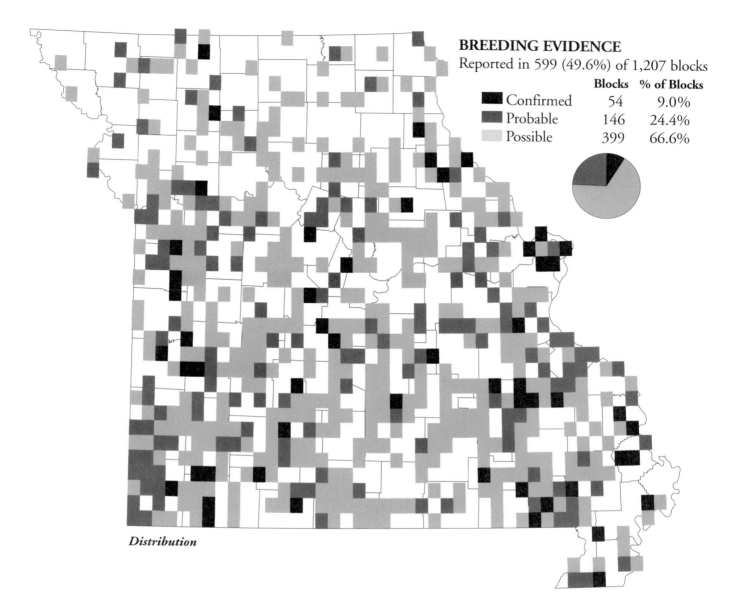

BREEDING EVIDENCE

Reported in 599 (49.6%) of 1,207 blocks

		Blocks	% of Blocks
■	Confirmed	54	9.0%
■	Probable	146	24.4%
■	Possible	399	66.6%

Distribution

begin as soon as the adults return to their territories. Evidence for the earliest onset of nesting was the observation of a bird on the nest on May 10 in Scott County in 1989. The earliest dates for nest with young and fledglings corre-spond with the dates described by Davis and Kushlan (1994). Green Herons are occasionally double-brooded in Missouri (Kaiser 1982) and late dates for nests with young and fledglings likely indicate a second brood.

Black-crowned Night-Heron
Nycticorax nycticorax

Rangewide Distribution: South central, southeastern Canada, United States to southern South America
Abundance: Fairly common but local
Breeding Habitat: Marshes, swamps, ponds, lake, lagoon & mangrove
Nest: Loose sticks, twigs & reeds in trees, shrubs & cattails
Eggs: 3–5 light bluish or greenish
Incubation: 24–26 days
Fledging: 42–49 days

Black-crowned Night-Herons are colonial breeders and gregarious throughout the year, often associating with other species of herons (Davis 1993). Breeding colonies are typically located near or over water in dense cover such as willows or herbaceous vegetation. Nests in trees are frequently at heights of less than 5 meters (Meyerriecks 1960). Most colony sites are occupied for a number of years.

Code Frequency

Rare breeders in Missouri (Robbins and Easterla 1992), Black-crowned Night-Herons were confirmed to breed in only one block during the seven-year Atlas Project. The numerous possible records shown on the map were likely not actual breeding locations for several reasons. Black-crowned Night-Herons characteristically disperse widely in late summer after breeding (Custer and Osborn 1978). They also travel up to 24 kilometers from breeding locations to feed (Hoefler 1979). Additionally, the Missouri Department of Conservation staff surveyed for breeding locations annually 1986–1990, and likely were aware of the majority of colonies outside of blocks.

Distribution

Evidence was found at too few locations to reveal any regional differences in Black-crowned Night-Heron distribution. They were found, outside of Atlas Blocks, in four mixed-heron/egret breeding colonies in Scott, Mississippi and Pemiscot counties. However, this information should not be construed to imply they are primarily residents of southern Missouri. Likewise, data are insufficient to reveal abundance details except that Black-crowned Night-Herons are rare breeders in Missouri.

Abundance

Surveys of known colonies outside of Atlas blocks indicated up to 50 individuals in two Scott County sites, and 15–100 and 55–60 in the Mississippi and Pemiscot county sites, respectively.

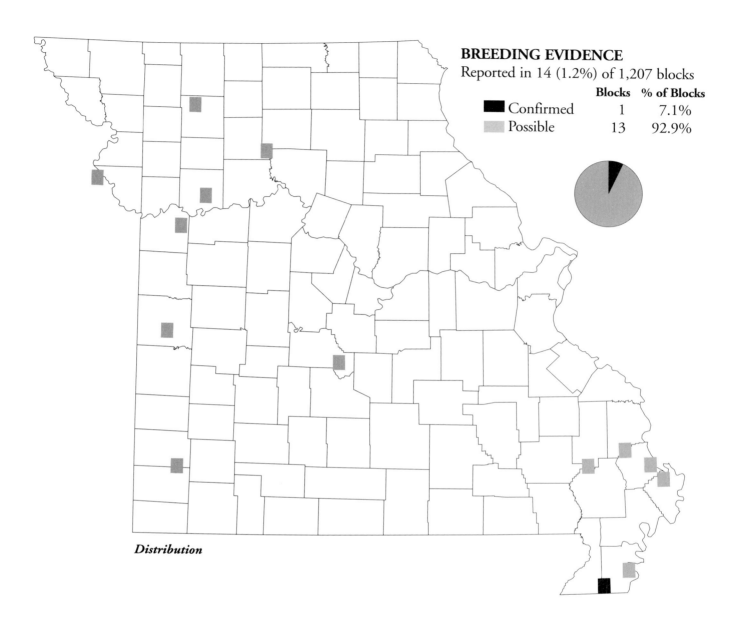

Yellow-crowned Night-Heron

Nyctanassa violaceus

Rangewide Distribution: Eastern United States, both
  coasts of Mexico
Abundance: Less common than Black-crowned
  Night-Heron
Breeding Habitat: Along streams, rivers & edges of
  wooded swamps & marshes
Nest: Twigs and sticks lined with roots & leaves in
  clumped trees or shrubs
Eggs: 4–5 light bluish-green
Incubation: 21–25 days
Fledging: 25 days

Yellow-crowned Night-Herons are mainly active during the night or the half-light of predawn and post-sunset. Their diets consist mainly of crustaceans such as crayfish, which are stalked among the riffles that connect the long pools of Ozark streams. During the day they retire back into the bushes along streams, returning again in the evening to forage in more open areas of the stream.

Code Frequency

The vast majority of records for this species indicated only possible breeding. Due to the difficulty in finding nests, Yellow-crowned Night-Herons were likely nesting in a number of the blocks in which they were seen. However, some sightings might have also resulted from post-breeding dispersal.

Distribution

Records from the Atlas Project suggest that Yellow-crowned Night-Herons are scattered but distributed statewide in appropriate habitat. However, of the 19 records for this species, 16 occurred south of the Missouri River. Fifty-three percent (10) of the records occurred in the Ozark Natural Division where the clear streams, abundant crayfish and streamside vegetation were a common mix. None were located in the northern 20 percent of the state from southern Linn County north, despite the fact that backwaters and wetlands in northern Missouri, especially in the floodplain marshes of the Big Rivers Natural Division, would be the most likely place to find this species. This secretive species is presumably much more widely distributed than indicated and specific habitat searches are needed to accurately determine breeding status.

Phenology

Records extended from May 20 to August 2, and an observation of fledglings provided the only confirmed record. Most nests were initiated in mid- to late May although the first spring arrivals may appear in early April (Robbins and Easterla 1992).

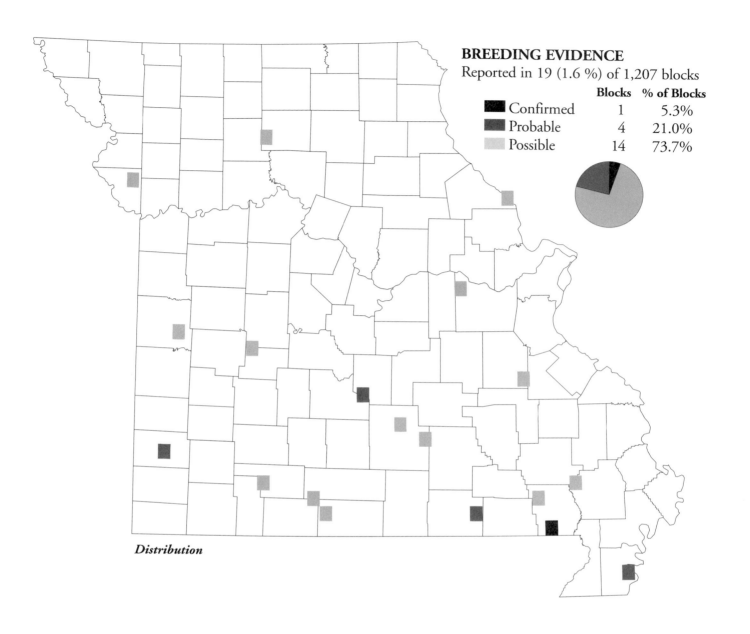

	Blocks	% of Blocks
Confirmed	1	5.3%
Probable	4	21.0%
Possible	14	73.7%

Distribution

Black Vulture
Coragyps atratus

Distribution

Black Vultures were reported in only two blocks: in Wayne and Taney counties. Reports from other than the Atlas Project were from along Missouri's southern border from Barry to Butler county. Historically, the Black Vulture was considered a regular, though not numerous, summer resident in the alluvial counties of the southeast (Widmann 1907). Obviously, Black Vultures today are an extremely rare breeding species in Missouri.

> **Rangewide Distribution:** United States southwest to East Coast; Mexico through South America
> **Abundance:** Common, with range expanding northeast
> **Breeding Habitat:** Bluffs, open woods, lowlands & swamps with debris
> **Nest:** Eggs on or in stump or on ground in dense vegetation without nest
> **Eggs:** 2 gray-green or blue-white, wreathed or marked with brown or lavender
> **Incubation:** 37–48 days
> **Fledging:** 80–94 days

Black Vultures barely range into the southern edge of Missouri where they are rare and local (Robbins and Easterla 1992). The few nests observed have been found in caves or on the forest floor at Mingo National Wildlife Refuge. Winter concentrations occur in Stone and Taney counties. A few individuals seen in summer suggest they may breed in that region as well.

Code Frequency

Both observations of possible breeding evidence were in appropriate breeding habitat, but because vultures can forage far from nest sites, these two blocks may not have contained nesting areas.

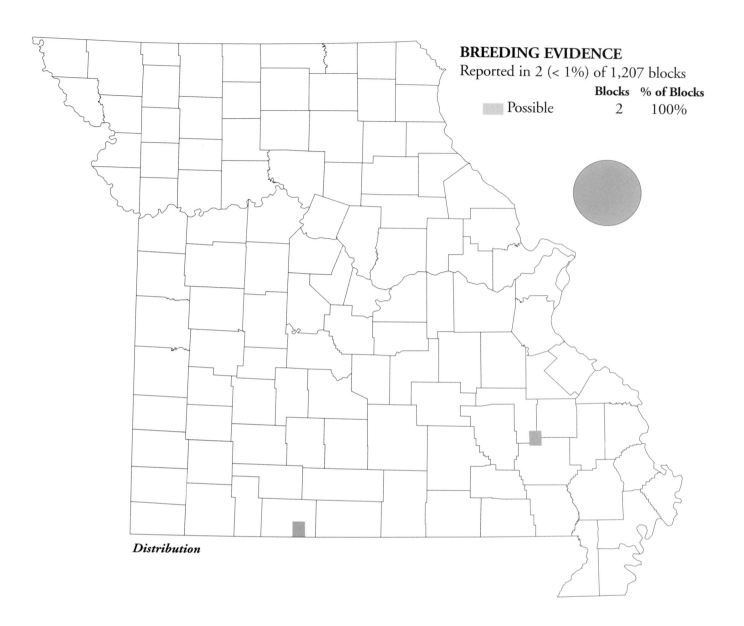

BREEDING EVIDENCE
Reported in 2 (< 1%) of 1,207 blocks

		Blocks	% of Blocks
	Possible	2	100%

Distribution

Turkey Vulture
Cathartes aura

Rangewide Distribution: Southern Canada, through-out United States, Central & South America
Abundance: Common, expanding north in the eastern United States
Breeding Habitat: Open areas in lowlands & mountains
Nest: In snags, caves & on stumps without a nest
Eggs: 2 white occasionally with brown marks
Incubation: 38–41 days
Fledging: 66–88 days

Turkey Vultures are named for their naked red heads that resemble those of Wild Turkeys. They are renowned in some areas for their clock-like return to roosts in spring. This species nests at cave entrances, in hollow logs within a forest or directly on the forest floor. Occasionally they select old barns and abandoned houses.

John James Audubon did not mention this species on his 1843 trip on the Missouri River. Widmann (1907) suggested the species had increased due to the presence of humans on the landscape.

Code Frequency

Nearly three-fourths of all records indicated possible breeding. Although many of these birds may have nested in or near the block where found, Turkey Vultures are renowned for their far-ranging daily travels. Confirmed records were scattered throughout the state, with 50 percent of them actual nests with eggs or young. Although not confirmed, Turkey Vulture presumably bred in or near many blocks.

Distribution

Turkey Vultures were one of the most widely distributed yet rarely confirmed species. They were found statewide except they were completely absent in the unforested regions of the Mississippi Lowlands. Possible records were obtained in forest islands in eastern New Madrid and southern Mississippi counties.

Abundance

No regional difference in abundance was apparent except a slight increase in the Ozark Natural Division.

Phenology

Given an incubation/fledging period that ranges from 3.5 to 4.3 months, nesting activity was probably well under-way by mid- to late April and was mostly completed by late August.

Abundance by Natural Division
Average Number of Birds / 100 Stops

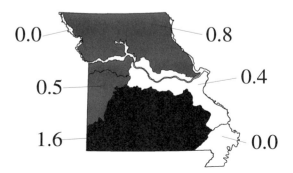

0.0   0.8

0.5   0.4

1.6   0.0

Breeding Phenology

EVIDENCE (# of Records)	MAR	APR	MAY	JUN	JUL	AUG	SEP
NB (2)				6/20	6/20		
NE (6)		4/27		6/15			
NY (15)			6/01			8/02	

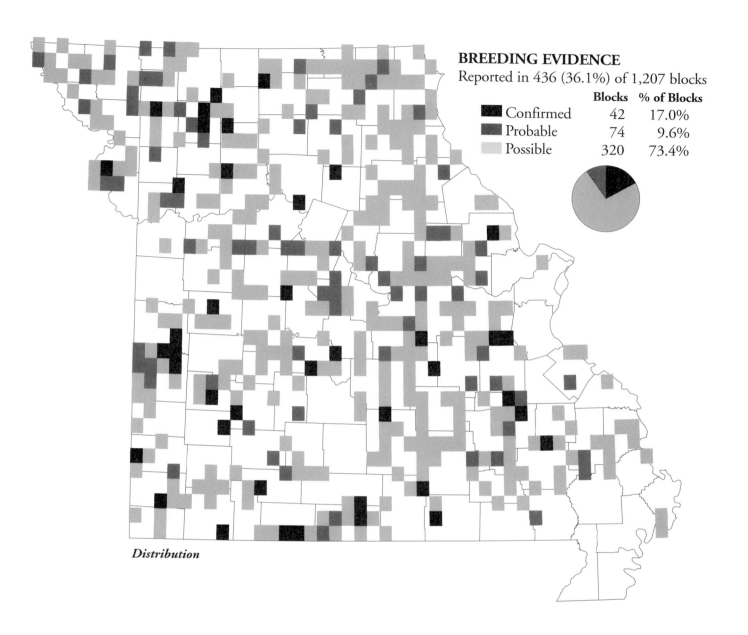

BREEDING EVIDENCE

Reported in 436 (36.1%) of 1,207 blocks

	Blocks	% of Blocks
Confirmed	42	17.0%
Probable	74	9.6%
Possible	320	73.4%

Distribution

Trumpeter Swan
Cygnus buccinator

Rangewide Distribution: South central Alaska, Idaho,
  Montana & Wyoming
Abundance: Locally common in a few breeding areas
Breeding Habitat: Usually freshwater with dense
  emergent vegetation such as inland waters & ponds
Nest: Emergent vegetation & feathers on ground
  surrounded with water
Eggs: 4–6 cream or white & nest stained
Incubation: 33–37 days
Fledging: 91–119 days

The Trumpeter Swan, the largest waterfowl species in
North America, apparently formerly bred in Missouri
marshes. According to Widmann (1907), they nested in
northeast Missouri and perhaps along the Missouri River in
northwestern Missouri. They have been absent from the
state as breeders and migrants throughout most of the 20th
Century.

From 1984 through 1986, family groups of swans were
transported by the Conservation Department from Lacreek
National Wildlife Refuge, South Dakota to Mingo National
Wildlife Refuge near Puxico, Missouri. This was an attempt
to establish a migratory pattern among Lacreek's wintering
birds. Of the 16 birds transported, only one bird was known
to return to Lacreek and it never returned to Mingo. Most
swans either dispersed from the refuge or died after a few
months. However, the adult pair of the family that was
transported in 1984 did remain at Mingo and they survived
for a number of years.

Code Frequency

Two goslings were observed at Mingo National Wildlife
Refuge on June 29, 1986, offspring of the original pair of
adult swans brought to the refuge in 1984. This was the
only documented breeding of Trumpeter Swans during the
Atlas Project.

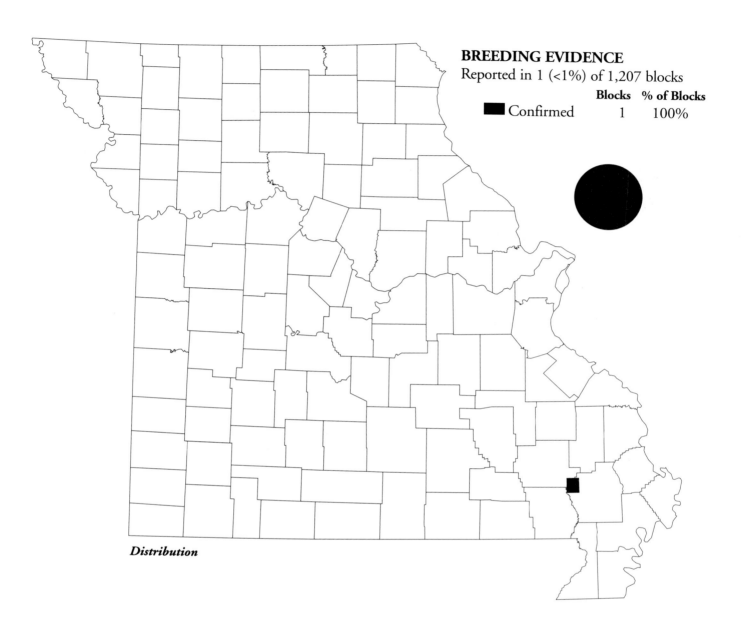

BREEDING EVIDENCE
Reported in 1 (<1%) of 1,207 blocks

	Blocks	% of Blocks
■ Confirmed	1	100%

Distribution

Canada Goose
Branta canadensis

Rangewide Distribution: Canada, Alaska & northern & central United States
Abundance: Widespread & abundant, expanding south
Breeding Habitat: Marsh, meadows, & small islands in ponds or reservoirs
Nest: Grass, forbs, moss, sticks, aquatic vegetation & feathers on ground or in nest box
Eggs: 4–7 white & nest-stained
Incubation: 25–30 days
Fledging: 40–73 days

The Giant Canada Goose, *Branta canadensis maxima*, is the only subspecies that nests in Missouri. It historically nested in the Midwest on elevated sites near water such as muskrat houses, islands, abandoned raptor nests and bluff faces. Giant Canada Geese discovered nesting along the lower Missouri River in the 1960s are believed to be a remnant native population, according to a telephone conversation with Conservation Department Waterfowl Research Biologist David Graber. Individuals descended from this remnant population likely contributed to Missouri's current population.

Code Frequency

Due to their size and conspicuous behavior, Canada Geese are perhaps the easiest of birds to find and identify. Therefore, the map likely provides an accurate representation of their distribution. They are also one of the easiest species to confirm, most often by observing goslings swimming behind or feeding with their parents. Therefore, where Canada Geese were not confirmed by the Atlas Project, they likely did not breed.

Distribution

The distribution of breeding confirmations indicated two primary concentrations of geese. One extended from central Missouri to the St. Louis area. This portion of the Missouri River was the historical breeding range for the species (McKinley 1961). A second breeding zone in the state's western counties was centered in the Kansas City area, and extended from Arkansas to Iowa. Scattered locations elsewhere indicate that the range is essentially statewide with the exception of the heavily forested regions of the Lower Ozark Natural Section and the highly agricultural Mississippi Lowlands.

Abundance

Because of the clumped nature of this species' distribution, valid abundance data were not obtained from the Atlas Project. From other sources, it is evident that Canada Geese have been increasing in number and range throughout much of the eastern United States and Missouri. The state's numerous farm ponds and lakes, and the provision of goose nesting tubs, have contributed to a thriving population. In some cities where geese have abundant food and are protected, flocks have become so large as to become nuisances.

Breeding Phenology

EVIDENCE (# of Records)	MAR	APR	MAY	JUN	JUL	AUG	SEP
NE (8)	3/15				7/04		
NY (13)	4/01						8/25

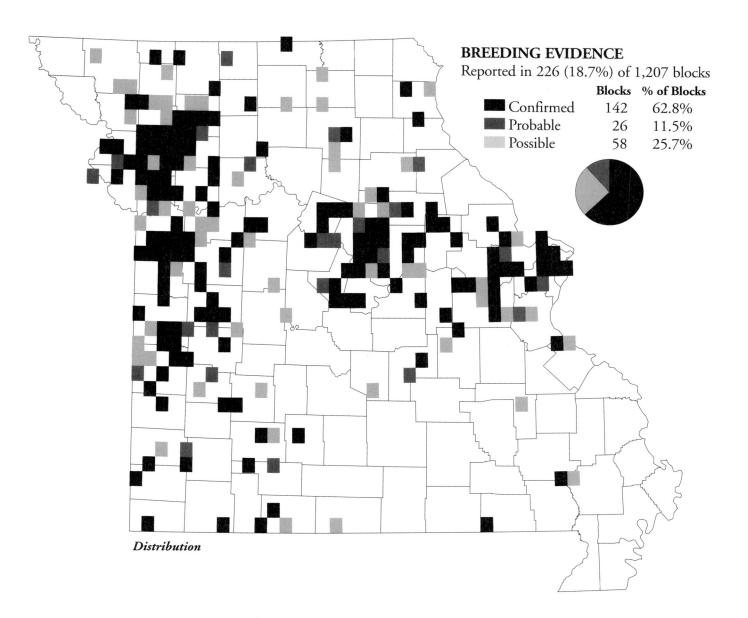

BREEDING EVIDENCE

Reported in 226 (18.7%) of 1,207 blocks

		Blocks	% of Blocks
■	Confirmed	142	62.8%
■	Probable	26	11.5%
▨	Possible	58	25.7%

Distribution

Phenology

Data indicating the onset of breeding were available from 127 blocks. Observation dates are consistent with nesting dates recorded during Canada Geese breeding surveys on conservation areas, according to comments by David Graber. Canada Geese are single-brooded but will re-nest if disturbed early in the nesting cycle. An adult on a nest on June 15 was presumably re-nesting.

Wood Duck
Aix sponsa

Rangewide Distribution: Southern Canada, north-
  western & eastern United States
Abundance: Fairly common
Breeding Habitat: Wooded swamps, sloughs, ponds,
  marshes & wet woods
Nest: Wood chips & down-lined tree cavity or nest
  box
Eggs: 10–15 creamy white
Incubation: 28–37 days
Fledging: 56–70 days

Wood Ducks frequent wooded, slow-moving streams and marshes where they forage for aquatic insects and seek potential nest cavity for nest sites among trees. For egg laying, they select large tree cavities within a few miles of wetlands. Although most nest cavities are near or over water, some are found up to two kilometers from water (Hepp and Bellrose 1995). Most suitable wetland areas are wooded or have shrubs and robust emergent plants. Once dependant on natural cavities, Wood Ducks have accepted artificial nest boxes. Although now a common species throughout Missouri, experts at the turn of the 20th century predicted their demise (Widmann 1907).

Code Frequency

Wood Ducks are more difficult to see than most ducks. The difficulty of locating natural cavities explains the low frequency of nest site observations. Locating ducklings following a female was much easier. Of the 200 confirmed records, 88 percent were observations of ducklings, leaving only 12 percent for all other codes. Of the 460 total records, 50 percent (237) were sightings of individuals or pairs.

Distribution

Wood Ducks were sighted in every county. The Osage Plains and Mississippi Lowlands natural divisions apparently lack the abundance of nesting cavities found in the other natural divisions. With the exceptions of a few areas in the Ozark Natural Division, and much of the Glaciated and Osage plains, their distribution was scattered evenly across the state.

Abundance

Roadside counts did not provide a meaningful assessment of Wood Duck abundance.

Phenology

Wood Ducks begin arriving in late February and are common after early April (Robbins and Easterla 1992). Most fledglings were observed in June (106) but 32 were logged in May, 34 in July, and 2 in August. This is the only regularly double-brooded duck in North America (Hepp and Bellrose 1995).

Notes

Given continued protection from illegal harvest, the use of nest boxes, the expansion of beaver populations and concurrent creation of Wood Duck preferred habitat, this species will probably continue to do well (Hepp and Bellrose 1995).

Abundance by Natural Division
Average Number of Birds / 100 Stops

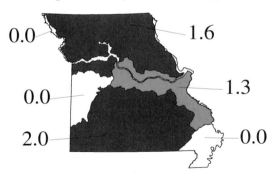

0.0   1.6

0.0   1.3

2.0   0.0

Breeding Phenology

EVIDENCE (# of Records)	MAR	APR	MAY	JUN	JUL	AUG	SEP
NE (2)				6/03	6/22		
NY (6)			5/30	6/12			

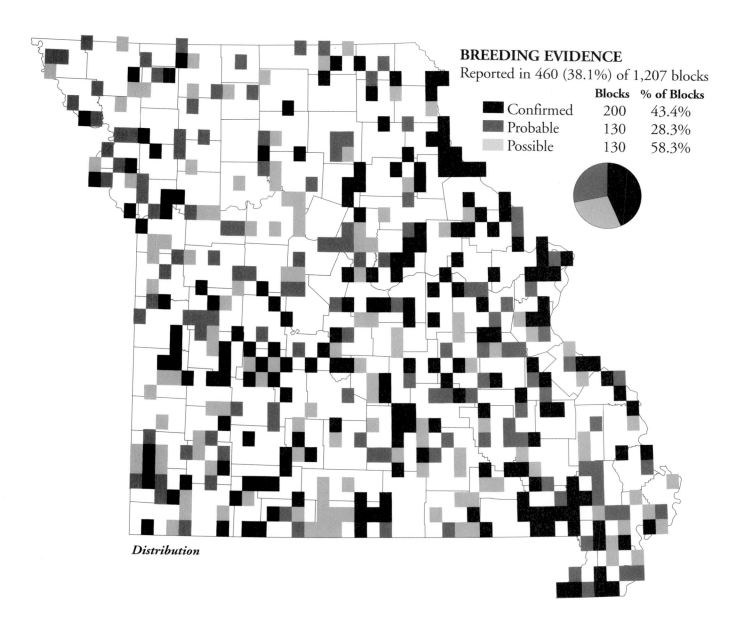

BREEDING EVIDENCE
Reported in 460 (38.1%) of 1,207 blocks

	Blocks	% of Blocks
Confirmed	200	43.4%
Probable	130	28.3%
Possible	130	58.3%

Distribution

Mallard

Anas platyrhynchos

> **Rangewide Distribution:** Canada, United States & Mexico, rare in Central America
> **Abundance:** Abundant & widespread
> **Breeding Habitat:** Grassy areas, hay fields, cattails & bulrush
> **Nest:** Cattails, reeds & grass in hollow logs near water or on ground
> **Eggs:** 7–10 greenish-buff, grayish-buff or whitish
> **Incubation:** 28 days
> **Fledging:** 42–60 days

Metallic green head feathers help identify the male Mallard. Mallards can be seen on most lakes and marshes in Missouri and they are abundant migrants spending the winter on available open water. During Widmann's time (1907), the hunting season ran until May 1, and he expected the breeding population would be extirpated.

Code Frequency

As with most waterfowl, a hen with ducklings was frequently the only clue to breeding. Thirty-seven percent of Mallard records were confirmed, 79 percent based on the observation of broods. Twenty-eight percent were possible sightings, and it is likely nesting was attempted in most blocks where this species was observed. It is assumed that Atlasers did not separate domestic from wild birds. Consequently, the distribution map may not accurately represent the breeding status of wild Mallards in Missouri

Distribution

Mallards were essentially distributed statewide although several counties in north central Glaciated Plains and southern central Ozark natural divisions lacked records. Records appear to be concentrated around urban areas and large lakes where small islands afford protection from nest predators. Many observations may have recorded domestic pairs nesting by farm ponds, in city parks or in yards along the shores of big lakes.

Abundance

Mallards are common breeders around large lakes and uncommon breeders in other wetlands statewide. Feeding by property owners around lakes has probably expanded the population.

Phenology

Nests with eggs and fledglings were recorded over four months, a long period for a single-brooded species. These records likely include second nesting attempts after a first one failed.

Breeding Phenology

EVIDENCE (# of Records)	MAR	APR	MAY	JUN	JUL	AUG	SEP
NE (6)		4/14					8/24
NY (3)			6/05		6/19		

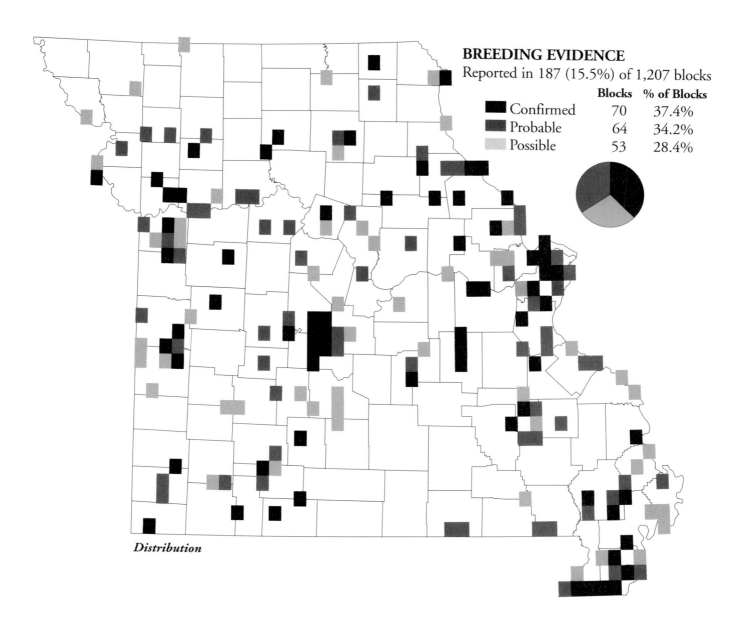

BREEDING EVIDENCE

Reported in 187 (15.5%) of 1,207 blocks

	Blocks	% of Blocks
Confirmed	70	37.4%
Probable	64	34.2%
Possible	53	28.4%

Distribution

Northern Pintail
Anas acuta

Rangewide Distribution: Greenland, Canada & most of North American & Europe
Abundance: Abundant & widespread, more common in western United States
Breeding Habitat: Grassland, fields, tundra, marsh & pond
Nest: Dry grass & leaves lined with feathers & hair, on the ground concealed in grass
Eggs: 6–9 olive-green & olive-buff
Incubation: 22–25 days
Fledging: 36–57 days

Northern Pintails are common migrants, returning when the ice melts in spring. They mainly forage in marshes on plant material but also ingest insects and other water-dwelling animals. Most individuals are absent from the state from May through late August (Robbins and Easterla 1992), although when the prairie pothole breeding grounds are drought stricken, Pintails may attempt to nest in Missouri. A few pairs may breed in northwestern Missouri on an irregular basis (Robbins and Easterla 1992). The normal breeding range is well north of Missouri. Iowa lists nine northern counties with nesting records and one in a west central county (Dinsmore et al. 1984).

Code Frequency

The Northern Pintail was recorded only once during the Atlas Project. Four flightless immatures followed an adult female around a one-hectare farm pond near Smithville, Clay County. The pond was adjacent to 45 hectares of ungrazed pasture, a likely nesting location.

Distribution

The single record provided no hint of this species' potential Missouri breeding range. However, three historical records were all from counties bordering the Missouri River north of Kansas City (Robbins and Easterla 1992). All historical and Atlas Project breeding records are in the Western Glaciated Plains Natural Section.

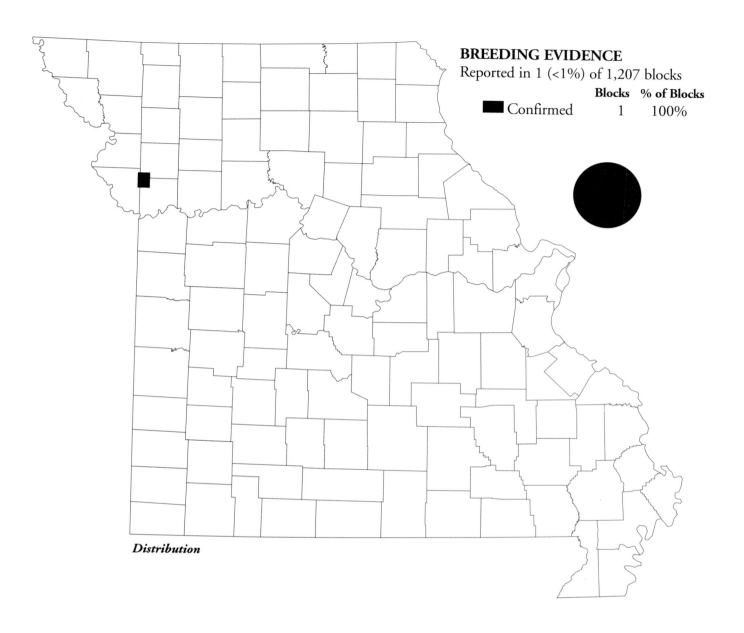

Reported in 1 (<1%) of 1,207 blocks

	Blocks	% of Blocks
■ Confirmed	1	100%

Distribution

Blue-winged Teal
Anas discors

Rangewide Distribution: Northwestern & southern
  Canada, southeastern Alaska, United States except
  coastal areas
Abundance: Locally common & widespread
Breeding Habitat: Prairie potholes, marsh, pond,
  stream & lake
Nest: Grass & cattail, lined with fine material, hidden
  on ground
Eggs: 8–11 creamy-white or olive-white
Incubation: 24 days
Fledging: 35–44 days

The Blue-winged Teal is one of four duck species that
breed regularly in Missouri. Typical breeding habitats are
grassy shorelines of lakes, ponds or other waterways that
possess emergent vegetation to provide cover for broods after
hatching. They typically nest in grassy meadows, and with
nests placed an average of 38 meters from water (Bellrose
1976).

Code Frequency

Blue-winged Teal are relatively easy to detect and identify
when on open water, but their well-concealed nests are difficult to detect. Therefore, the possible breeding locations
shown on the map may have involved breeding individuals.
Interestingly, the probable records likely were not breeding
locations because eight of the nine were observations of
pairs. Breeding Blue-winged Teal females separate from
males in late spring and summer to stay close to nests or
broods. In fact, because females tend to remain hidden,
males seen alone late in the season may be a better indicator
of breeding than pairs.

Distribution

Blue-winged Teal appeared to have a sparse distribution
in several areas. None were recorded in the southeastern
Ozark highlands and only one block recorded them in the
Osage Plains.

Phenology

Confirmed breeding evidence was rare, so Atlas Project
data reveal little of the Blue-winged Teal's breeding
phenology.

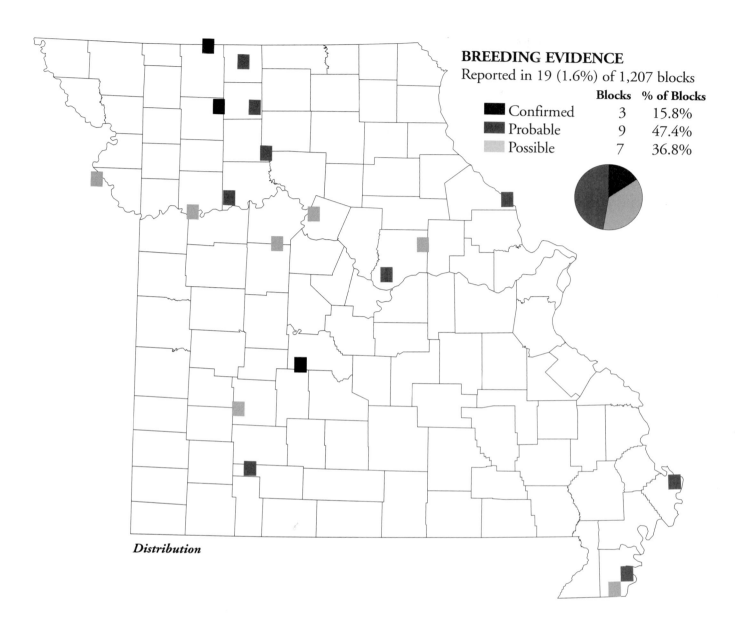

BREEDING EVIDENCE
Reported in 19 (1.6%) of 1,207 blocks

	Blocks	% of Blocks
Confirmed	3	15.8%
Probable	9	47.4%
Possible	7	36.8%

Distribution

Northern Shoveler
Anas clypeata

> **Rangewide Distribution:** Western Canada, western United States & Eurasia
> **Abundance:** Common in western & increasing in eastern United States
> **Breeding Habitat:** Shallow fresh water, muddy meadow or grassland with emergent vegetation
> **Nest:** Depression of grass & vegetation in short grass near boggy water
> **Eggs:** 9–12 olive-buff to greenish-gray
> **Incubation:** 22–25 days
> **Fledging:** 38–66 days

Northern Shovelers are at the southern periphery of their breeding range in Missouri and breed irregularly in the state in very small numbers. When drought prevails in the Northern Midwest prairie pothole region, where this species normally nests, these dabblers probably use available wetland farther south. Also, a few Northern Shovelers linger into the summer in most years, with 30 individuals observed at Swan Lake on June 1, 1987 (Robbins and Easterla 1992). During migration they are abundant in suitable wetland habitat.

Code Frequency

In addition to a single bird and a pair recorded within blocks, two confirmed records independent of the Atlas Project indicate Northern Shovelers might breed irregularly to annually in extremely small numbers.

Distribution

Atlas records were obtained in Squaw Creek and Mark Twain national wildlife refuges. Confirmed breeding records outside blocks were from Holt County (Robbins and Easterla 1992).

Phenology

Most migrants are absent from the state from June through August and most fledglings have been seen in May (Robbins and Easterla 1992). Atlas records were obtained May 12 and July 27, within the range of dates when breeding is expected to occur.

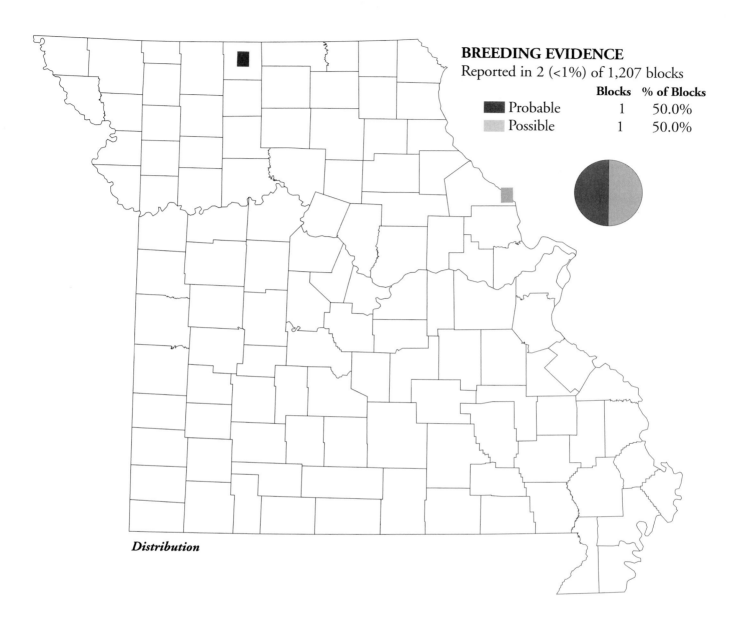

Reported in 2 (<1%) of 1,207 blocks

	Blocks	% of Blocks
Probable	1	50.0%
Possible	1	50.0%

Distribution

Hooded Merganser
Lophodytes cucullatus

Rangewide Distribution: Southwestern & southeastern Canada, northwestern & eastern United States

Abundance: Uncommon, occurs exclusively in North America

Breeding Habitat: Woods near water, lakes, swamps & marshes

Nest: Tree cavity near water lined with grass, leaves & down

Eggs: 10–12 white, often nest-stained

Incubation: 32–33 days

Fledging: 71 days

The Hooded Merganser is one of four duck species that regularly breed in Missouri. Like Wood Ducks, they nest in tree cavities near sloughs, ponds and streams.

Code Frequency

Hooded Mergansers are relatively easy to see and identify when with ducklings on open water. Their infrequent detection during the Atlas Project therefore suggests they are indeed a rare nesting species. Females with broods of ducklings accounted for most breeding confirmations. Because of difficulty in locating nest cavities, the four blocks in which Hooded Mergansers were recorded as possible breeders may have actually been nesting localities.

Distribution

Although almost half the blocks in which Hooded Mergansers were recorded were in Pike and Ralls counties, the remaining blocks were scattered widely enough to indicate that these birds may have a statewide distribution. Hooded Mergansers can breed wherever suitable cavity trees or artificial nest boxes are near appropriate brood-rearing wetlands (Duggar et al. 1994). Such nesting localities discovered during the Atlas Project included the Ted Shanks and Fountain Grove conservation areas and the Clarence Cannon National Wildlife Refuge.

Phenology

Considering the early date that fledglings were observed, the breeding season likely began before April 1. Because migration through Missouri continues during early April (Robbins and Easterla 1992), resident breeders can be expected on nesting grounds well before the last migrants have departed.

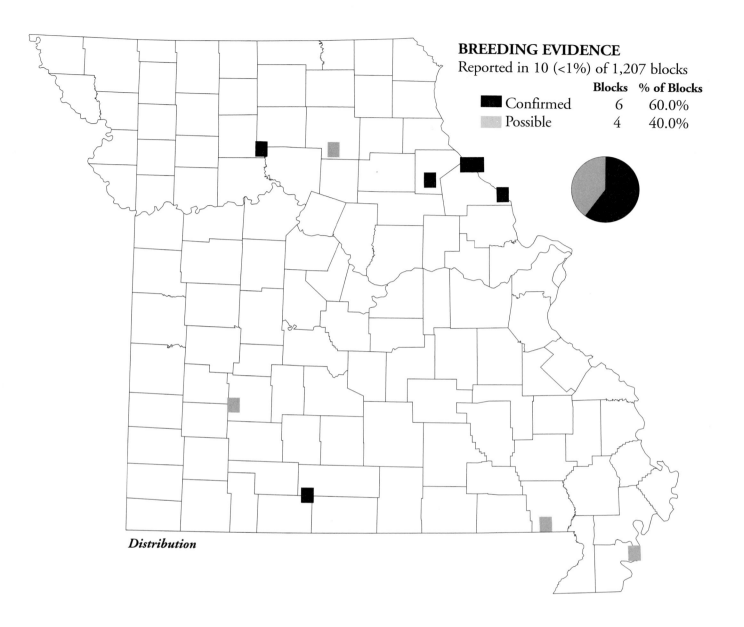

Reported in 10 (<1%) of 1,207 blocks

	Blocks	% of Blocks
Confirmed	6	60.0%
Possible	4	40.0%

Distribution

Mississippi Kite
Ictinia mississippiensis

> **Rangewide Distribution:** Southeastern United States
> **Abundance:** Common in southern Great Plains, less common southeast
> **Breeding Habitat:** Forested waterways & open rangeland
> **Nest:** Flat construction of twigs & sticks lined with leaves on upper branches
> **Eggs:** 1–2 white or bluish-white, unmarked or faintly spotted
> **Incubation:** 32–32 days
> **Fledging:** 34 days

Perhaps Missouri's most graceful bird of prey, the Mississippi Kite glides, swoops and turns in the aerial search for insects. Never stopping to hover like its cousin the White-tailed Kite, it continually pursues insects. Throughout much of their range, Mississippi Kites nest in the canopies of large riparian forests.

Code Frequency

The few nests observed in Missouri have typically been in the tops of tall trees in dense forests. Therefore, although the species is conspicuous and easily identified, locating nests is difficult. Mississippi Kites presumably nested in some of the blocks where sighted.

Distribution

The majority of sightings occurred along the Mississippi River south of Cape Girardeau and in the St. Louis area, with most nests near rivers. Results from the Atlas Project echo the historical range described by Widmann (1907). Scattered sightings in southwestern Missouri also followed the pattern Widmann described. In years following the Atlas Project, nests were located in the Joplin and Kansas City areas.

Abundance

Most records occurred along the Mississippi River and in public wetland areas in the Mississippi Lowlands. A few nest sites probably occurred farther north along the Mississippi River to Pike County where both adults and young have been seen. A report of 300 birds over the Mississippi River in New Madrid and Mississippi counties on May 16, 1991 (Wilson, 1991a) illustrates the gregarious nature of this species, which seems to form loose nesting colonies.

Phenology

Information from the Atlas Project was insufficient to define the phenology of this species. According to Robbins and Easterla (1992) migrants arrive by late April to early May and depart by September.

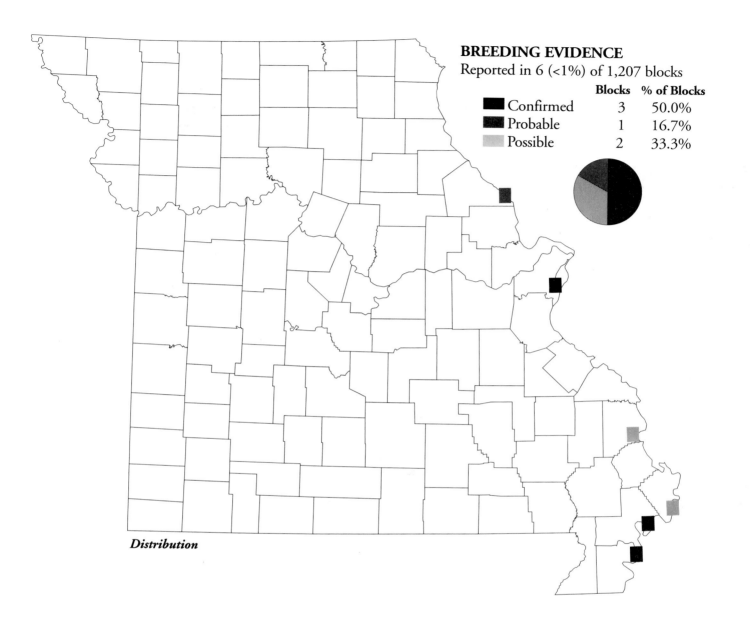

Bald Eagle
Haliaeetus leucocephalus

Rangewide Distribution: Canada, Alaska & United States

Abundance: Common in Alaska, uncommon elsewhere

Breeding Habitat: Cliffs with large trees near water

Nest: Large sticks & vegetation, deeply lined with fine materials, in trees on cliffs

Eggs: 2 bluish-white, often nest-stained

Incubation: 34–36 days

Fledging: 70–98 days

Although regular and locally abundant in Missouri in winter, Bald Eagles are rare breeders in the state. Griffin's (1978) summary of the history of Bald Eagle breeding in Missouri suggests that prior to 1900, Bald Eagles bred throughout much of the state and were especially numerous in southeastern Missouri swamps. Throughout the early 20th century, the few nests reported were primarily along the Mississippi River and in the Ozarks. A failed nest in Camden County in 1962 (Easterla 1964) was the last known nesting attempt until 1982 (Wilson 1985).

From 1981-90, Bald Eagle restoration projects were conducted at Mingo National Wildlife Refuge in Wayne County and at the Schell-Osage Conservation Area in Vernon County. Seventy-four young Bald Eagles were released and no doubt supplemented a rebounding breeding population. Beginning in 1985, Bald Eagles resumed breeding in the state regularly and since then the number of productive territories has generally increased (Wilson 1995).

Code Frequency

This conspicuous and readily-identified species is relatively easy to confirm as breeding by its large, prominently located nest. The main factor limiting the number of reports was the Atlas Project sampling procedure. Because Bald Eagles were listed as Endangered by Missouri and the United States Fish and Wildlife Service, independent nesting territory surveys were undertaken and several nests were found outside of Atlas blocks.

Distribution

Breeding Bald Eagles displayed no definitive regional distribution within the state. Most nests were near major rivers, lakes and wetlands as expected for a fish-eating species.

Abundance

Bald Eagles remain extremely rare as breeders in Missouri. Fortunately, the number of productive nests has increased including those found outside Atlas blocks. From 1986 to 1992, the number of nests increased from two to 10.

Phenology

Observations made during 1991–1994 independent of the Atlas Project determined the fledging of Bald Eagles in Missouri ranged between May 15–July 15 with a median of June 16. Considering the length of the Bald Eagle's incubation and rearing periods, this would require nesting to be initiated as early as mid-January.

Breeding Phenology

EVIDENCE (# of Records)	MAR	APR	MAY	JUN	JUL	AUG	SEP
NB (2)	3/21	3/21					
NY (1)		4/20	4/20				
FY (3)		4/10	5/06				

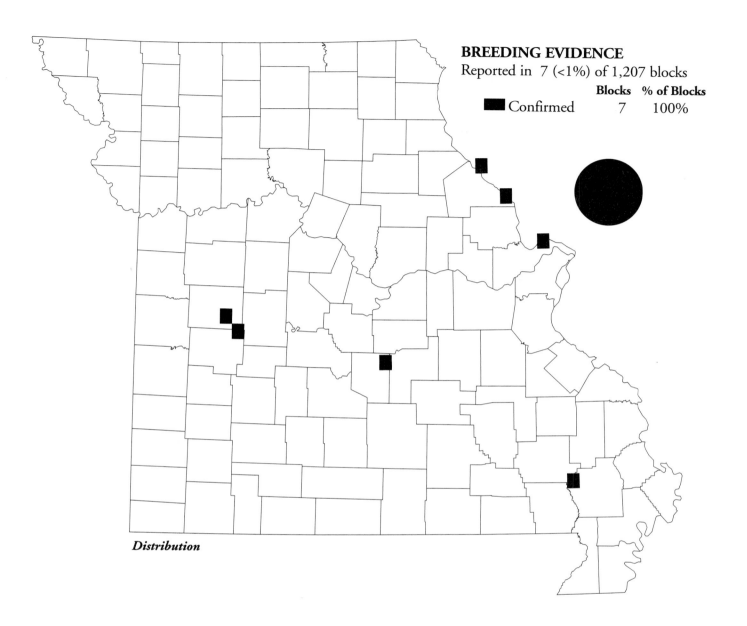

Northern Harrier
Circus cyaneus

> **Rangewide Distribution:** Most of Canada, entire United States
> **Abundance:** Fairly common
> **Breeding Habitat:** Open areas, ravines, native grasses & meadows
> **Nest:** Sticks & grass on elevated ground in thick vegetation
> **Eggs:** 5 bluish-white, usually unmarked, 10 percent spotted with brown
> **Incubation:** 31–32 days
> **Fledging:** 30–35 days

Although a common migrant in open grassland habitat, the Northern Harrier is listed in Missouri as an Endangered breeding species. This species forages and nests in both wet and upland grasslands with some woody vegetation intermixed (Palmer 1988). Widmann (1907), encouraged farmers "to give, at least on his own grounds, the fullest protection to a benefactor that removes the pest which eats his grain and girdles his fruit trees." During the mid-1900s, Northern Harrier populations suffered great losses due to pesticide-related egg shell thinning and losses of wetland nesting habitat (Terres 1987).

Code Frequency

Wandering, non-breeding Northern Harriers occasionally appear throughout the summer months (Robbins and Easterla 1992). Therefore, a sight observation of a bird does not necessarily indicate breeding. Observation of harriers usually led to a more intensive search to confirm breeding. For example, when an adult male and an adult female were present in the same vicinity, some Atlasers intensified nest searches in an attempt to confirm breeding. Confirmed breeding reports occurred in 1989, 1990 and 1992 within Atlas blocks.

Distribution

Observations were made statewide, however, most confirmations were in the Osage Plains Natural Division and the Western Glaciated Plains Natural Section. Several public prairies in the Osage Plains Natural Division hosted pairs of Northern Harriers. Most Atlasers reported finding this species in grasslands, both native prairie and non-native pastures, and hayfields.

Phenology

Northern Harriers are common migrants in Missouri from February to May, and again from August to November (Robbins and Easterla 1992). They nest fairly late in the season. As a ground-nesting species, they require new growth to conceal their nest location, which may explain the late nesting season.

Notes

During the Atlas Project, land enrolled in the Conservation Reserve Program provided additional nesting and foraging habitat for Northern Harriers.

Breeding Phenology

EVIDENCE (# of Records)	MAR	APR	MAY	JUN	JUL	AUG	SEP
NY (2)				6/16	7/04		
FY (2)				6/19	6/21		

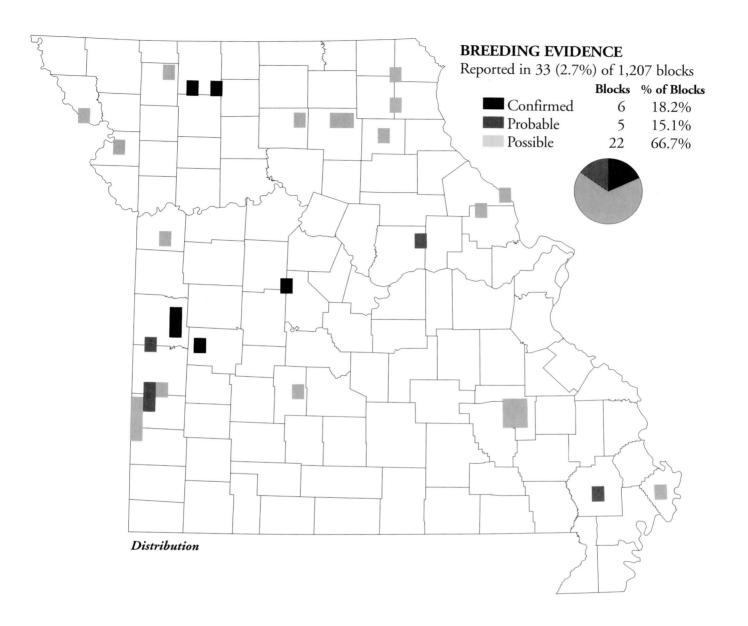

BREEDING EVIDENCE
Reported in 33 (2.7%) of 1,207 blocks

	Blocks	% of Blocks
Confirmed	6	18.2%
Probable	5	15.1%
Possible	22	66.7%

Distribution

Sharp-shinned Hawk
Accipiter striatus

Rangewide Distribution: Western & central Alaska, southern Canada, northwestern United States from Great Lakes to East Coast, resident population in Mexico through South America

Abundance: Fairly common

Breeding Habitat: Large mountainous forests, especially Shortleaf Pine

Nest: Broad, flat construction of sticks & twigs, lined with fine material, in trees near trunk

Eggs: 4–5 white or bluish-white wreathed with brown marks

Incubation: 32–35 days

Fledging: 24–27 days

The smaller of the Missouri accipiters, the Sharp-shinned Hawk has a more buoyant flight than the Cooper's Hawk. In Missouri (Kritz 1989) and throughout much of their range, Sharp-shinned Hawks select dense even-aged pine stands for nesting (Reynolds et al. 1982). All but two of the 17 nests that Kritz (1989) found were in Short-leaf Pine (*Pinus echinata*) plantations. Nests were often situated in a whorl of branches near the top of a tree.

Code Frequency

Sharp-shinned Hawks are secretive and harder to find during the breeding season than during migratory periods because they select dense forest cover for nesting. Therefore, they were no doubt more plentiful and more widely distributed than Atlas Project data indicate. On the other hand, it is conceivable that some of the possible records resulted from late-departing migrants. This species is often confused with the similar Cooper's Hawk.

Distribution

Sharp-shinned Hawks were mainly found in the Osage Plains, Ozark, and Ozark Border natural divisions with two scattered records from the Glaciated Plains. The Atlas Project indicated they may be more widespread than previously thought, with records from 29 counties in the Ozark, Ozark Border and Osage Plains natural divisions. Kritz (1989) located 17 nests in 1985–1986 in the eastern Ozark and Ozark Border natural divisions, finding them in 11 of the most continuously forested counties in the Ozarks. However, he searched only areas in which he believed he had the best opportunity of finding breeders.

Abundance

While detected on Miniroutes in both the Osage Plains and Ozark natural divisions we know very little about the abundance of the species. Based on Kritz's (1989) work and the Atlas Project findings, there are likely fewer that 50 nests statewide.

Phenology

In Kritz's 1989 study, egg laying occurred between May and July, hatching between June and August and young fledged between July and September.

Breeding Phenology

EVIDENCE (# of Records)	MAR	APR	MAY	JUN	JUL	AUG	SEP
FY (4)				6/12		7/30	

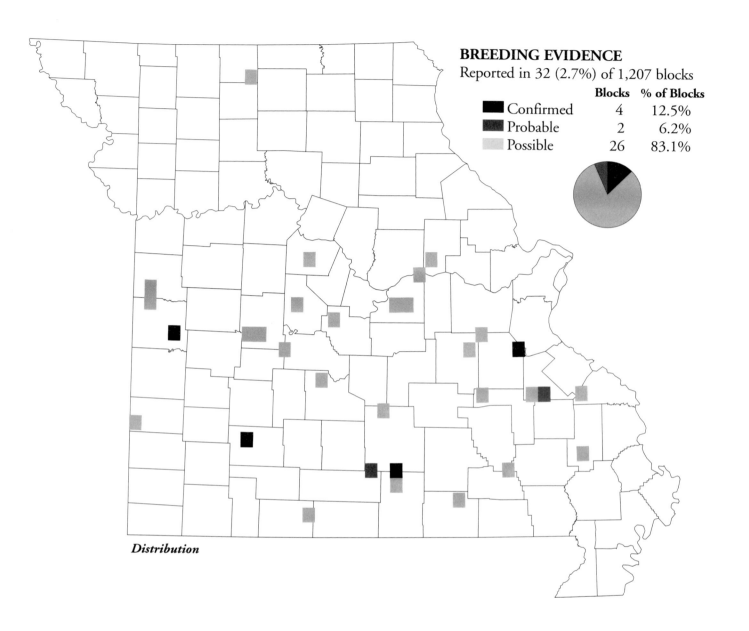

BREEDING EVIDENCE

Reported in 32 (2.7%) of 1,207 blocks

	Blocks	% of Blocks
Confirmed	4	12.5%
Probable	2	6.2%
Possible	26	83.1%

Distribution

Cooper's Hawk
Accipiter cooperii

Rangewide Distribution: Southern Canada, entire United States, northern Mexico
Abundance: Uncommon & may be declining
Breeding Habitat: Mature deciduous & pine forests & riparian zones
Nest: Flat & deep construction of sticks & twigs, lined with chips & bark, in trees
Eggs: 4–5 blue or green-white, usually nest stained with brown spots
Incubation: 32–36 days
Fledging: 27–34 days

In Missouri, Cooper's Hawks nest in mature, even-aged forests with moderate canopy closure, frequently consisting of Short-leaf Pine *(Pinus echinata)* (Kritz 1989). According to Kritz, they tolerate human disturbance and habitat fragmentation and often nest near woodland edges, in small plantations and near human habitation. Compared with Sharp-shinned Hawks, Cooper's Hawks select dense forest habitats with more vertical stratification and open canopies for nesting (Kritz 1989). Apparently this species was once a widespread and common breeder in Missouri. Widmann (1907) described them as "a fairly common summer resident in all parts of the state."

Code Frequency

Cooper's Hawks are difficult to locate because of their secrecy at the nest site and their forested breeding habitat. In most blocks they were recorded as possible breeders and, although some of these records may have resulted from transient individuals, Cooper's Hawks may well have bred in many of these blocks.

Distribution

Cooper's Hawks found in the southern half the state were usually associated with oak-pine forests. North of the native pine range, the Cooper's Hawks found were associated with deciduous forests and, on two occasions, pine plantations. A grouping of reports occurred in Harrison, Grundy and Mercer counties with breeding confirmed in the latter county. Other observations in northern Missouri, in which breeding was probable or confirmed, indicated Cooper's Hawks can nest anywhere in the state where an appropriate habitat occurs. The only record from the Mississippi Lowlands was at Donaldson Point Conservation Area in New Madrid County in a floodplain forest between the levee and the Mississippi River.

Phenology

The few breeding confirmations recorded during the Atlas Project provide a conservative picture of Cooper's Hawk nesting phenology. An adult observed on the nest on May 24 and a nest with young on June 2 correspond closely with the nesting events observed by Kritz (1989), who first observed eggs in the nest on May 15 in 1985 and May 24 in 1986.

Breeding Phenology

EVIDENCE (# of Records)	MAR	APR	MAY	JUN	JUL	AUG	SEP
NB (1)				6/07	6/07		
NY (6)			6/02		6/26		
FY (7)			6/04		7/14		

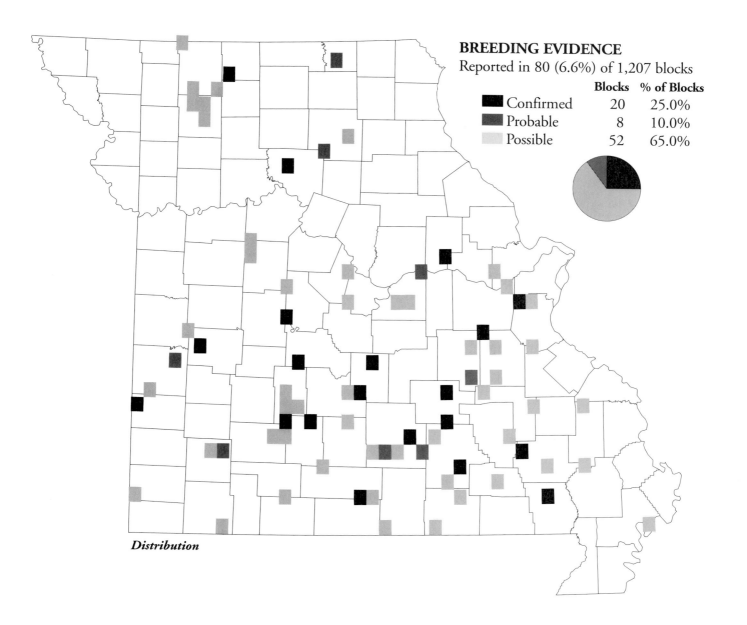

BREEDING EVIDENCE
Reported in 80 (6.6%) of 1,207 blocks

	Blocks	% of Blocks
■ Confirmed	20	25.0%
■ Probable	8	10.0%
■ Possible	52	65.0%

Distribution

Red-shouldered Hawk
Buteo lineatus

Rangewide Distribution: California coast, eastern United States
Abundance: Fairly common
Breeding Habitat: Moist forests & mixed woods near streams
Nest: Sticks, twigs & bark lined with leaves & moss in tree near trunk
Eggs: 3 white or bluish-white with brown marks, often nest stained
Incubation: 28 days
Fledging: 39–45 days

When seen at close range, Red-shouldered Hawks are one of Missouri's most beautiful raptors. Their rusty shoulders, surrounded by black-and-white checkered back and wings, cast a spectacular pattern in the sunlight as they fly among the trees. At certain times of year, they forage along streams and backwaters for crayfish and also eat birds, mammals, reptiles and amphibians (Crocoll 1994). Midmorning is usually the best time to find them as they soar into the sky over their territories. Like the Red-tailed Hawk, this species hunts from a perch. It is frequently seen by those canoeing down a quiet Ozark stream. More than any other factor, loss of bottomland hardwood forests has contributed to the decline in numbers for this species (Hands et al. 1989).

Code Frequency

Winter and spring nest searches might have resulted in an increase in the number of confirmed reports. Despite inadequate nest detection, Red-shouldered Hawks likely nested in some or most of the blocks where possible observations were recorded.

Distribution

Most records for this species were from the Ozark and Ozark Border natural divisions. Reports were scattered statewide except the Western Glaciated Plains and most of the Grand River natural sections. Rivers and swamps with extensive bottomland hardwood forests supported isolated populations of this species. Open lands and woodland plains seemed to support fewer Red-shouldered Hawks, but more Red-tailed Hawks and Great Horned Owls. Owls are a known nest predator of this species (Hands et al. 1989).

Abundance

Roadside point counts tallied this species at 1/2–2 birds per 100 stops in the Ozarks and Ozark Border natural divisions. Other survey techniques, such as canoeing, are needed to better estimate abundance.

Phenology

While some birds are present in the winter, migrants first arrive in early March and leave the state by mid-November (Robbins and Easterla 1992). Observations of birds on nests fell within the early April to early June range of egg-laying dates reported by Hands et al. (1989).

Abundance by Natural Division
Average Number of Birds / 100 Stops

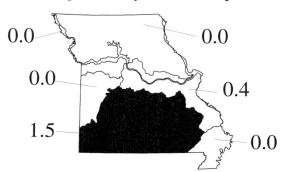

0.0   0.0
0.0   0.4
1.5   0.0

Breeding Phenology

EVIDENCE (# of Records)	MAR	APR	MAY	JUN	JUL	AUG	SEP
NY (6)		5/01		6/07			
FY (3)			5/20	6/21			

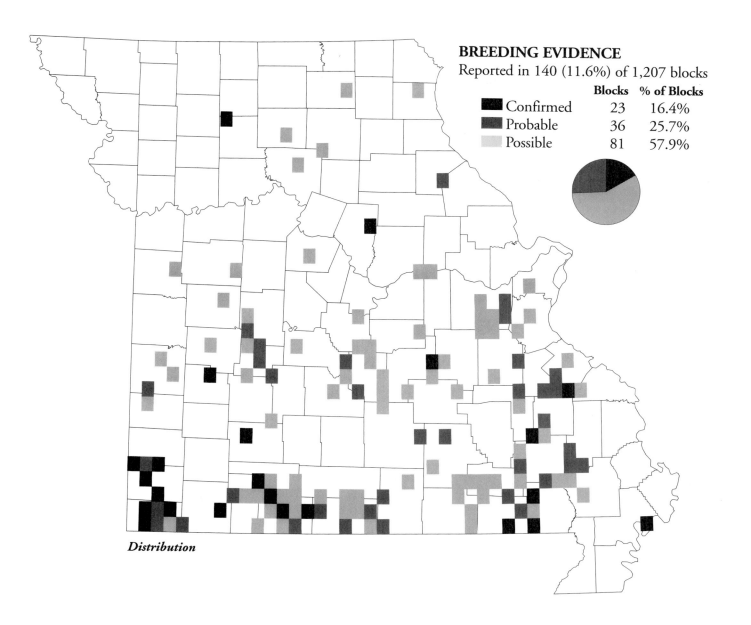

BREEDING EVIDENCE
Reported in 140 (11.6%) of 1,207 blocks

	Blocks	% of Blocks
Confirmed	23	16.4%
Probable	36	25.7%
Possible	81	57.9%

Distribution

Broad-winged Hawk

Buteo platypterus

Rangewide Distribution: Central & southern Canada;
eastern United States

Abundance: Fairly common

Breeding Habitat: Continuous dense forests often near water

Nest: Loose sticks, twigs & leaves lined with bark, lichen,
evergreen sprigs & green leaves in crotches of deciduous
or coniferous trees

Eggs: 2–3 white or blue-white, wreathed, with brown marks
or unmarked

Incubation: 28–32 days

Fledging: 35 days

The smallest of the buteos to breed in Missouri, Broad-winged Hawks are especially noticeable during migration, in April, September and October, when several hundred can sometimes be seen at once. In contrast, during the breeding season Broad-wings can be difficult to find. They often nest deep within the forest and may easily go undetected unless sighted soaring above the canopy, or unless birders recognize their shrill, monotone whistle.

Code Frequency

Because Broad-winged Hawks are difficult to detect, they may have occurred more widely than indicated by the map. Evidences beyond those indicating possible breeding were apparently difficult for Atlasers to detect, presumably because of the species' secrecy around nest sites.

Distribution

Few species' distribution maps are aligned with the Ozarks to the extent of the Broad-winged Hawk's. The map indicates this species is essentially absent from western central and northern Missouri during the breeding season. The two locations at the northern edge of the state may have been migrants. Broad-winged Hawks may have been over-looked in forested regions that received less Atlaser effort, such as the heart of the Ozarks and the Lincoln Hills in northeastern Missouri.

Phenology

The peak of the spring migration occurs in late April (Robbins and Easterla 1992). The dates that young were observed in the nest define the nesting season for this single-brooded raptor.

Breeding Phenology

EVIDENCE (# of Records)	MAR	APR	MAY	JUN	JUL	AUG	SEP
NY (3)			5/29	6/14			

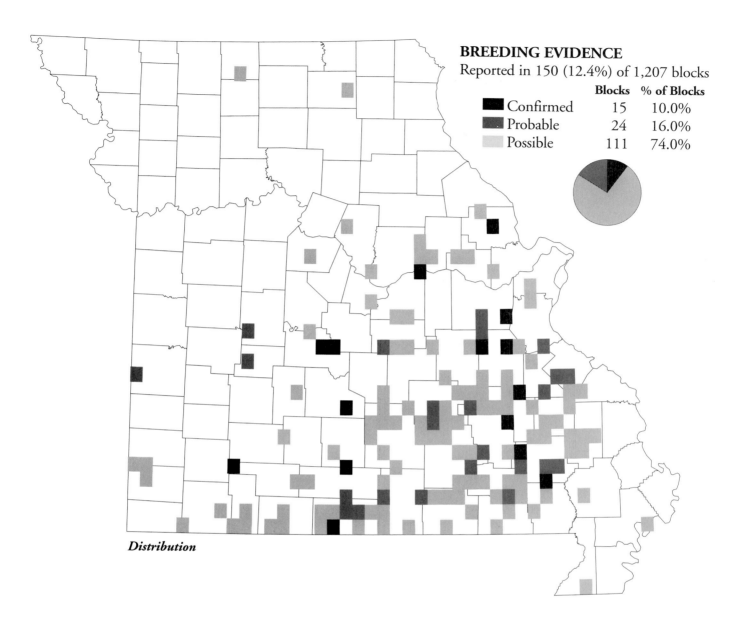

BREEDING EVIDENCE
Reported in 150 (12.4%) of 1,207 blocks

	Blocks	% of Blocks
Confirmed	15	10.0%
Probable	24	16.0%
Possible	111	74.0%

Distribution

Swainson's Hawk

Buteo swainsoni

> **Rangewide Distribution:** Southwestern & south central Canada, entire western United States
> **Abundance:** Common on Great Plains in the West
> **Breeding Habitat:** Pasture & prairie with groves or scattered trees
> **Nest:** Large lined construction of sticks, twigs, brambles & grass, in trees
> **Eggs:** 2–3 bluish or greenish white with sparse brown marks
> **Incubation:** 34–35 days
> **Fledging:** 30 days

Large flocks of Swainson's Hawks migrate between their winter grounds in southern South America and breeding grounds in the western North American Plains. While their normal breeding habitat is open prairie land and forested riparian corridors, they also nest in trees near farmsteads and towns. Because they are very tame, and frequently perch near roadsides, uninformed target shooters often shoot them (Terres 1987). They usually build their medium-sized, shallow nests on smaller branches away from the main tree trunk.

In 1907 Widmann reported one Lawrence County site and several in adjacent states. The Swainson's Hawk is currently listed as Endangered in Missouri.

Code Frequency

Swainson's Hawks can easily be detected and identified as they soar on slightly raised wings. Nests are easily located in late April before becoming shrouded by leaves. Because most Atlasing effort occurred later, Swainson's Hawks may have bred undetected in some blocks where found.

Distribution

In addition to the locations shown on the map, several nests located outside blocks produced young between 1986 and 1992. These nests were located in suburban Springfield near houses and schools, and were scattered throughout the surrounding counties in tree rows between mixed croplands and pastures. Summering Swainson's Hawks have been reported sparsely from western Greene County and southern Vernon County in southwest Missouri. Robbins and Easterla (1992) reported them in several counties in west central Missouri with Barton County being the most frequented area.

Abundance

Scattered records from southwestern Missouri indicate perhaps as many as 5–10 nesting pairs within the state annually.

Phenology

The Swainson's Hawk arrives in March and April and departs in late August and September (Robbins and Easterla 1992). Adults on nests have been observed by the authors in late April and early May.

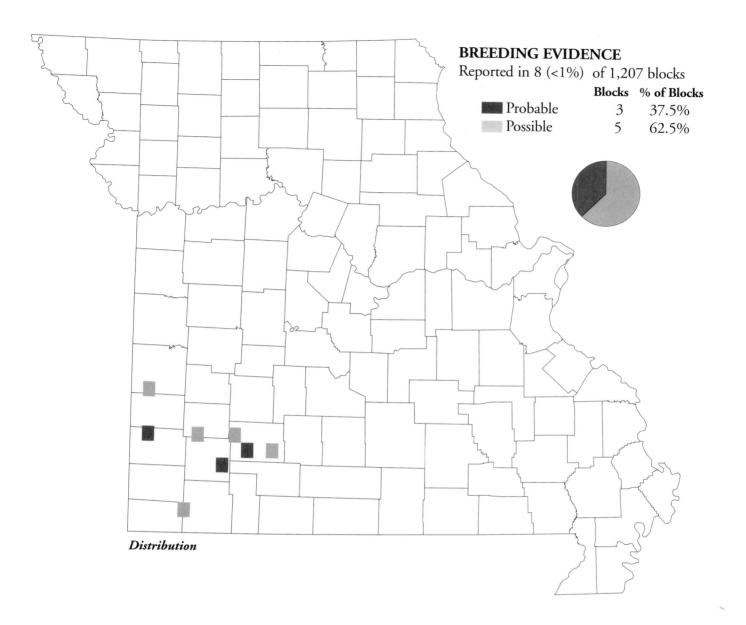

BREEDING EVIDENCE
Reported in 8 (<1%) of 1,207 blocks

	Blocks	% of Blocks
■ Probable	3	37.5%
▨ Possible	5	62.5%

Distribution

Red-tailed Hawk
Buteo jamaicensis

Red-tailed Hawks nest in conspicuous locations around the countryside. Sometimes nests are situated in a lone tree amid rows of corn, soybeans and pasture lands, or in urban areas where small mammals are numerous. Usually, however, nest sites are associated with fragmented rural forests and can be well hidden within the protective canopy of a wood-lot or forest edge. The highly-visible rusty-red tail feathers cannot be confused with those of any other species in the state. Great Horned Owls, which do not build a structured nest in trees, frequently usurp nests of Red-tailed Hawks (Ehrlich et al. 1988).

Code Frequency

Due to their conspicuousness, Red-tailed Hawks were likely recorded by Atlasers wherever they occurred. Where not recorded, therefore, they probably were absent or in low numbers. While most reports concerned possible and proba-ble sightings, 115 nests were located. The use of the UN code may have been erroneous. Because Red-tailed Hawks' nests may last for several years, these nests may not have been used during the 1986-92 survey period.

Distribution

Red-tailed Hawks were found statewide with only scat-tered records for the Mississippi Lowlands. Also, slightly fewer birds were located in Shannon, Carter and Reynolds counties where the landscape approaches continuous forest, a habitat less frequented by this species. Most confirmed observations occurred in the open western one-third of the state and parts of the Ozark Border Natural Division. The apparent scarcity of records in the north central Glaciated Plains is puzzling.

Abundance

The relative abundance in the Osage Plains was highest for the state at 5.7 birds per 100 stops followed by the Ozark region at 2.3 birds per 100 stops. This difference could result from the difficulty of observation in forested Ozark roadside areas, or this species may frequent more open habitats than are generally available in the Ozark and Ozark Border natural divisions.

Phenology

Atlas Project observations span more than enough time for a second brood, although Ehrlich et al. (1988) question this possibility.

Abundance by Natural Division
Average Number of Birds / 100 Stops

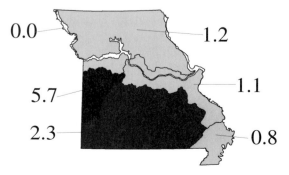

0.0   1.2

5.7

1.1

2.3   0.8

Breeding Phenology

EVIDENCE (# of Records)	MAR	APR	MAY	JUN	JUL	AUG	SEP
NB (9)	3/01				7/02		
NY (41)	3/31					8/01	
FY (22)		4/19			7/24		

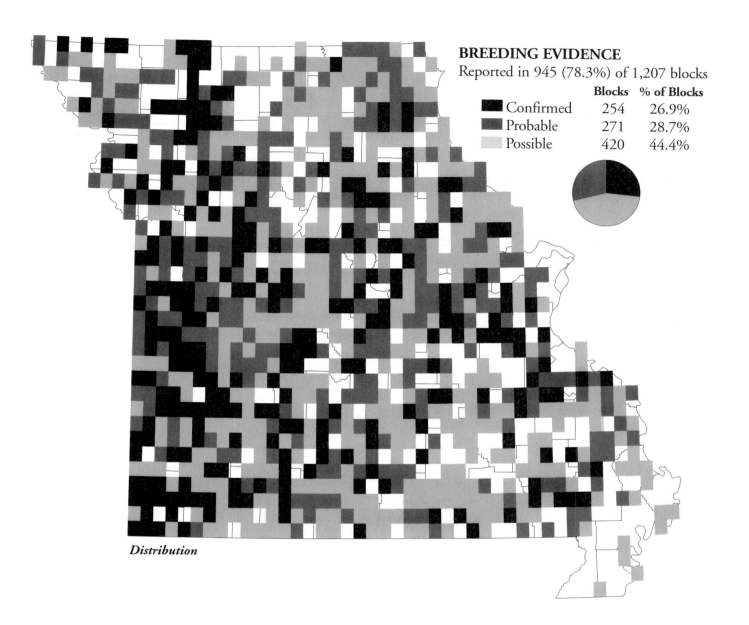

BREEDING EVIDENCE
Reported in 945 (78.3%) of 1,207 blocks

	Blocks	% of Blocks
Confirmed	254	26.9%
Probable	271	28.7%
Possible	420	44.4%

Distribution

American Kestrel
Falco sparverius

Rangewide Distribution: Southern Canada, North, Central & South America
Abundance: Common & widespread
Breeding Habitat: Cultivated & urban open or part open areas with trees
Nest: Little, if any nesting material in holes or hollows of trees or snags
Eggs: 4–5 white or pink-white with brown or lavender marks or unmarked
Incubation: 29–31 days
Fledging: 30–31 days

American Kestrels are adaptable in their choice of nesting habitats. Their requirements–moderately short grass in which to find prey and secure cavities in which to raise their young–are usually met in open, rural areas. However, these small falcons occasionally nest in towns and even city centers. Breeding pairs may be found in all types of habitat except deep woods (Palmer 1988). Nest sites are often situated in tree cavities, but they can also occur in utility poles excavated by woodpeckers, in the walls of buildings, and in specially-designed kestrel boxes (Nagy 1972).

Code Frequency

Because American Kestrels are relatively conspicuous while hunting and perching, the blocks in which they were found may accurately represent their true breeding distribution in Missouri. Evidence of probable breeding, such as pairs and territoriality, were more detectable for kestrels than most species. Because kestrels nest in cavities in open country, Atlasers were frequently able to confirm breeding. Nonbreeding individuals may have been observed in blocks where kestrels were recorded as possible breeders.

Distribution

The most obvious characteristic of the American Kestrel's breeding distribution was its general avoidance of the most forested parts of the state. Kestrels were recorded in greatest density from west central Missouri north to the Iowa line. Overall, their distribution appeared to reflect the occurrence of open agricultural land. In north central and northeastern Missouri, there were fewer blocks where the species was recorded despite a prevalence of seemingly appropriate habitat. Although the Mississippi Lowlands appeared more suitable for kestrels than for most bird species, the number of blocks in which this species was reported appeared low considering the abundance of agricultural land in this natural division. Additionally, there were only two blocks in this division where breeding was confirmed. Perhaps nest site availability or appropriate foraging habitat limits the kestrel's breeding distribution in the Mississippi Lowlands. Confirmed breeding was recorded for most blocks in the Kansas City and St. Louis areas.

Phenology

The earliest confirmation of breeding for American Kestrels was on April 22 when a bird was seen entering a nest cavity in circumstances indicating nesting. According to Toland (1983) Missouri kestrels often raise two broods during a season. He observed double broods in 14 of 53 nest boxes and reported the first brood was initiated in March

Breeding Phenology

EVIDENCE (# of Records)	MAR	APR	MAY	JUN	JUL	AUG	SEP
NB (1)	4/10	4/10					
NE (1)					7/12	7/12	
FY (10)		5/12			7/03		

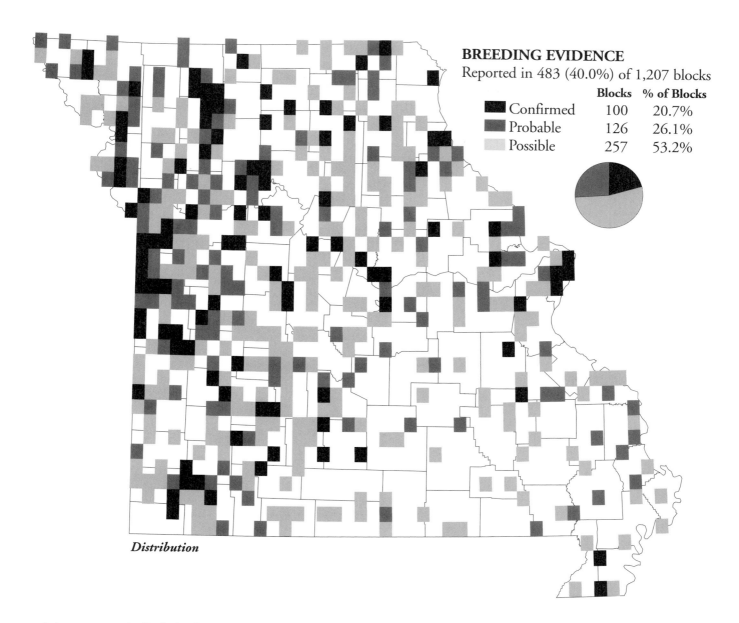

	Blocks	% of Blocks
Confirmed	100	20.7%
Probable	126	26.1%
Possible	257	53.2%

Distribution

and the young typically fledged in early June. The second nesting began in mid- to late June and fledging occurred in mid- to late August, Toland said. The only definitive evidence of a second brood obtained from the Atlas Project was a nest with eggs that was observed July 12.

Peregrine Falcon
Falco peregrinus

> **Rangewide Distribution:** Patchy areas in Canada, Alaska & western United States
> **Abundance:** In the United States, rare & local in the west & extirpated in the east
> **Breeding Habitat:** Mountains, tall buildings & open forest
> **Nest:** Scrape with accumulated debris on ledges, cliffs & old tree cavities
> **Eggs:** 3–4 white or pink-cream, occasionally with brown or red marks
> **Incubation:** 29–32 days
> **Fledging:** 35–42 days

Historically, Peregrine Falcons nested in small numbers on bluffs along the Mississippi, Missouri and Gasconade rivers. By the 1880s and 1890s only a few pairs remained in the state (Robbins and Easterla 1992). The last known probable breeding activity occurred in 1911 (Robbins and Easterla 1992). Currently, reintroduction projects in St. Louis and Kansas City have released young birds to the wild. A small population of urban-dwelling peregrines has been established, with the released birds selecting tall buildings as substitutes for cliff nesting sites. It is hoped that, with time, pairs may return to former nest sites on suitable bluffs.

Code Frequency

Because of interest in restoration projects, the Peregrine Falcon has been included here despite the fact that no evidence of breeding was found in Atlas Project blocks. This species is currently listed as Extirpated in Missouri.

Distribution

Nesting activity took place outside of Atlas blocks. The Missouri Department of Conservation Restoration Project in Kansas City resulted in several nesting attempts. Although birds released in Kansas City failed to produce young in that city, they have successfully nested in Omaha, Des Moines and Wichita. The World Bird Sanctuary Peregrine Falcon Restoration Project in St. Louis has experienced more success. Four nests sites were located between 1991 and 1996 in the St. Louis area. The first nest was found in 1991 when a pair of Peregrine Falcons nested on the Bell Center building, according to a personal conversation with Mike Cooke of the World Bird Sanctuary. Since then, St. Louis-released Peregrines have occupied and successfully raised young in three other nest sites in the St. Louis area, located on buildings and a bridge.

Abundance

Four nesting attempts occurred in St. Louis between 1991 and 1996. One nest was active from 1991 to 1994, one from 1992 to present, one in 1995 and one in 1996. About 17 young were reared from eggs laid by the attendant females, who also reared several fostered young.

Phenology

St. Louis birds set up nesting territories in February. Egg laying typically occurred from April through June, however, on one occasion, birds hatched in early April (Mike Cooke, personal communication.)

Ring-necked Pheasant
Phasianus colchicus

Rangewide Distribution: Southwestern to northern Canada, north central & southwestern to south central United States
Abundance: Introduced in United States & locally common
Breeding Habitat: Open country, cultivated & grassy areas
Nest: Depression in grass or weeds with some lining of leaves & grass
Eggs: 10–12 brownish-olive, occasionally pale blue & unmarked
Incubation: 23–25 days
Fledging: 12 days

This species, an Asian native, has been introduced throughout many parts of the United States along with many other pheasant species, most of which have failed to establish populations. Past attempts to introduce Ring-necked Pheasants to Missouri either failed or were mildly successful on the local level. Relocating naturalized Missouri birds resulted in successful introductions beginning in 1987. There has been considerable success in target areas where the habitat was suitable for pheasants' annual needs.

Code Frequency

Pheasants were easy to detect, especially in spring when crowing is more common. Most records were observations probably based on crowing males. The observation of broods accounted for 82 percent of confirmed records.

Distribution

The Atlas Project distribution map closely matches the known range diagramed in *The Ring-necked Pheasant in Missouri*, in which Hallett (1990) reported 47 counties with populations. Two primary populations of the Korean subspecies occur in the state, one in the Glaciated Plains of northern Missouri and one in the Mississippi Lowlands (Hallett 1990). Possible records scattered through much of the state may have observed game farm escapees. The sampling process apparently missed a small, persistent population in St. Charles County. Missouri is at the southern edge of this species' Midwestern range, which includes most of the northern Great Plains (Hallett 1990).

Abundance

Birds were 1.5 times as abundant in the Glaciated Plains as in the Mississippi Lowlands. The Glaciated Plains' population is centered in the northern parts of the Western Glaciated Plains and Grand River natural sections.

Phenology

Atlas Project observations of fledglings agree with annual breeding activity reports by pheasant biologists (Hallett 1990). Atlasers sighted 72 percent of the fledglings in June and July. Egg laying for a few birds was initiated in early April in the Farmers City and Rock Port blocks in Atchison County.

Abundance by Natural Division
Average Number of Birds / 100 Stops

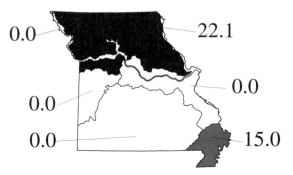

0.0   22.1

0.0   0.0

0.0   15.0

Breeding Phenology

EVIDENCE (# of Records)	MAR	APR	MAY	JUN	JUL	AUG	SEP
NB (1)		5/02	5/02				
NE (3)			6/04			8/14	
NY (4)			5/19		7/27		

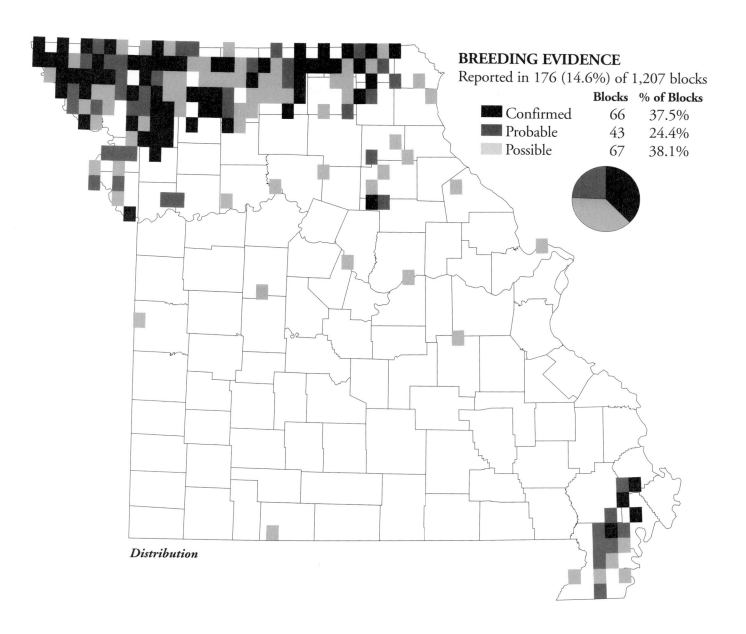

BREEDING EVIDENCE

Reported in 176 (14.6%) of 1,207 blocks

	Blocks	% of Blocks
Confirmed	66	37.5%
Probable	43	24.4%
Possible	67	38.1%

Distribution

Ruffed Grouse
Bonasa umbellus

Rangewide Distribution: Southern Canada; central Alaska, northwestern and northeastern United States
Abundance: Fairly common in forests
Breeding Habitat: Deciduous & coniferous woods with dense understory
Nest: Deep hollow lined with feathers at base of tree or log
Eggs: 9–12 buff with brown spots
Incubation: 23–24 days
Fledging: 10–12 days

The courtship drumming of the Ruffed Grouse is less well known to Missouri residents today than in the mid-1800s. According to Widmann (1907) this species was numerous in most wooded parts of Missouri until the 1880s. Frequent burning and grazing of forest lands have created more open land and less brushy young forests, the habitat of this species (Thompson et al. 1988). By the 1950s most of the Ruffed Grouse in Missouri were limited to small pockets of habitat. Thompson et al. (1988) suggested that today much more habitat is available than grouse to fill it. Reintroduction by the Missouri Department of Conservation in the second half of this century has restored Ruffed Grouse to many forested areas of the state.

Code Frequency

Early season observers documented drumming individuals and two reports of fledglings. Special Conservation Department drumming counts provide a more accurate picture of the species' distribution (Thompson et al. 1988). Many Atlasers were not in the field during the "drumming season," so it is likely much evidence that would have resulted in potential breeding records was not observed.

Distribution

An accurate distribution for the state was not determined by Atlas Project results. Most records do fall into the known range as described in University of Missouri Extension agricultural guide: *Ruffed Grouse in Missouri: Its Ecology and Management* (Thompson et al. 1988). The Pike County record is within the known continuous range of this species. The Ray and Nodaway records are near known release sites (Thompson et al. 1988).

Abundance

The Atlas Project did not collect enough information to determine the relative abundance of this species. Drumming counts conducted in March and April are used to establish estimates of relative abundance (Thompson et al. 1988).

Phenology

Male drumming normally occurs in March and April. Atlas Project fledgling sightings are consistent with hatching dates reported by Thompson et al. (1988).

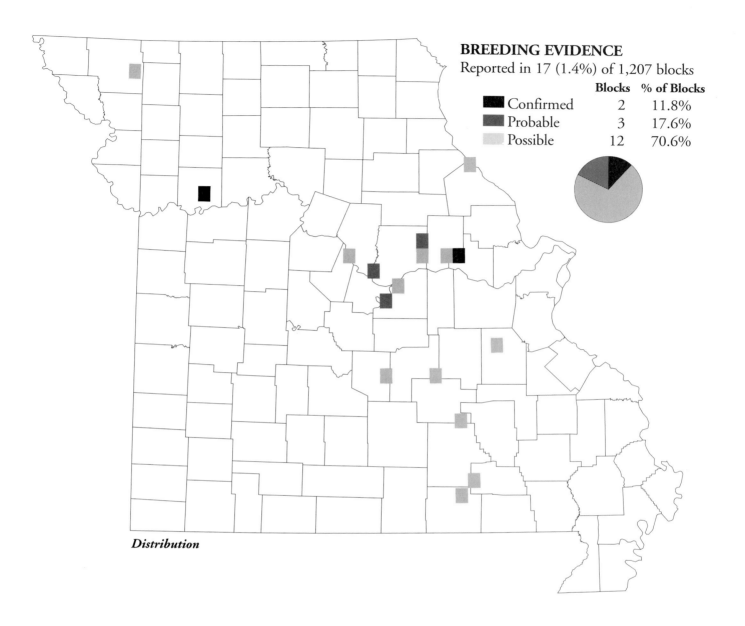

BREEDING EVIDENCE
Reported in 17 (1.4%) of 1,207 blocks

	Blocks	% of Blocks
Confirmed	2	11.8%
Probable	3	17.6%
Possible	12	70.6%

Distribution

Greater Prairie-Chicken
Tympanuchus cupido

Rangewide Distribution: South central Canada, central United States and southeastern Texas
Abundance: Uncommon locally & declining
Breeding Habitat: Open tall-grass prairie, pasture & hayfield
Nest: Depression on ground lined with leaves, feathers & grass
Eggs: 10–12 olive, spotted with dark brown
Incubation: 23–24 days
Fledging: 7–10 days

The population decline of Greater Prairie-Chickens reflects human modification of the landscape. Because they were associated with native prairies (Christisen 1985), Prairie-chickens were once numerous in Missouri. Their decline is related to the replacement of grasslands with croplands and widespread use of fescue (*Festuca* spp.) on remaining grasslands (Missouri Department of Conservation et al. 1991; Skinner et al. 1984). Native prairie once occupied 34 percent of Missouri but now occupies less than 0.5 percent (Schroeder and Robb 1993). Presumably the decline in prairie-chicken numbers has paralleled or exceeded the decline in native prairie acreage, especially because some remaining prairies are too small to support viable populations (Schroeder and Robb 1993).

Code Frequency

Greater Prairie-Chickens are easy to detect during spring courtship. In forty-three percent of blocks in which prairie-chickens were recorded, Atlasers observed courtship behavior, likely by viewing birds on a display ground. Even blocks in which possible evidence was recorded were likely true breeding areas as prairie-chickens do not migrate and rarely venture far from breeding areas. Because prairie-chickens are rare and clumped in a few areas of favorable habitat, Atlas Project sampling would be expected to miss many populations. Therefore, the map displays a conservative picture of prairie-chicken distribution in Missouri.

Distribution

All but three blocks with breeding evidence were in the Osage Plains, the last remaining foothold for this declining species. By chance, the Atlas Project detected outlying remnant populations in Audrain County and a recently-discovered population in the Missouri River bottoms in Carroll County. The Missouri Department of Conservation has carefully surveyed Prairie Chickens and can provide a more thorough picture of their range in Missouri.

Phenology

Courtship was first detected March 2 and extended beyond May 29. Nest initiation dates are variable, ranging from mid-April to early June (Schroeder and Robb 1993).

Breeding Phenology

EVIDENCE (# of Records)	MAR	APR	MAY	JUN	JUL	AUG	SEP
NE (2)				6/20	6/23		

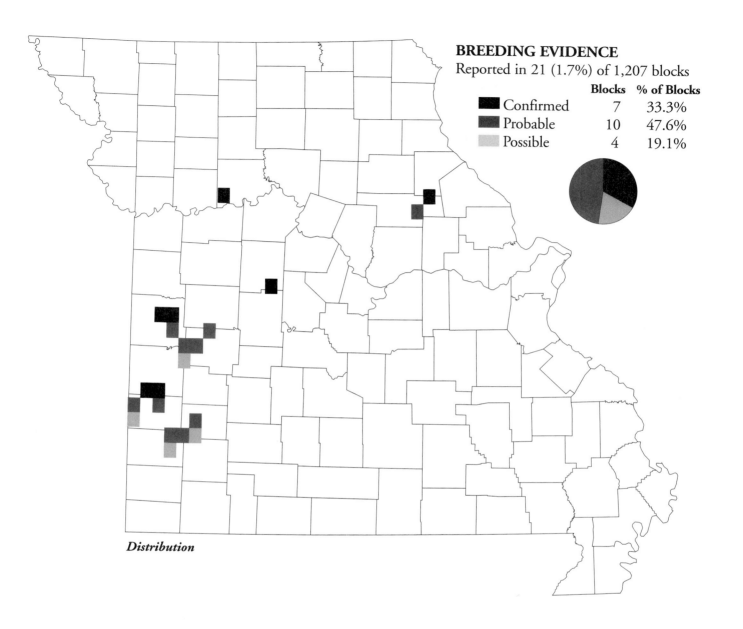

BREEDING EVIDENCE
Reported in 21 (1.7%) of 1,207 blocks

	Blocks	% of Blocks
Confirmed	7	33.3%
Probable	10	47.6%
Possible	4	19.1%

Distribution

Gray Partridge
Perdix perdix

Rangewide Distribution: Southern Canada, northern
   United States
Abundance: Widely introduced from Europe
Breeding Habitat: Pastures with fencerows, hayfields
   & grainfields
Nest: Depression on ground lined with leaves, straw &
   grass
Eggs: 15–17 olive, occasionally white & unmarked
Incubation: 23–25 days
Fledging: 13–15 days

Gray Partridges were introduced into North America in
the early 1900s (Carroll 1993). They are apparently well
adapted to areas of intensive agriculture such as those growing cereal grains and row crops. These permanent residents
breed in pastures, grain stubble and hedgerows (Carroll
1993).

A population, established in northwestern Iowa between
1905 and 1913, expanded rapidly following the 1960s
(Dinsmore et al. 1984). Trapped wild birds were also
released in southeastern Iowa in the 1970s (Dinsmore et al.
1984). Presumably some of these birds moved southwest
into northern Missouri.

The first record of a Gray Partridge in Missouri was in
the extreme northwestern corner in 1987 or 1988 (Robbins
and Easterla 1992). Subsequently, coveys have been found
in Atchison, DeKalb, Holt, and Nodaway counties in northwestern Missouri (Robbins and Easterla 1992) and, according to a personal report by retired Conservation Department
writer Joel Vance, in Knox County in northeastern
Missouri.

Code Frequency

Atlasers obtained only three records of Gray Partridge.

Distribution

Individuals sighted were in appropriate breeding habitat
southwest of Maryville in Nodaway County, and northwest
of Rockport in Atchison County. The report by Vance,
mentioned above, occurred near the Scotland and Knox
county line.

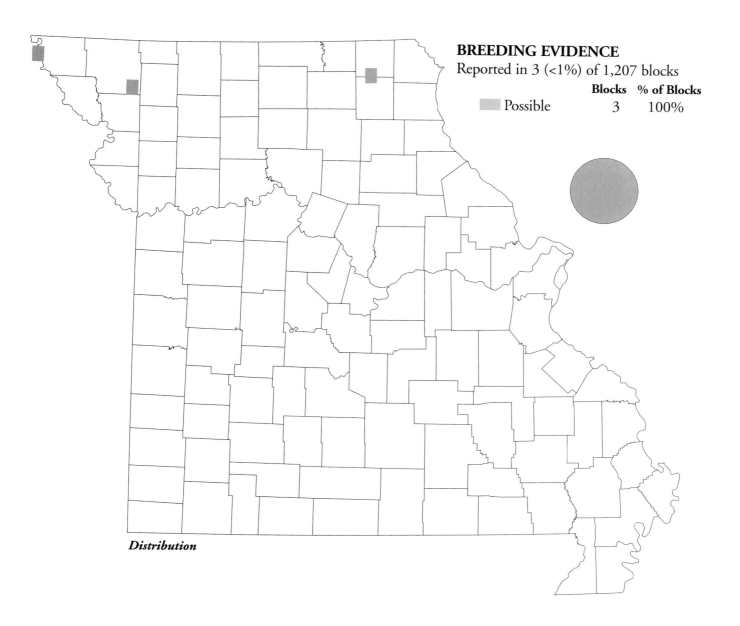

Reported in 3 (<1%) of 1,207 blocks

	Blocks	% of Blocks
Possible	3	100%

Distribution

Wild Turkey
Meleagris gallopavo

> **Rangewide Distribution:** Widespread through North America
> **Abundance:** Locally common where introduced, disappearing from original range
> **Breeding Habitat:** Open areas at edge of woods & mountains
> **Nest:** Depression on ground, lined with dead leaves & grass
> **Eggs:** 10–12 buff to white with dull brown marks
> **Incubation:** 27–28 days
> **Fledging:** 6–10 days

Each spring, Missourians have a great opportunity to enjoy the elaborate strutting displays and loud gobbling vocalizations of Wild Turkeys. By 1907, Widmann noted the Wild Turkey was gone from most of northern Missouri and only remained in small flocks in the Ozark and Mississippi Lowlands natural divisions.

Management efforts have restored this once widely-extirpated species to good numbers in Missouri and many other states. Forty-nine states have enough birds to have a hunting season. Eaton (1992) claimed by the year 2000, the Wild Turkey will be restored in all suitable habitats nationwide.

Code Frequency

Because of their large size, Wild Turkeys are easily detected compared with most bird species. Turkey poults, occasionally disturbed by Atlasers who were walking in forests and fields, were by far the most frequent confirming observation. Nineteen percent of confirmations represented actual nests. Early season visits to listen for Wild Turkeys usually documented the presence of this species.

Distribution

The Atlas project found birds in all counties except Dunklin and Pemiscot of the Mississippi Lowlands. In many areas in the Mississippi Lowlands Natural Division and Springfield Plateau and Elk River natural sections this species was reported in fewer blocks than the rest of the state. Wild Turkeys use forests, grasslands and woodlands as foraging areas for seed, grass and insects. Consequently, areas of the state that have less forest cover, fewer grasslands or fewer mast-producing trees probably have reduced winter survival, which results in fewer birds.

Abundance

Between 1954 and 1979, a Department of Conservation reintroduction program used locally-acquired birds and successfully established Wild Turkeys in 101 of 114 counties (Missouri Department of Conservation 1993). Today all Missouri counties have sufficient numbers to allow hunting. Wild Turkeys' relative abundance was highest in the Glaciated Plains, Ozark Border and Ozark natural divisions. These areas supply a better mix of grasslands and forests than the Osage Plains.

Abundance by Natural Division
Average Number of Birds / 100 Stops

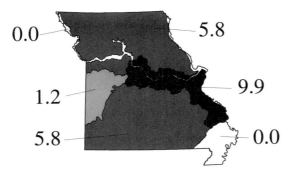

0.0   5.8

1.2   9.9

5.8   0.0

Breeding Phenology

EVIDENCE (# of Records)	MAR	APR	MAY	JUN	JUL	AUG	SEP
NB (1)		5/02	5/02				
NE (57)	4/09						8/29
NY (10)		5/01			7/06		

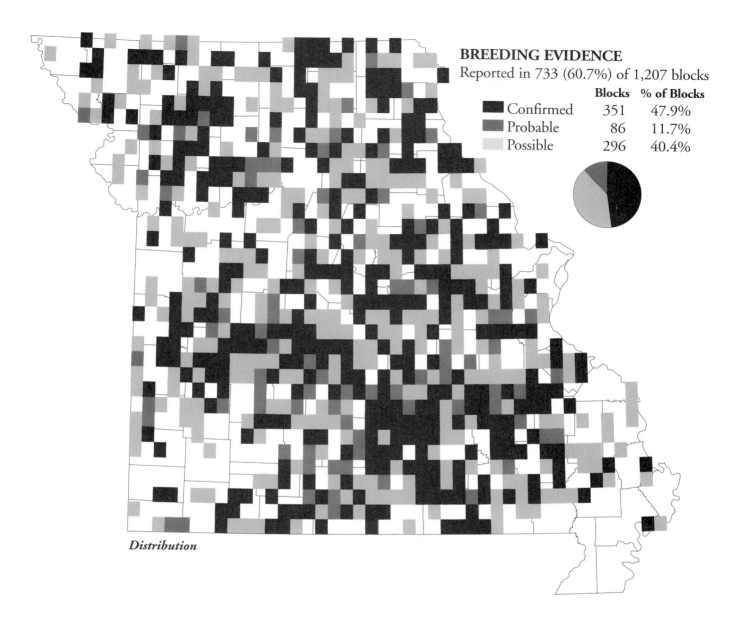

BREEDING EVIDENCE
Reported in 733 (60.7%) of 1,207 blocks

	Blocks	% of Blocks
■ Confirmed	351	47.9%
■ Probable	86	11.7%
■ Possible	296	40.4%

Distribution

Phenology

Wild Turkeys are easily located in April and May when they are displaying and calling. Most nests and eggs were found in May (22) and June (16) but four were found in April, six in July and three in August. Atlasers did not determine whether hens abandoned or attended the late August nests.

Northern Bobwhite
Colinus virginianus

Rangewide Distribution: Native to eastern United
  States & introduced in northwest
Abundance: Common
Breeding Habitat: Tall grassland, fields & open
  woodland
Nest: Shallow depression lined with grass, hidden by
  vegetation
Eggs: 10–16 white to creamy/buff & unmarked
Incubation: 23–24 days
Fledging: 6–7 days

The Northern Bobwhite's call is familiar to, and easily
recognized by most Missourians. It clearly whistles "bob bob
white" with a rising inflection on the "white." A careful
observer driving through the countryside often sees the
Bobwhite calling from a fence post. Because they do not
migrate, quail, as Bobwhite are frequently called, suffer the
ravages of unpredictable ice and snow in Missouri when
good cover is not available. Spring rains can negatively affect
the success of this single-brooded, ground-nesting species
(Dailey 1996). Their large clutch size, however, allows for a

rapid rebound in population numbers following a few productive years. Agricultural land, brushland and open grassland throughout the state insure the continued availability of
this species' habitat. One of their natural haunts, grassland
with widely-scattered trees, has mostly been converted to
other uses or succeeded into forest, losing the grass compo-
nent this species requires.

Code Frequency

The Northern Bobwhite was the sixth most-commonly
recorded species during the Atlas Project. This is no doubt
due to their persistent vocalizations that commence in
spring and continue to midsummer. Singing males, pairs,
territoriality and sightings of fledged birds accounted for 77
percent of documentations. While Atlasers readily observed
these breeding season behaviors, they found it difficult to
locate nests and eggs. These birds likely bred in most blocks
in which they were sighted.

Distribution

Northern Bobwhites were found statewide. They were
recorded in 95 percent of blocks in the Ozark and
Mississippi Lowlands natural divisions and in 100 percent in
the Glaciated Plains and Osage Plains natural divisions. This
is one of few species distributed so widely in forested regions
as well as in the most heavily agricultural regions.

Abundance

Noticeably lower relative abundances were observed in
regions where clean agricultural practices prevail, such as the
Mississippi Lowlands, and heavily forested areas, such as the
Ozark and Ozark Border natural divisions. However,
Bobwhites were still counted at many stops on relative
abundance surveys and are considered a common resident
even in these low-abundance areas. However, roadside quail

Abundance by Natural Division
Average Number of Birds / 100 Stops

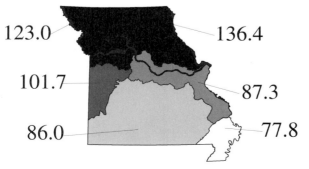

123.0   136.4
101.7   87.3
86.0   77.8

EVIDENCE (# of Records)	MAR	APR	MAY	JUN	JUL	AUG	SEP
NE (22)		4/15				8/14	
NY (11)			5/14			8/13	

Breeding Phenology

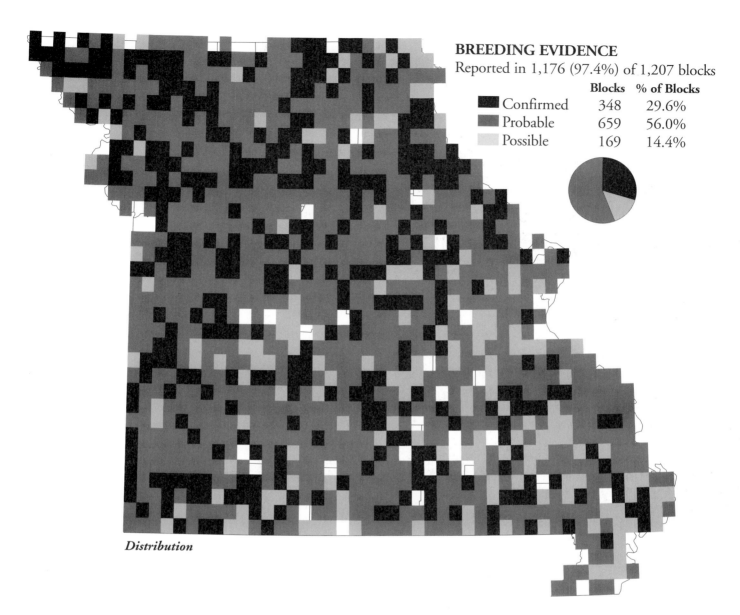

BREEDING EVIDENCE
Reported in 1,176 (97.4%) of 1,207 blocks

	Blocks	% of Blocks
Confirmed	348	29.6%
Probable	659	56.0%
Possible	169	14.4%

Distribution

counts, conducted in 1983–1995 by the Conservation Department, show continuing declines over the long run (Dailey 1996).

Phenology

This single-brooded, gallinaceous, ground-nesting species may renest repeatedly when weather, depredation or agricultural activity cause brood failure. This likely explains the mid-August nests with eggs and young.

King Rail
Rallus elegans

Rangewide Distribution: Eastern United States, Cuba & interior Mexico

Abundance: Common in freshwater, swamps & marshes

Breeding Habitat: Shallow water with grassy vegetated wet soil

Nest: Deep basket of dry aquatic vegetation 6-8" above ground

Eggs: 10–12 buff with brown spots

Incubation: 21–24 days

Fledging: 63+ days

The secretive nature of King Rails, and the dense cover in which they nest, make them as hard to see as other rails. Their rusty coloration, long slender bill and blackish barred flanks closely resemble the Virginia Rail, however, larger size usually identifies the King Rail. This is usually the only definitive characteristic observed in short encounters with this species. The best time to hear and view this species is dawn or dusk when its "rusty bedsprings" call can be heard or when it ventures into the open. To flush rails, people need to walk closely together during a daylight wading survey.

Code Frequency

As many wetland areas are outside blocks, Atlasers logged only three records, all resulting from sightings of fledglings. Fledgling sightings and voice are probably the easiest detection methods for this species.

Distribution

Ted Shanks Conservation Area and Mark Twain National Wildlife Refuge at Annada are the only locations where this species was recorded during the Atlas Project. Outside the Atlas Project, King Rails have also been observed in other marshes at Schell-Osage and Duck Creek conservation areas and Mingo and Squaw Creek national wildlife refuges.

Abundance

No abundance information was gleaned from the Atlas Project. However, King Rails were formerly a common breeder in fresh water marshes along the big rivers (Widmann 1907). The King Rail is now a rare and local breeder (Robbins and Easterla 1992) and it is listed as Endangered in Missouri. The United States Fish and Wildlife Service listed King Rails as a Species of Management Concern in 1986 due to habitat loss and pesticide susceptibility.

Of Missouri's native wetlands, only 10 percent–670,000 acres–remain (Auckley 1996). Only a fraction of this remainder is suitable habitat for King Rails. Restoration of scarce wetland communities would greatly enhance breeding populations in Missouri.

Phenology

Dates of fledgling observations suggest that nesting commenced in May. Widmann (1907) observed young on June 1. Meanley (1969) reported Arkansas nests mainly between April and June.

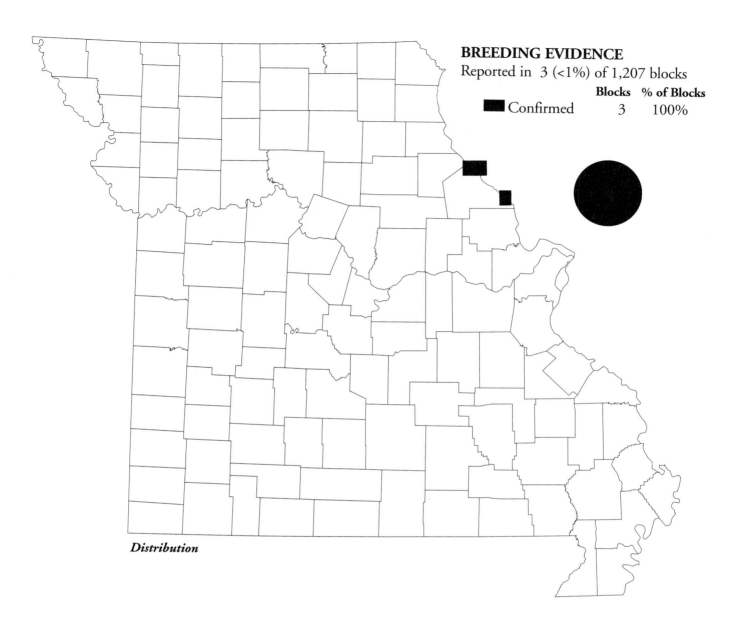

BREEDING EVIDENCE
Reported in 3 (<1%) of 1,207 blocks

	Blocks	% of Blocks
■ Confirmed	3	100%

Distribution

Sora

Porzana carolina

> **Rangewide Distribution:** Southern Canada, northern & southwestern United States
>
> **Abundance:** Common & widely distributed in North America
>
> **Breeding Habitat:** Freshwater marshes, swamps & wet meadows
>
> **Nest:** Baskets of dead aquatic vegetation built up to 6" above water
>
> **Eggs:** 10–12 brown & buff, with brown marks
>
> **Incubation:** 18–20 days
>
> **Fledging:** 21–25 days

Soras are secretive, marsh-dwelling rails with loud, descending, whinnying calls. The little information that birders have about the Missouri breeding population has been gathered from scanty past records. The Atlas Project produced little new information on this species.

Code Frequency

Because the habitat where they reside is difficult to survey, Soras may actually be more widespread and numerous than the three Atlas Project records indicated. The Jasper County record represented a possible June nesting attempt.

Distribution

Atlas Project information was insufficient to accurately assess the true distribution of the species. Soras are likely distributed statewide, but may not be present every year. They are likely more widely distributed in wet years when marshy areas provide a greater expanse of nesting habitat.

Abundance

Information was insufficient to assess the abundance of Soras. They are assumed to be a casual nester when optimal conditions are present. They may also be more abundant in Missouri when breeding grounds farther north are under drought conditions.

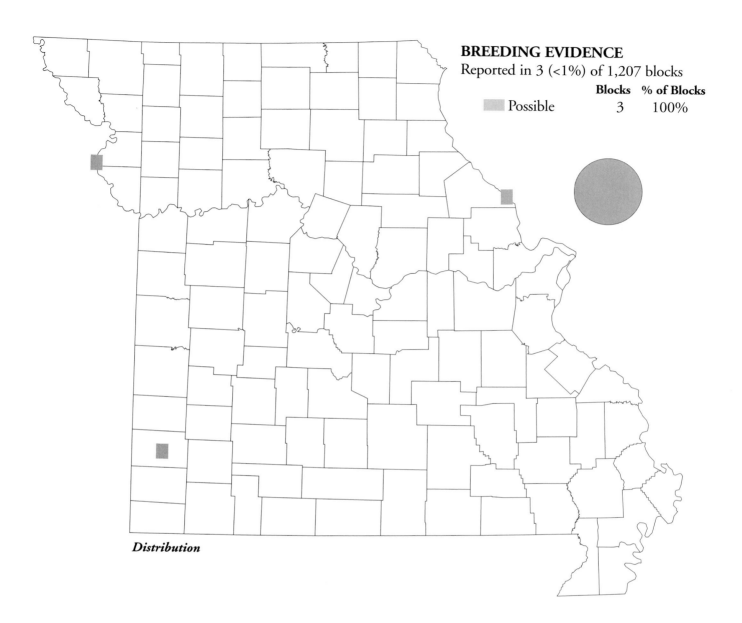

BREEDING EVIDENCE
Reported in 3 (<1%) of 1,207 blocks

		Blocks	% of Blocks
Possible		3	100%

Distribution

Common Moorhen

Gallinula chloropus

Rangewide Distribution: Eastern & southwestern United
   States, Gulf Coast through most of South America
Abundance: Common
Breeding Habitat: Marshes, swamps & lakes with emer-
   gent vegetation or grassy edges
Nest: Rimmed cup of bleached aquatic plants, lined with
   grass, usually over water or on ground or shrubs
Eggs: 5–8 cinnamon or buff with reddish-brown or olive
   marks
Incubation: 19–22 days
Fledging: 40–50+ days

The Common Moorhen breeds in freshwater marshes
vegetated with cattails (*Typha* spp.), bulrushes (*Scirpus* spp.)
and willows (*Salix* spp.). They do not require a large marsh
and may reside in small patches of emergent vegetation at
the edge of lakes or rivers (Terres 1987). Their nests, which
are usually anchored in vegetation at the water's surface, are
similar in appearance to an American Coot's nest but smaller in diameter (Fredrickson 1971).

Code Frequency

Common Moorhens are extremely hard to locate because
their habitat is difficult to survey and they nest in dense
emergent vegetation. Breeding evidence for the Common
Moorhen was found in only one block during the seven-year
Atlas Project.

Distribution

An individual in breeding habitat sighted at the Ted
Shanks Conservation Area in Pike County may have nested
undetected and the species may have been present but undetected in other marshlands.

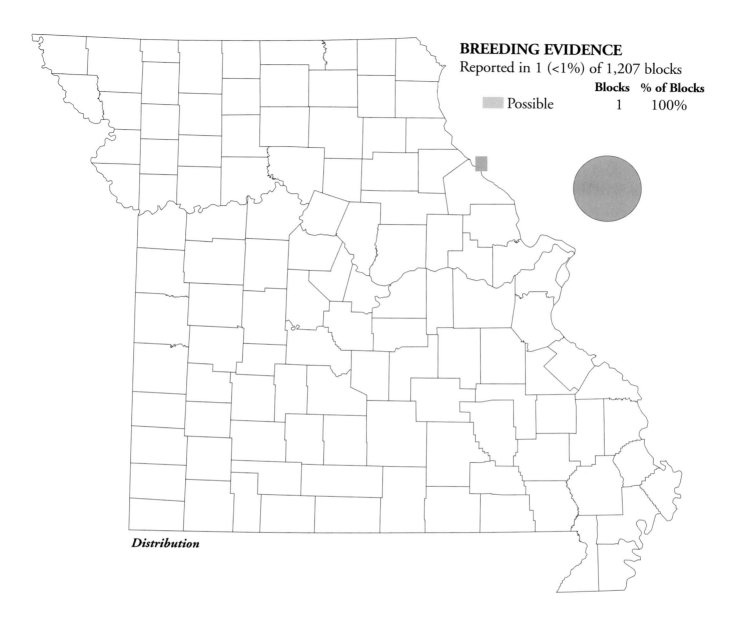

	Blocks	% of Blocks
Possible	1	100%

Distribution

American Coot
Fulica americana

> **Rangewide Distribution:** Southern & south central Canada, entire United States except central eastern states, through Mexico & Central America
> **Abundance:** Common to abundant
> **Breeding Habitat:** Ponds, lakes, rivers, marshes with cattails & bulrushes
> **Nest:** Large floating construction of dead stems anchored to vegetation & lined with fine materials
> **Eggs:** 8–12 pinkish-buff with blackish-brown marks
> **Incubation:** 21–25 days
> **Fledging:** 49–56(?) days

American Coots are common transients and rare summer residents in Missouri (Robbins and Easterla 1992). They tend to breed in large, undisturbed wetlands interspersed with patches of emergent vegetation. Typical breeding areas are approximately half open water and half emergent vegetation (Sanderson 1977). Coots are fairly difficult to locate when breeding because they nest in densely vegetated marsh habitat. Nests, which are mounds of dead vegetation that project above the water level, are typically well hidden. Their presence is often best revealed by their courtship vocalizations and searches of marsh vegetation are usually required to obtain further evidence of breeding.

Coots were apparently more common in the past. Widmann (1907) described them as "not very rare" breeders in suitable locations, not only in the floodplains of the large rivers, but in the prairie and Ozark regions.

Code Frequency

American Coots were recorded in only 15 blocks during the seven-year Atlas Project. The small number of reports was likely due to the scarcity of breeders and difficulty in searching marsh habitat. In addition, non-nesting individuals occasionally summer in Missouri, and this further complicates attempts to document which are breeders. Although non-breeders may have been counted as some of the possible and probable breeding records, breeding was likely more widespread and common than indicated by the four blocks in which breeding was confirmed. Some of the possible and probable breeding locations were conceivably breeding areas that could not be explored thoroughly enough to obtain higher evidence.

Distribution

The data indicate American Coots breed throughout Missouri and support the contention (Robbins and Easterla 1992) that they are rare breeders. The loss of suitable wetland habitat during this century has apparently reduced the number of American Coots breeding in Missouri.

Phenology

Due to a scarcity of confirmed breeding evidence, few substantial data were obtained on the breeding phenology of American Coots in Missouri. The observation of hatching-year young on June 19 indicates nesting would have begun in late April or early May. Considering these dates, reproduction may occasionally be imperiled by the drying of marsh habitats in late spring and summer.

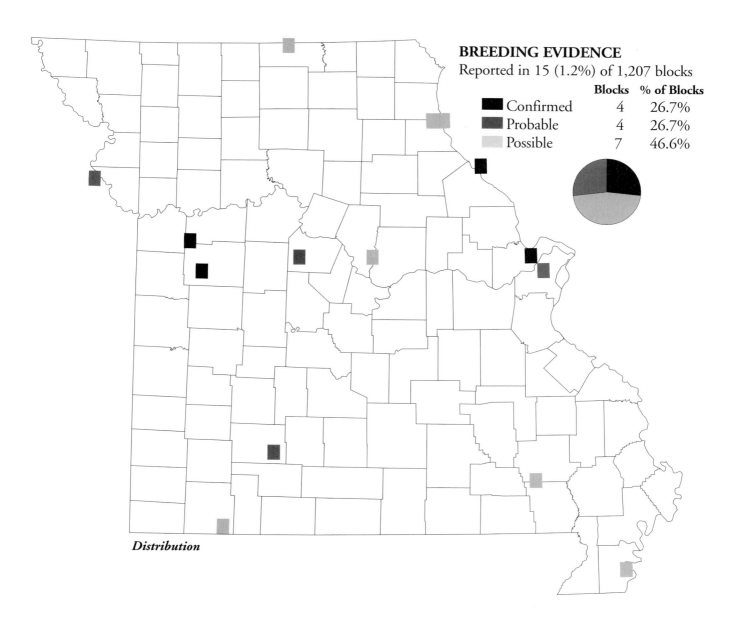

	Blocks	% of Blocks
Confirmed	4	26.7%
Probable	4	26.7%
Possible	7	46.6%

Distribution

Killdeer
Charadrius vociferus

> **Rangewide Distribution:** Alaska & lower half of Canada, throughout United States, northern South America
> **Abundance:** Common in most area
> **Breeding Habitat:** Pastures, meadows & open areas
> **Nest:** In depression on ground with little or no vegetation
> **Eggs:** 3–5 buff colored with blackish or brown markings
> **Incubation:** 24–28 days
> **Fledging:** 25–days

Although the Killdeer is a common shorebird in Missouri, it can be found in nonshore habitats throughout the state. They are common in open graveled areas such as driveways, stream bed islands and even flat, gravel urban rooftops. In forested regions with occasional pastureland, Killdeer will nest in exposed soil around farm ponds, cattle wallows and creek beds. Its "kill deer, kill deer kill deer" call frequently gives away its presence and its distraction display leads the intruder away from its nest of cryptically-colored eggs.

Code Frequency

In 87 percent of blocks this noisy, striking-colored species was recorded. Confirmed evidence was recorded in a surprisingly low 47 percent. Because the camouflaged eggs are hard to locate in their scrape nest, 78 percent of confirmations resulted from a distraction display or direct observation of young. Only 16 percent of confirmed records were of nests and eggs or young.

Distribution

Killdeer were found statewide. They were reported less frequently in the Ozark Natural Division, especially in heavily-forested Shannon County. Even though easily detected, they were not reported in a number of blocks scattered throughout the state.

Abundance

Killdeer were about six times more abundant in the open agricultural landscape of the Mississippi Lowlands and the Big Rivers natural divisions than in other regions of the state. About half as many were tallied in the Osage Plains with the remaining three sections averaging 10 birds/100 stops (7.9, 8.3 and 14.8).

Phenology

Most Killdeer return to Missouri in February and March with some overwintering in mild years (Robbins and Easterla 1992). Killdeer sometimes rear two broods, which may account for their extended breeding season.

Abundance by Natural Division
Average Number of Birds / 100 Stops

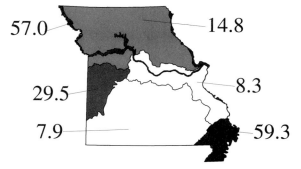

57.0   14.8

29.5   8.3

7.9   59.3

Breeding Phenology

EVIDENCE (# of Records)	MAR	APR	MAY	JUN	JUL	AUG	SEP
NE (66)	4/07				7/09		
NY (13)		5/12			7/03		

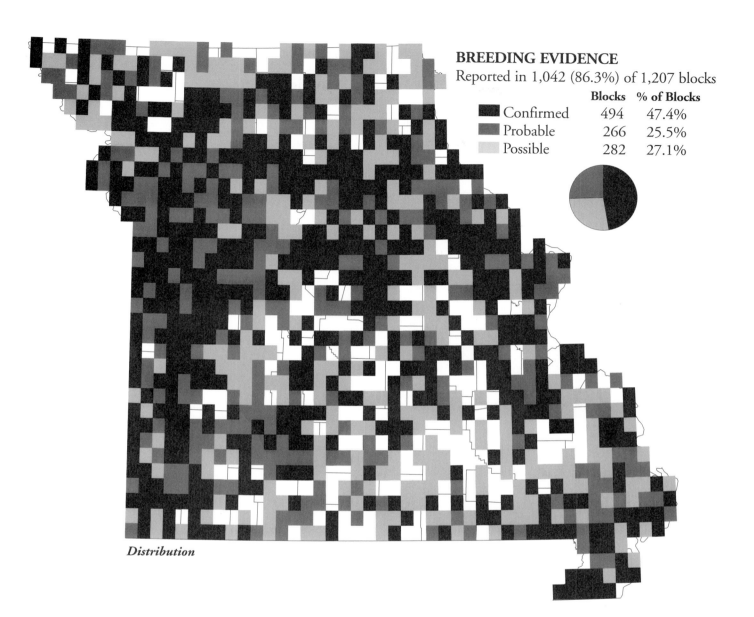

BREEDING EVIDENCE
Reported in 1,042 (86.3%) of 1,207 blocks

	Blocks	% of Blocks
Confirmed	494	47.4%
Probable	266	25.5%
Possible	282	27.1%

Distribution

Black-necked Stilt
Himantopus mexicanus

Rangewide Distribution: Patchy on United States coasts & western interior
Abundance: Rare to uncommon & local
Breeding Habitat: Marshes, wet savannas, mud flats & flooded fields
Nest: Variable; eggs often laid on wet ground; occasional nest of mud, sticks, shells & debris, lined with pebbles, shell bits & sticks
Eggs: 4 buff with dark brown or black marks, often nest stained
Incubation: 22–26 days
Fledging: 28–32+ days

Black-necked Stilts appear to be expanding their breeding range to the east and north (Jacobs 1991). The first record of Black-necked Stilts breeding in Missouri was the observation of three nests with eggs in Stoddard County on June 28, 1990 as a direct result of Atlas fieldwork (Jacobs 1991). Black-necked Stilt nests discovered during the Atlas Project and others found in Missouri have all been located on low contour rice field levees.

Code Frequency

Black-necked Stilts were located in only two blocks during the course of the Atlas Project and were confirmed to breed in both. Those nests were easily observed on newly constructed levees bare of vegetation. Those discoveries prompted a more thorough search for Black-necked Stilts throughout the rice-producing district of southeastern Missouri, during which several individuals were located outside of Atlas blocks.

Phenology

The two nests discovered indicate that incubation occurred in June and fledging in early July. However, because of the scarcity of nests, a complete picture of the Black-necked Stilt nesting phenology in Missouri is presently unavailable.

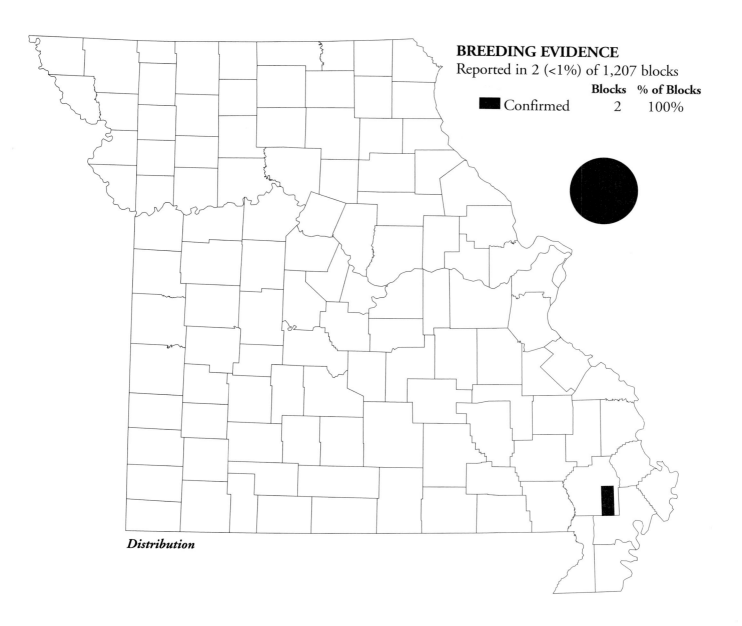

Distribution

Spotted Sandpiper
Actitis macularia

> **Rangewide Distribution:** Throughout northern two-thirds of North America
> **Abundance:** Common & widespread
> **Breeding Habitat:** Grass, woods & shrubs near water
> **Nest:** Lined with grass or moss
> **Eggs:** 1–4 brownish, greenish or pinkish-buff, marked with dark brown
> **Incubation:** 20–24 days
> **Fledging:** 17–21 days

Spotted Sandpipers are perhaps the most easily recognized among the difficult-to-identify small shorebirds. When standing, they teeter continuously. In the air, their stiff-winged flight near the surface of water is characteristic and a certain identification key. Along with the Upland Sandpiper, Killdeer and the American Woodcock, they are the only shorebirds that regularly breed in Missouri. They can be observed foraging around ponds and along streams, but are difficult to track to a nest hidden in the grass or vegetation away from the shoreline.

Code Frequency

Most codes were possible breeding evidence, however, 19 percent of all records were confirmations. The four fledglings and two nests found suggest that it is difficult but not impossible to find nests of this species. Confirmations were made in every natural division except the Mississippi Lowlands.

Distribution

Spotted Sandpipers were located in every natural division, although blocks were widely scattered. Widmann (1907) described Spotted Sandpipers as fairly common along big rivers throughout the state. Because access to long stretches of big rivers is difficult, this species may be more common than Atlas Project data indicates.

Phenology

Few records were available to outline the timing of nesting. Observations of one nest with eggs and one with young fell within the typical May to July rangewide nesting season (Terres 1987). The fledglings sighted in August could easily have been migrant birds (Robbins and Easterla 1992) or a late nesting by this two- or three-brooded species.

Notes

Atlasers observed no Brown-headed Cowbird parasitism. Terres (1987) noted one record for this species: a Brown-headed Cowbird egg found with four Spotted Sandpiper eggs. As the young cowbird would not be fed by adult Spotted Sandpipers, this species would be an ineffective host.

Breeding Phenology

EVIDENCE (# of Records)	MAR	APR	MAY	JUN	JUL	AUG	SEP
NE (1)			5/20	5/20			
NY (1)			5/30	5/30			

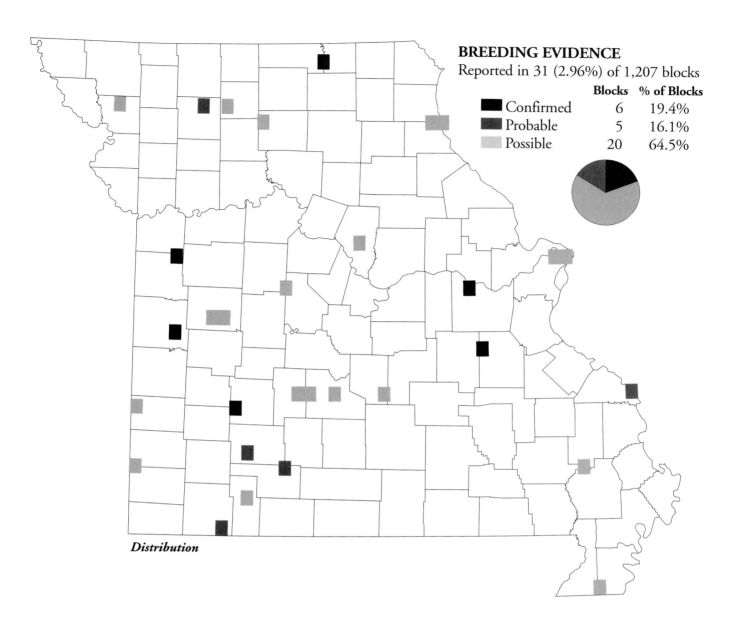

BREEDING EVIDENCE
Reported in 31 (2.96%) of 1,207 blocks

	Blocks	% of Blocks
Confirmed	6	19.4%
Probable	5	16.1%
Possible	20	64.5%

Distribution

Upland Sandpiper
Bartramia longicauda

Rangewide Distribution: Central Canada, south central Alaska, eastern Washington through northeastern United States
Abundance: Fairly common, declining in east
Breeding Habitat: Short-grass fields, pastures & meadows
Nest: Depression in grass on ground, lined with dry grass
Eggs: 4 cream or pinkish-buff with red-brown marks
Incubation: 21–27 days
Fledging: 30–31 days

Urbanization and the conversion of grasslands to agriculture has decreased available habitat for Upland Sandpipers in several areas. In the eastern United States, remnant populations of Upland Sandpipers are restricted to airports, military bases and other extensive grassland areas. They are still common in the western part of the range. Missouri lies on a midline between the two regions.

Known as the Field Plover during Widmann's time (1907), the breeding and transient migratory population of Upland Sandpipers was severely reduced by an open hunting season that historically lasted until May 1. Now that laws protect them, they are fairly common in some areas.

Code Frequency

The somewhat ventriloquistic, whistling calls of Upland Sandpipers can be heard over long distances which makes them easy to detect in May or June when they are calling most. Possible breeding codes were primarily observations by call or by sighting a bird. Probable breeding evidence was often obtained by finding a bird present on a revisit, which suggests territoriality. Nests were hard to find. Most confirmation codes were observations of fledglings or distraction displays near a suspected nest site.

Distribution

This species was recorded in only 139 blocks with most records located in the grasslands of the Glaciated Plains and Osage Plains natural divisions. Scattered records occurred in this species' former range in the Ozark and Ozark Border natural divisions. Interesting records from Douglas, Wright and Texas counties were associated with open grassland and former prairie remnants within the Ozark Natural Division. A May 31 record near Dexter, Stoddard County, in Crowley's Ridge Natural Section of the Mississippi Lowlands Natural Division, was a possible breeding site. Upland Sandpipers were unexpectedly found in primarily agricultural lands along the Missouri River in Holt and Atchison counties.

Abundance

Abundance was nearly equal between the Glaciated Plains and Osage Plains natural divisions. They were by far most common in the Western Glaciated Plains and Grand River natural sections.

Phenology

Territorial birds were located between May 10 and July 7. Considering a 21–27 day incubation period, Upland Sandpipers likely laid most eggs in May and most young were off the nest in June.

Abundance by Natural Division
Average Number of Birds / 100 Stops

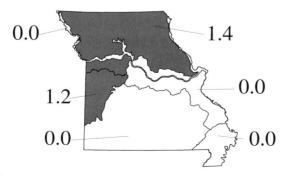

0.0   1.4

1.2   0.0

0.0   0.0

Breeding Phenology

EVIDENCE (# of Records)	MAR	APR	MAY	JUN	JUL	AUG	SEP
NY (1)				7/12	7/12		

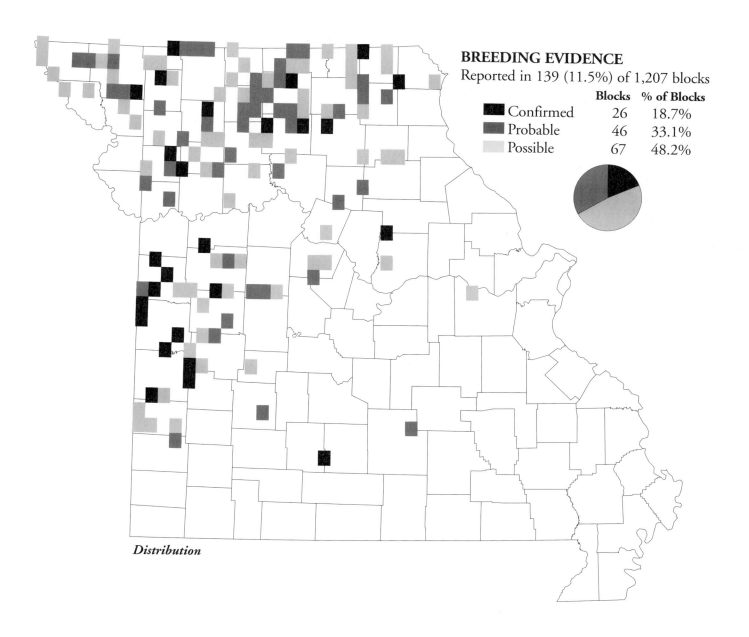

BREEDING EVIDENCE

Reported in 139 (11.5%) of 1,207 blocks

	Blocks	% of Blocks
Confirmed	26	18.7%
Probable	46	33.1%
Possible	67	48.2%

Distribution

American Woodcock
Scolopax minor

> **Rangewide Distribution:** Southeastern Canada & eastern United States
> **Abundance:** Fairly common, but local
> **Breeding Habitat:** Forests & thickets by streams or moist woodlands
> **Nest:** Ground with abundant leaves, lined with dead leaves & conifer needles, occasionally rimmed with twigs
> **Eggs:** 4 pink-buff or cinnamon with brown marks, somewhat wreathed
> **Incubation:** 20–21 days
> **Fledging:** 14 days

American Woodcocks arrive en masse when the spring thaw enables them to probe for insects and worms in the damp soil. Males initiate their elaborate courtship flights and displays within days of their arrival. Display areas and nesting grounds are often scrub/shrub successional habitats and the brushy edges between woods and fields. They normally avoid the interiors of mature forest and breeding territories often include dry areas for nesting and damp areas for feeding (Sanderson 1977).

Code Frequency

American Woodcocks are uncommon breeders (Robbins and Easterla 1992) that were difficult for Atlasers to confirm. Because this early-season breeder is most active during twilight and after dark, most Atlaser effort likely took place too late in the season and at the wrong time of day. Most woodcock records fell in the possible breeding category. Probable breeding records resulted primarily from observations of territorial and courtship displays. American Woodcocks, however, often display while migrating (Peterjohn and Rice 1991). Therefore, the observation of courtship does not necessarily imply nesting in the block or even in the state.

Distribution

Although woodcock records were obtained nearly statewide, there were fewer along the western border. Most of the probable or confirmed breeding records occurred in eastern Missouri. Regional variations may result from the American Woodcock becoming less common as it approaches the western edge of its range. Two records associated with Crowley's Ridge were the only reports of American Woodcocks in the Mississippi Lowlands, a natural division that lacks breeding habitat.

Phenology

American Woodcocks are early migrants and breeders as suggested by confirmed breeding dates. They produce only one brood per year (Ehrlich et al. 1988).

Breeding Phenology

EVIDENCE (# of Records)	MAR	APR	MAY	JUN	JUL	AUG	SEP
NE (6)	3/07		5/03				
NY (1)		4/20	4/20				

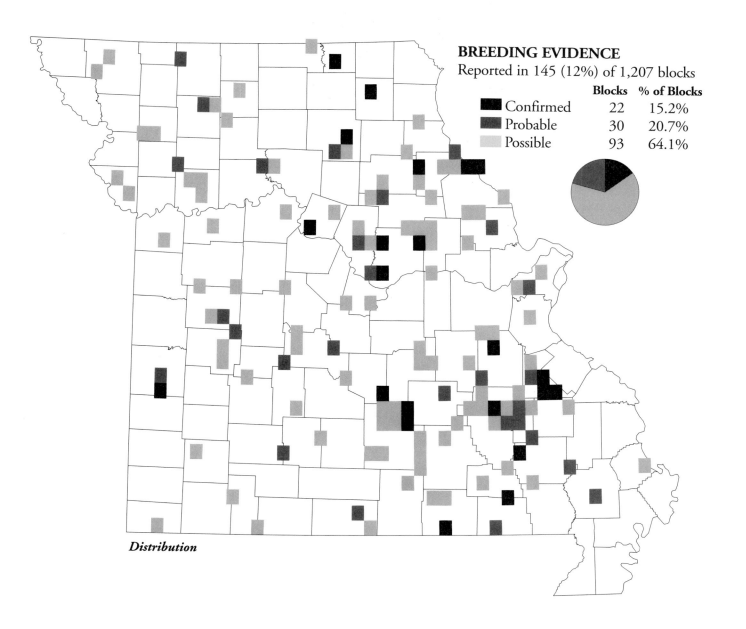

BREEDING EVIDENCE
Reported in 145 (12%) of 1,207 blocks

	Blocks	% of Blocks
Confirmed	22	15.2%
Probable	30	20.7%
Possible	93	64.1%

Distribution

Least Tern
Sterna antillarum

Rangewide Distribution: Western & eastern Gulf
   coasts, central United States
Abundance: Fairly common & local, declining inland
   & west
Breeding Habitat: Sand or gravel areas with scarce
   vegetation surrounded by water
Nest: Usually colonies with shallow ground depres-
   sions, usually unlined
Eggs: 2 olive-buff colored with dark brown markings
Incubation: 20–22 days
Fledging: 19–20+ days

The Least Tern is the smallest tern and the only one that presently nests in Missouri. A former breeding species on sandbars throughout the Big Rivers Natural Division (Widmann 1907), Least Terns are now restricted to several islands on the lower Mississippi River. The interior population of this species is listed as Endangered by the United States Fish and Wildlife Service.

Code Frequency

Courtship behavior indicating probable breeding was recorded in two blocks in Pemiscot County. Possible codes were listed for New Madrid and Mississippi counties.

Distribution

All Atlas Project records occurred in counties next to known nesting colonies on lower Mississippi River islands. In August and early September following the nesting season, the Least Tern can be found statewide in small numbers (Robbins and Easterla 1992).

Phenology

First arrivals occur by mid-May (Robbins and Easterla 1992), with breeding colonies forming on sandbars as flood waters subside. After breeding, individuals disperse statewide. Most birds depart Missouri by mid-September (Robbins and Easterla 1992).

Notes

In 1993, some Least Terns bred in agricultural fields rather than on sand bars in the Mississippi River. This was likely due to continued flood water levels on the river. Brad Jacobs observed several pairs on nests in a field that was surrounded by water. On a later visit nests were observed to have been depredated by unknown predators when the flood waters had subsided and the nest sites were reconnected to the adjacent land.

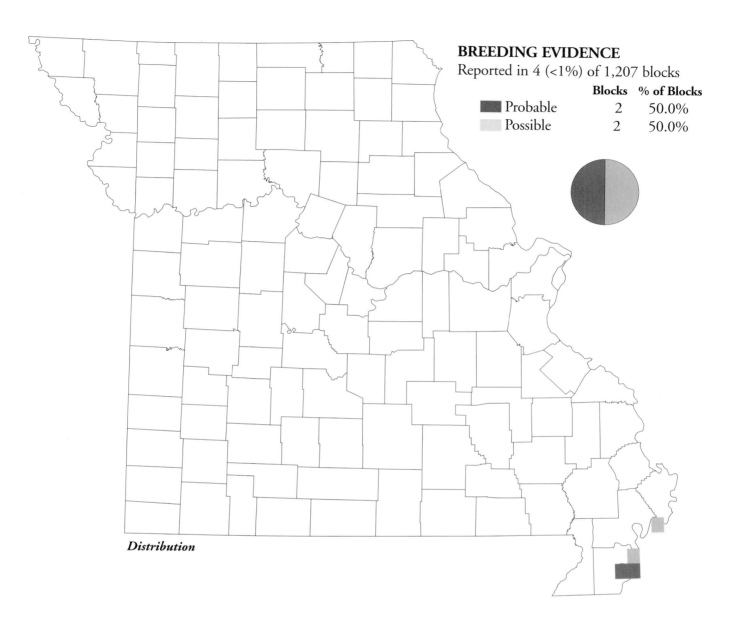

BREEDING EVIDENCE
Reported in 4 (<1%) of 1,207 blocks

	Blocks	% of Blocks
Probable	2	50.0%
Possible	2	50.0%

Distribution

Rock Dove
Columba livia

Rangewide Distribution: Native to Eurasia, introduced & established through most of the world
Abundance: Widespread & common
Breeding Habitat: Cliffs & ledges, buildings, bridges & caves
Nest: Unlined saucer of roots, stems & leaves on building ledges, eaves or bridges
Eggs: 2 white & unmarked
Incubation: 16–19 days
Fledging: 25–26 days

The first Rock Doves, or pigeons, are thought to have arrived in North America in the early 1600s (Terres 1987) and much later in Missouri. Widmann (1907) did not list them, however, perhaps because of their domesticated status. Today, feral Rock Doves are found throughout the lower 48 states in association with farms and in urban and suburban areas.

Code Frequency

Rock Doves nest where sheltered sites are near grain food sources, such as farm buildings and grain elevators. Warehouses, bridges and ornate city buildings are commonly used as nest sites. Occasionally they nest far from human habitation in cavities along bluff faces. Rock Doves are easily seen as they fly across the sky, typically in flocks. Therefore, where not recorded in a block, they likely were few or not present. Although Rock Doves may range several kilometers from nest sites, they may have nested in most blocks where they were recorded.

Distribution

Breeding Rock Doves were sparingly distributed statewide away from urban and suburban areas. They were most notably not detected in many parts of the Ozark Natural Division and Eastern Glaciated Plains Natural Section, where forest cover predominates. Their habit of nesting in cave entrances and on cliffs as well as in barns and silos may have resulted in some Ozark records. They were also recorded in few blocks within the northeastern sector of the state.

Abundance

Rock Doves were more than twice as abundant in the Big Rivers Natural Division than in the division where it was next-most abundant, the Osage Plains. The grainfields along the Missouri and Mississippi rivers, as well as the barns and grain elevators, may provide both secure nesting sites and an abundance of food.

Abundance by Natural Division
Average Number of Birds / 100 Stops

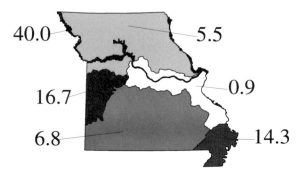

40.0   5.5

16.7   0.9

6.8   14.3

Breeding Phenology

EVIDENCE (# of Records)	MAR	APR	MAY	JUN	JUL	AUG	SEP
NB (8)	4/06				6/30		
NE (12)	4/10				6/27		
NY (24)	3/04					8/01	
FY (3)		4/28		6/07			

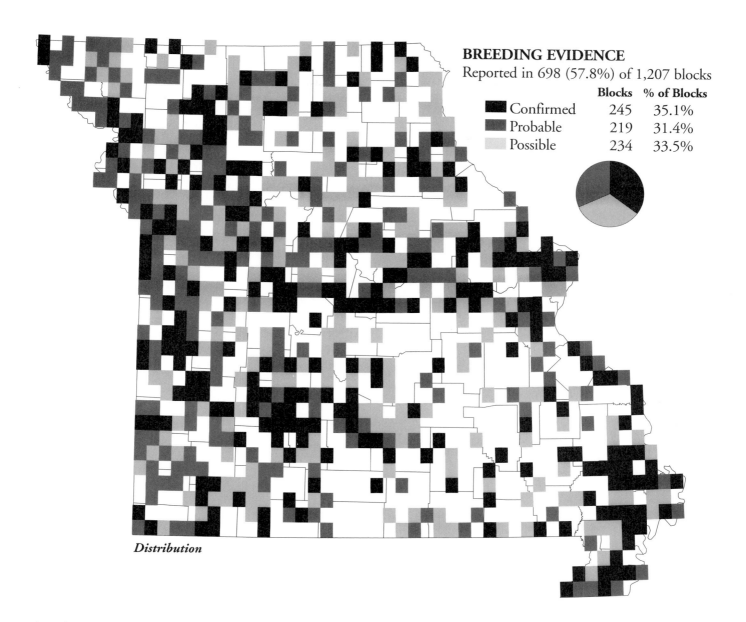

BREEDING EVIDENCE
Reported in 698 (57.8%) of 1,207 blocks

	Blocks	% of Blocks
Confirmed	245	35.1%
Probable	219	31.4%
Possible	234	33.5%

Distribution

Phenology

Rock Doves raise several broods each year and lay eggs every month of the year (Terres 1987). Most egg laying occurred from March through June and August through November (Terres 1987). Atlas surveyors found evidence of nesting March through August. Because little Atlasing was conducted after August, there were few opportunities to determine if the species nests as late as November in Missouri.

Mourning Dove
Zenaida macroura

Rangewide Distribution: Southern Canada & entire United States
Abundance: Abundant & widespread
Breeding Habitat: Forests, grasslands, urban areas & croplands
Nest: Lined twigs, sticks & grass in trees or on ground
Eggs: 2 white & unmarked
Incubation: 13–14 days
Fledging: 12–14 days

The cooing of Mourning Doves can be heard in spring and summer throughout Missouri towns, cities and countryside. They nest primarily at forest edges, farm groves, suburban yards and city parks. In areas with few trees, they will sometimes nest directly on the ground. During late summer and early fall, flocks of thousands congregate in fields to feed on waste grain and sunflower *(Helianthus spp.)* seeds.

Code Frequency

Mourning Doves are easy to detect. Recorded in 1186 blocks, the species was the Atlas Project's fourth most-commonly recorded bird. Atlasers recorded it in only 16 fewer blocks than the most widely recorded species, the Northern Cardinal. Mourning Doves and their nests are easily observed with minimal effort. Although they likely bred in most blocks where found, they were apparently present in low numbers where recorded only as possible breeders.

Distribution

Mourning Doves were found statewide. The only definitive distributional pattern that emerged was the lack of higher nesting evidence from the more forested parts of the state. Fewer confirmations in the Ozark Natural Division were perhaps due to that region's lack of forest edge and open grasslands and the greater difficulty of locating nests in more wooded conditions. Birds in heavily forested areas may have had to travel greater distances to forage, thus making it even more difficult to locate nests.

Abundance

The greatest abundances were recorded in the Big Rivers, Osage Plains, Glaciated Plains and Mississippi Lowlands natural divisions. The Ozark and Ozark Border natural divisions had less than half as many birds per route, perhaps due to greater forest cover in these sections. This species generally avoids heavily-forested regions (Mirarchi and Baskett 1994). Abundance values suggest roadside point counts can be useful in monitoring population trends in this species. In Missouri, both point and transect counts are conducted by the Conservation Department to determine Mourning Dove relative abundance levels.

Abundance by Natural Division
Average Number of Birds / 100 Stops

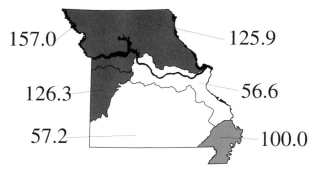

157.0   125.9

126.3   56.6

57.2   100.0

Breeding Phenology

EVIDENCE (# of Records)	MAR	APR	MAY	JUN	JUL	AUG	SEP
NB (34)	4/09				7/14		
NE (90)	4/02						8/08
NY (32)	4/05						8/04
FY (11)		5/13			7/30		

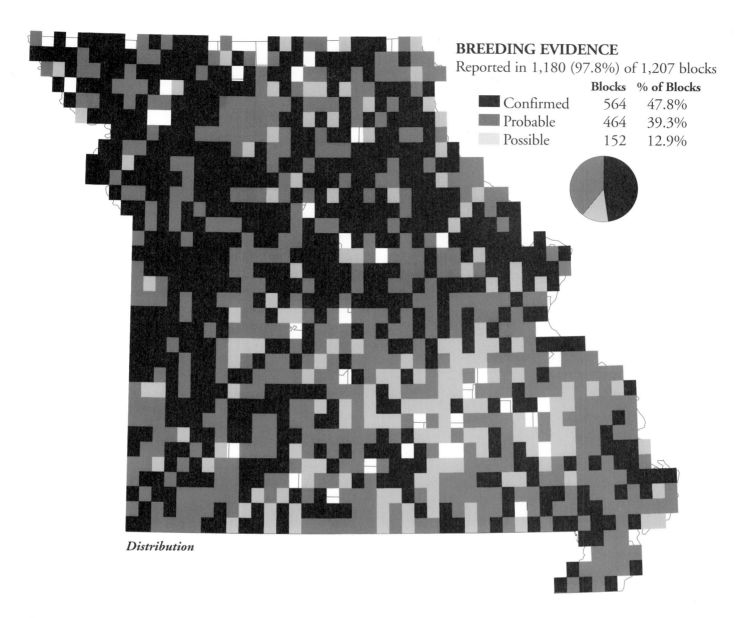

BREEDING EVIDENCE
Reported in 1,180 (97.8%) of 1,207 blocks

	Blocks	% of Blocks
Confirmed	564	47.8%
Probable	464	39.3%
Possible	152	12.9%

Distribution

Phenology

This species nests from late March through early September (Hallett 1990). Although nests have been observed into mid-October, most atlasing was completed by this date. Due to the ability of Mourning Doves to raise 2–6 broods a season (Ehrlich et al. 1988), an abundance of young birds is often observed during late summer.

Notes

Only one record of Brown-headed Cowbird parasitism was documented out of the 119 nests that were observed with eggs or young. Mourning Doves are considered rare hosts of Brown-headed Cowbirds (Ehrlich et al. 1988).

Black-billed Cuckoo

Coccyzus erythropthalmus

Rangewide Distribution: South central to southeastern Canada, north central & central United States to East Coast
Abundance: Uncommon to fairly common
Breeding Habitat: Groves of trees, forest edges & moist thickets
Nest: Twigs lined with ferns, grass, burrs, catkins & roots in trees near trunk, or on logs or vines on the ground
Eggs: 2–3 blue-green, occasionally marked with same
Incubation: 10–13 days
Fledging: 7–9 days

Black-billed Cuckoos nest in groves, forest edges and thickets, especially thickly wooded areas near streams (Harrison 1975). In Missouri, they commonly use willow stands in marshes. Because the primary breeding range of Black-billed Cuckoos lies to the north of Missouri Black-billed Cuckoos are expected in Missouri primarily during migration.

Code Frequency

Because of the late and variable timing of the spring migration of cuckoos (Robbins and Easterla 1992), some late migrants may have been recorded as possible breeders in a number of the blocks. Consequently, data from probable and confirmed blocks may provide the best representation of the Black-billed Cuckoo's distribution in Missouri. Misidentification of Yellow-billed versus Black-billed cuckoos may also have occurred. However, in 82.5 percent of the blocks where Black-billed Cuckoos were recorded, Yellow-billed Cockoos were also recorded, suggesting that most Atlasers had no difficulty separating the songs and calls of the two species.

Distribution

As expected, this northerly-distributed species was more common in northern Missouri. About half of the probable and confirmed records were in the northwestern quarter of the state. None-the-less, Black-billed Cuckoos were unexpectedly scattered throughout southern Missouri. There is no general habitat pattern to explain this distribution.

Phenology

The Black-billed Cuckoo's breeding season apparently extends late into the year as the observation of a nest with eggs indicates. The Black-billed Cuckoo's brief 11-day incubation and 10-day rearing periods enable them to rear multiple broods within a single season.

Notes

Although Black-billed Cuckoos have been reported to lay eggs in the nests of other birds (Ehrlich et al. 1988), no evidence of the behavior was observed during the Atlas Project.

Breeding Phenology

EVIDENCE (# of Records)	MAR	APR	MAY	JUN	JUL	AUG	SEP
NB (1)				7/05	7/05		
NE (1)						8/18	8/18
FY (1)				7/16	7/16		

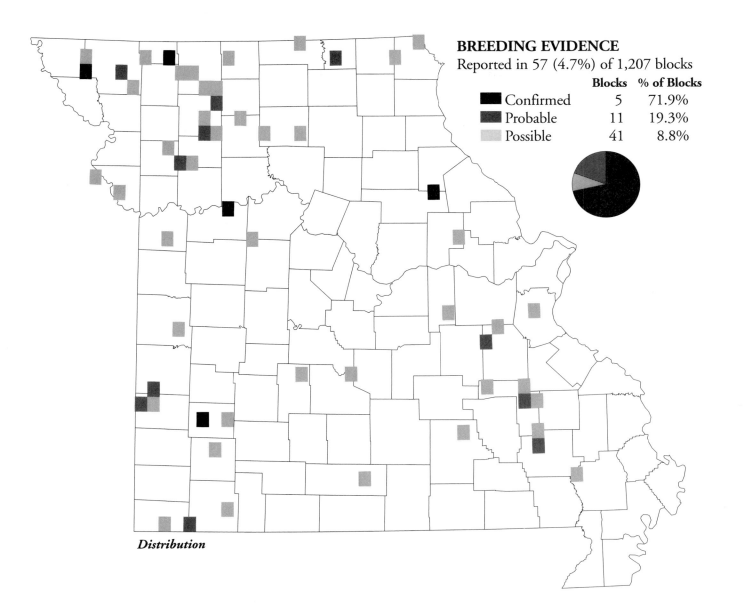

BREEDING EVIDENCE
Reported in 57 (4.7%) of 1,207 blocks

	Blocks	% of Blocks
■ Confirmed	5	71.9%
■ Probable	11	19.3%
□ Possible	41	8.8%

Distribution

Yellow-billed Cuckoo
Coccyzus americanus

> **Rangewide Distribution:** Interior California across
>   United States to East Coast
> **Abundance:** Common to widespread
> **Breeding Habitat:** Woods with dense undergrowth &
>   thickets
> **Nest:** Loose twigs lined with roots, leaves & pine
>   needles in shrubs
> **Eggs:** 4 light blue fading to light greenish-yellow
> **Incubation:** 9–11 days
> **Fledging:** 7–8 days

Also known as Raincrows, Yellow-billed Cuckoos are forest and open woodland birds that feed mainly on large caterpillars. They become especially abundant during years of caterpillar outbreaks, resulting in large year-to-year fluctuations in population size (Kaufman 1996). In addition to consuming caterpillars in large quantities, they eat fruit, frogs, lizards and berries (Ehrlich et al. 1988).

Code Frequency

Yellow-billed Cuckoos are easy to detect because of their vocalizations. Confirming breeding is sometimes difficult, perhaps because foliage is too dense for nests to be found. Sixty-four percent of all confirmed records resulted from the observation of food being carried to young.

Distribution

This species was documented in about 86 percent of the blocks in the Big Rivers, Mississippi Lowlands, Osage Plains and Ozark natural divisions. Due primarily to the 68 percent frequency reported in the Eastern Glaciated Plains Natural Section, fewer blocks (78 percent) were reported in the Glaciated Plains Natural Division. Most noticeable was a total absence of reports from several blocks in northeastern Missouri. Whether this distributional pattern was due to shifts in caterpillar abundance is unknown.

Abundance

Although Yellow-billed Cuckoos were found statewide, the greatest abundance was recorded in the Ozark Natural Division, with lowest abundance in the Glaciated Plains. This pattern may relate to prey abundance during the Atlas Project.

Phenology

Yellow-billed Cuckoos raise one or, possibly, two broods per season (Ehrlich et al. 1988). The extended breeding season discovered by the Atlas Project could relate to differences in breeding among individuals as determined by food abundance or it may represent multiple broods.

Abundance by Natural Division
Average Number of Birds / 100 Stops

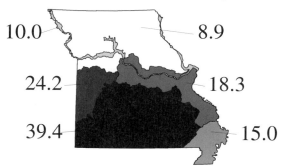

10.0   8.9
24.2   18.3
39.4   15.0

Breeding Phenology

EVIDENCE (# of Records)	MAR	APR	MAY	JUN	JUL	AUG	SEP
NB (13)			5/24			8/02	
NE (11)			5/25			8/02	
NY (5)			5/26		7/16		
FY (105)			6/01				9/03

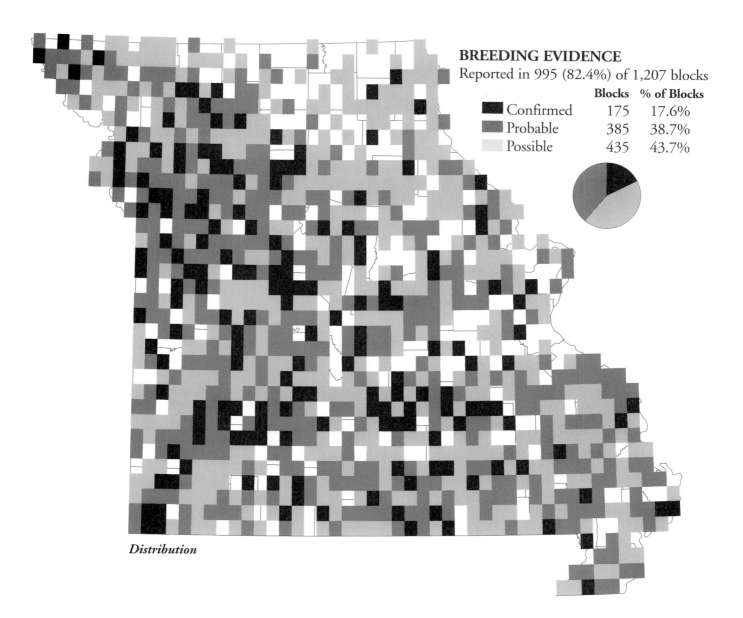

Distribution

Notes

Like their Old World counterparts, Yellow-billed Cuckoos occasionally lay eggs in the nests of other birds although they also build their own nests and raise their young. Widmann (1907) listed the Black-billed Cuckoo, Gray Catbird, Brown Thrasher, Wood Thrush, Cedar Waxwing, Northern Cardinal and Rose-breasted Grosbeak as repeated hosts, although none of these species has been observed to successfully raise Yellow-billed Cuckoos.

Greater Roadrunner

Geococcyx californianus

Rangewide Distribution: Southern United States to north central Louisiana & Mexico

Abundance: Common

Breeding Habitat: Mixed brush, open dry land & pine/oak woods

Nest: Sticks lined with fine materials, placed low in tree or thicket

Eggs: 4–6 white with chalky yellowish coat

Incubation: 20 days

Fledging: 18 days

This southwestern species occurs in Missouri glades, old fields and occasionally around residences and urban areas. Greater Roadrunners have generally been expanding their range in Missouri during the later half of the 20th century. The first record of a roadrunner in Missouri was in Taney County in 1956 (Brown 1963). By the mid-1970s it had been recorded in all southwestern counties with reports as far north as the Missouri River (Norris and Elder 1982). Following a series of hard winters in the late 1970s, roadrunners disappeared from much of their Missouri range, although they had noticeably expanded their range again by the late 1980s (Robbins and Easterla 1992).

Code Frequency

Greater Roadrunners are unlikely to be seen in Missouri because of their rarity and because they occur primarily in rocky, remote situations. Thus, they may have occurred in a few more blocks than shown on the map. Because locating roadrunner nests is extremely difficult, they may have bred in many of the blocks in which they were recorded. Additionally, because Missouri is on the northern fringe of the breeding range, some of the birds may have been unmated individuals. The single probable breeding record resulted when a roadrunner was sighted twice in the same locality with at least ten days intervening, indicating territoriality.

Distribution

Atlasers detected Greater Roadrunners in seven blocks located in Barry, Newton, Stone and Taney counties. The latter county contained half of the blocks and the highest level of breeding evidence. Greater Roadrunners occurred at several location during 1986–1992 outside of Atlas blocks. At one of these locations, near Hardenville in Ozark County, breeding was documented in June by the observation of recently-fledged young.

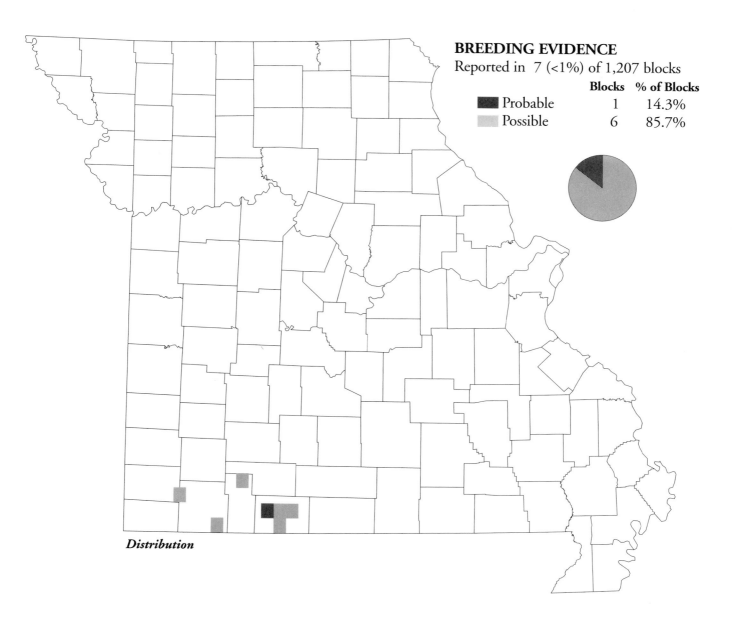

Reported in 7 (<1%) of 1,207 blocks

	Blocks	% of Blocks
Probable	1	14.3%
Possible	6	85.7%

Distribution

Barn Owl
Tyto alba

Rangewide Distribution: Southern Canada, United States to South America & worldwide

Abundance: Common in the West, uncommon & declining in eastern United States

Breeding Habitat: Tree cavities, barns, sheds & man-made structures

Nest: No nest structure, occasionally sticks & chips in a crevice or building

Eggs: 5–7 white & often nest-stained, elliptical in shape

Incubation: 30–34 days

Fledging: 52–56 days

Barn Owls are most often found in association with open grasslands. In some regions they occupy areas with crops such as cotton or grains. Buildings or hollow trees are used as roosting and nesting sites in most breeding territories. Waste grain in fields and around farmsteads is apparently important in establishing sufficient rodent populations to support nesting. Barn Owls are believed to have been most numerous and widespread in the Midwest during the early 1900s (Colvin 1985). They apparently benefitted from diversified farming practices and an abundant rodent prey base associated with less efficient grain harvest and storage. They became increasingly scarce during the mid- to late 1900s in response to cleaner farming methods, the removal of barns and other open buildings and the use of rodenticides (Colvin 1985).

Code Frequency

The nocturnal, secretive nature of Barn Owls, coupled with their extreme rarity, resulted in only 11 records during the seven-year Atlas Project. Because nests are usually hidden and difficult to find, the eight blocks in which they were classified as possible breeders may have contained nests.

Distribution

Barn Owls were very sparingly recorded. Over half of the blocks in which Barn Owls were recorded were in the Mississippi Lowlands Natural Division. That division's openness and the cotton and cereal grains that are grown there apparently offer suitable habitat for rodents that are prey for Barn Owls.

Abundance

Barn Owls are indeed a rare species. Data are insufficient to evaluate any variation in abundance across Missouri.

Phenology

Barn Owls are apparently capable of initiating nesting early in the year. Barn Owls are recognized to have a lengthy nesting season elsewhere at this latitude (Colvin 1985).

Breeding Phenology

EVIDENCE (# of Records)	MAR	APR	MAY	JUN	JUL	AUG	SEP
NY (2)			5/29	6/09			

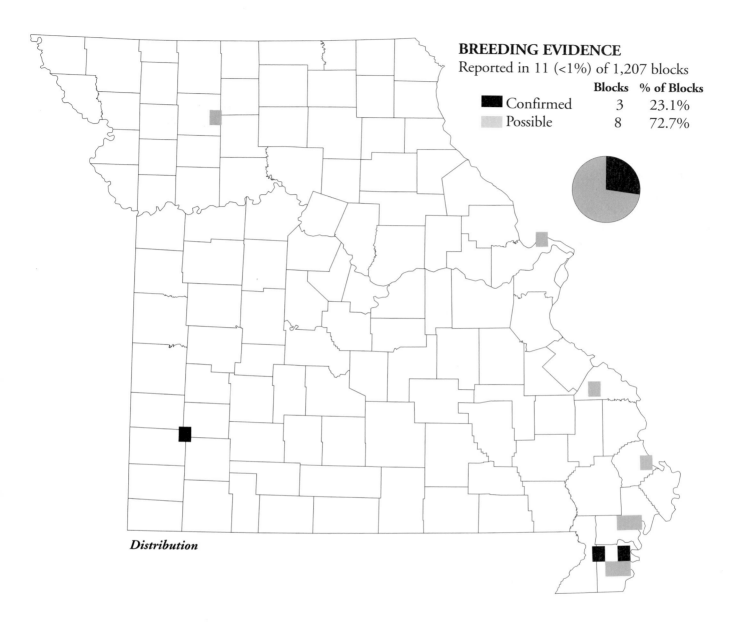

BREEDING EVIDENCE
Reported in 11 (<1%) of 1,207 blocks

	Blocks	% of Blocks
Confirmed	3	23.1%
Possible	8	72.7%

Distribution

Eastern Screech-Owl

Otus asio

Rangewide Distribution: Southeastern Canada & eastern United States
Abundance: Common
Breeding Habitat: Open deciduous woods & scrub, parks, towns & riparian zones
Nest: Tree cavity or stump lined with remnants of feathers, fur & debris
Eggs: 4–5 white
Incubation: 26 days
Fledging: 27 days

Eastern Screech-Owls occur and breed in open woodlands, deciduous and coniferous forests and residential areas in towns. They have the broadest ecological niche of any owl within their range (Gehlbach 1995). A prerequisite for nesting is a suitable tree cavity or substitute such as a man-made nest box. They often allow people to approach closely.

Code Frequency

Highly nocturnal, Eastern Screech-Owls are detected primarily by their whinnying calls. Because of their nocturnal behavior and their early breeding season, Eastern Screech-Owls were probably missed by the majority of Atlasers. For the same reasons, Eastern Screech-Owls were rarely classified as probable or confirmed breeders. Breeding was confirmed primarily by evidence detectable during the day, such as fledglings. Although this owl was reported in only 31 percent of blocks, it is reasonable to assume that it occurred in many blocks where it was not recorded.

Distribution

Eastern Screech-Owls ranged throughout Missouri. Although they may have bred undetected in the majority of blocks, observations of breeding behavior were concentrated in the southern and western regions. This species was scarce in the north, in the Mississippi Lowlands and along the eastern border of the state. In general, blocks where Eastern Screech-Owls were recorded corresponded to greater forest cover (fig. 9) or greater Atlaser effort (fig. 5).

Phenology

Atlas data suggest that egg laying can begin as early as late February. Late dates for eggs and nestlings were unexpected because screech-owls are reported by Ehrlich et al. (1988) to be single-brooded. However, Gehlbach (1995) reported great variation in nesting dates, with eggs in the nest from December 15 to August 15. He believed the later dates were due to renesting.

Breeding Phenology

EVIDENCE (# of Records)	MAR	APR	MAY	JUN	JUL	AUG	SEP
NE (2)		4/20			7/08		
NY (7)	4/04			6/17			

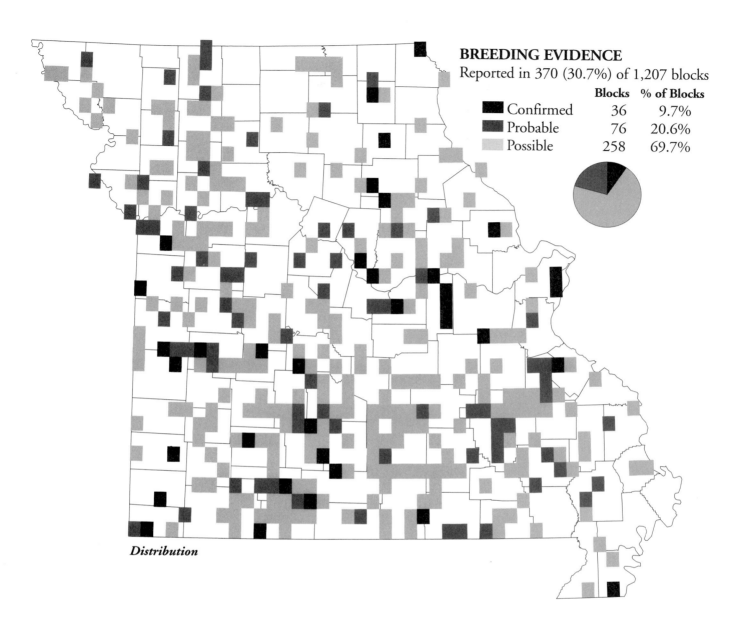

BREEDING EVIDENCE
Reported in 370 (30.7%) of 1,207 blocks

	Blocks	% of Blocks
Confirmed	36	9.7%
Probable	76	20.6%
Possible	258	69.7%

Distribution

Great Horned Owl
Bubo virginianus

Rangewide Distribution: Canada, Alaska & entire
   United States to southern South America
Abundance: Widespread & common
Breeding Habitat: Various moist or arid coniferous &
   deciduous forests
Nest: Sticks, moss, hair, bark & roots, lined with
   feathers or down, in tree cavities or abandoned nests
Eggs: 2–3 dull white
Incubation: 26–35 days
Fledging: 35 days

These large, nocturnal birds are ubiquitous, occurring in
deep forests, open areas interspersed with small woodlots
and even in urban areas. In Ohio, their numbers seem to be
expanding within more open habitats (Peterjohn and Rice
1991). Great Horned Owls prefer to nest in tree cavities,
but they will occupy abandoned hawk, crow and heron
nests when suitable cavities are unavailable (Price 1940).
They will occasionally nest in barn lofts, abandoned buildings and on bluffs (Williams 1950).

Code Frequency

Great Horned Owls appear to initiate nesting earlier in
the year than any other native bird. Courtship begins in
December and incubating adults have been observed in
January. Because the majority of Atlas Project field effort
occurred later in the season, and during daylight when this
nocturnal species was not active, it can be assumed that
Great Horned Owls occurred and bred undetected in many
blocks. Most breeding was confirmed by the observation of
fledglings, the evidence easiest to detect during the day and
late in the breeding season.

Distribution

Because evidence indicating probable or confirmed
breeding is relatively difficult to detect, most of the possible
records were likely from actual breeding areas, suggesting
that Great Horned Owls bred statewide. Their relative
scarcity in the Mississippi Lowlands may relate to a shortage
of suitable nest trees. Prey would seem to be plentiful in
that natural division. A relative scarcity of Great Horned
Owls in portions of the Ozarks might have been due to the
lack of open and fragmented habitats in that region. For
unknown reasons, this species was encountered less frequently in portions of central, eastern and northern
Missouri.

Abundance

Great Horned Owls were recorded too infrequently on
Miniroutes and Breeding Bird Surveys to establish meaningful abundance data.

Breeding Phenology

EVIDENCE (# of Records)	MAR	APR	MAY	JUN	JUL	AUG	SEP
NE (2)	3/01	3/01					
NY (13)	3/15			5/20			
FY (3)	4/05			6/17			

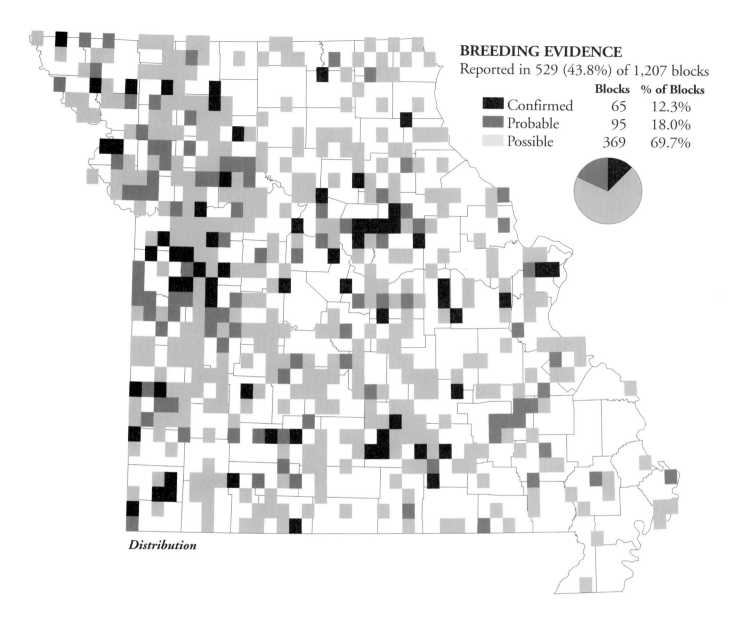

BREEDING EVIDENCE
Reported in 529 (43.8%) of 1,207 blocks

		Blocks	% of Blocks
■	Confirmed	65	12.3%
■	Probable	95	18.0%
■	Possible	369	69.7%

Distribution

Phenology

Great Horned Owls are permanent residents that initiate nesting early, with young fledging at the same time that rodent prey becomes available in spring (Marti 1974). Earlier breeding confirmation dates might have been record-ed by more early-season Atlasing. Late dates for nestlings revealed a surprisingly protracted breeding season for this single-brooded species. Renesting may have accounted for some of the late nesting.

Barred Owl
Strix varia

> **Rangewide Distribution:** Central & southeastern Canada, eastern United States & central Mexico
> **Abundance:** Common, expanding south from northwestern range
> **Breeding Habitat:** Dense coniferous & mixed forests & river valleys
> **Nest:** May add green sprigs to abandoned nests but adds no material to tree cavity
> **Eggs:** 2–3 white
> **Incubation:** 28–33 days
> **Fledging:** 42 days

Barred Owls associate with mature open woods containing numerous old trees and large cavities for nesting (Voous 1988). They are more frequently encountered in relatively moist woodlands than are Great Horned Owls. They seem somewhat dependent on forest size, preferring woodlands of greater than 40 hectares and avoiding small woodlots and narrow riparian corridors (Peterjohn and Rice 1991).

Code Frequency

Although difficult to locate because of their nocturnal habits, Barred Owl vocalizations are easily-recognized and can at times be heard during the day. Their nest sites and other confirmed breeding evidence, however, are extremely difficult to find. Thus, most blocks in which Barred Owls were recorded as probable and possible breeders were likely indeed nest site locations. Additionally, Barred Owls may have occurred and bred in many blocks in which they were not recorded at all. The most common confirmed breeding evidence during the Atlas Project was fledged young. This presumably was due to the young owls' tendency to leave the nest and perch nearby before learning to fly well.

Distribution

Blocks in which Barred Owls were recorded appeared to have a fairly uniform statewide distribution, except they were not recorded in several adjoining blocks in the Mississippi Lowlands and in the northwestern corner of the state. In general, low Barred Owl distribution corresponded to those blocks receiving the least amount of Atlas Project coverage (fig. 5). In northwestern Missouri, where coverage was more thorough, Barred Owl distribution may be fairly accurately portrayed.

Phenology

Barred Owls breed early with incubation initiated as early as February (Bent 1938). Because most Atlas Project efforts began no earlier than April, much early nesting evidence was missed. The nesting season was apparently somewhat protracted. Because Barred Owls are single-brooded (Harrison 1975), there appears to be great variation in the dates of nest initiation.

Breeding Phenology

EVIDENCE (# of Records)	MAR	APR	MAY	JUN	JUL	AUG	SEP
NY (5)		4/20			7/15		
FY (3)		4/17		6/01			

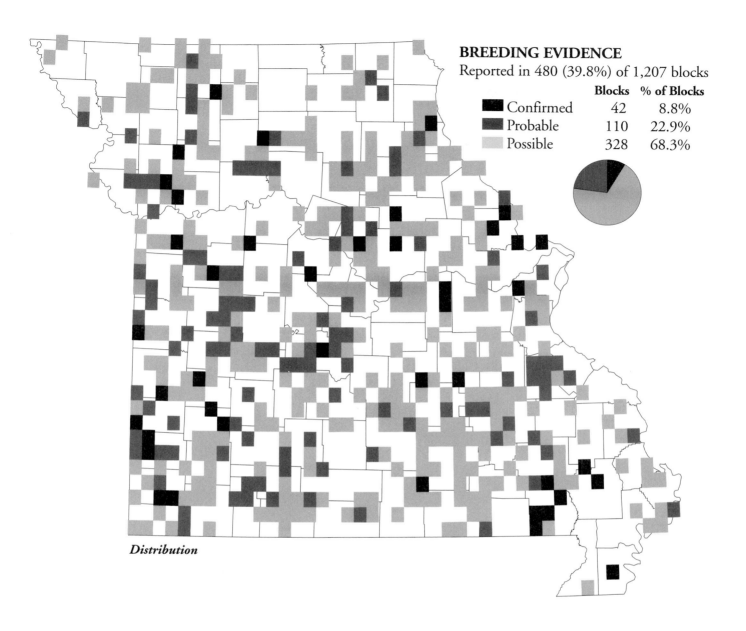

BREEDING EVIDENCE
Reported in 480 (39.8%) of 1,207 blocks

	Blocks	% of Blocks
Confirmed	42	8.8%
Probable	110	22.9%
Possible	328	68.3%

Distribution

Short-eared Owl

Asio flammeus

> **Rangewide Distribution:** Canada, Alaska, northern United States & worldwide
> **Abundance:** Fairly common & widespread
> **Breeding Habitat:** Open prairies, meadows, savannahs & marshes
> **Nest:** Unlined or sparsely lined with grass, weeds & feathers, on the ground
> **Eggs:** 4–7 white/cream-white
> **Incubation:** 26–28 days
> **Fledging:** 31–36 days

Short-eared Owls are most active in the early morning and evening as they search for food or perform courtship displays. They select grasslands for nesting, so early season mowing of hayfields may disrupt nesting attempts. Short-eared Owls are rare, sporadic breeders in Missouri. Prior to the Atlas Project, there were only three records of possible breeding during the latter half of the 20th century (Robbins and Easterla 1992). Short-eared Owls are known to move nomadically to locate areas with high rodent populations, then settle and breed (Ehrlich et al. 1988). They are apparently more likely to nest in Missouri during years in which rodent populations are high.

Code Frequency

All three confirmed observations of Short-eared Owls involved carrying food to young. Because of their dawn and dusk activity periods, Short-eared Owls may have been missed by Atlasers in some blocks.

Distribution

Breeding locations discovered during the Atlas Project occurred in the Glaciated Plains and Osage Plains natural divisions, apparently regions of the state where they formerly nested more extensively. Widmann (1907) listed nesting records in the northern and western prairie regions at the turn of the century. The Conservation Reserve Program (CRP), a government cropland set-aside program, appeared to greatly enhance habitat for Short-eared Owls during the course of the Atlas Project. This and other grassland species apparently nested on CRP acreages across the state, but especially in the Glaciated Plains Natural Division.

Phenology

Food carried to young was observed between June 17 and July 2. Although Terres (1987) indicated that eggs of this single-brooded species are typically laid between March and June in prairie states, April is apparently the most likely month for egg laying in Missouri.

Breeding Phenology

EVIDENCE (# of Records)	MAR	APR	MAY	JUN	JUL	AUG	SEP
FY (3)				6/17	7/02		

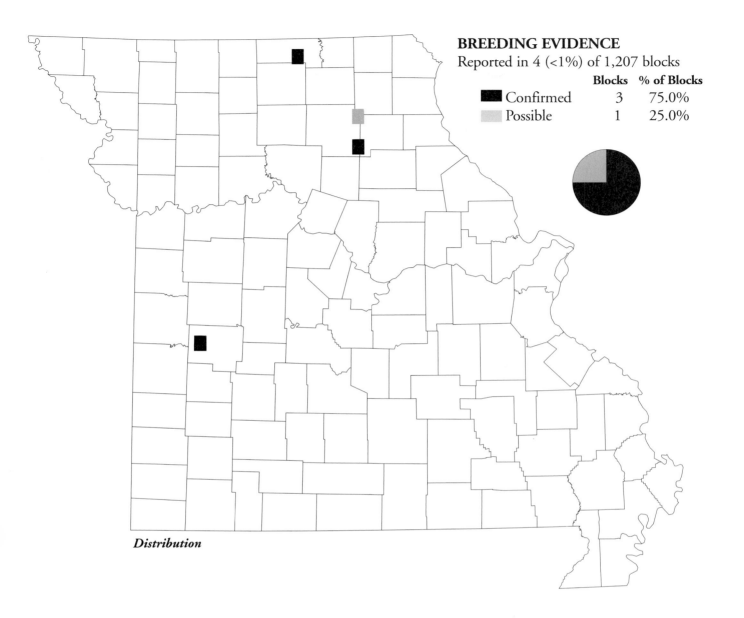

Reported in 4 (<1%) of 1,207 blocks

	Blocks	% of Blocks
Confirmed	3	75.0%
Possible	1	25.0%

Distribution

Common Nighthawk

Chordeiles minor

Rangewide Distribution: Southern Canada, entire
    United States south through Panama
Abundance: Common & widespread
Breeding Habitat: Open & semi-open areas with little
    vegetation
Nest: Depression on flat ground
Eggs: 2 white/olive with olive mottling
Incubation: 19 days
Fledging: 21 days

Historically, this goatsucker probably nested on open barrens and burnt-over tracts left by forest and prairie fires (Terres 1987) and they can still be found nesting on glades far from human habitation. Beginning in the late 1800s (Widmann 1907), this species has increasingly nested on flat, gravel-covered roofs. Therefore, most nesting today is associated with cities and towns, and the increase in population of this species has coincided with urban development. In addition to the availability of rooftop nest sites, nighthawks may find foraging more successful in cities than in the wild because city lights attract large quantities of insects.

Code Frequency

Common Nighthawks are easily detected as they hunt insects overhead and they are readily identified by white wing bars and their "peent" calls. Even at night, they often can be seen overhead if illuminated by city lights. They also forage during daylight hours and thus were more likely to be detected by daytime Atlasers than more strictly nocturnal birds, such as Whip-poor-wills and owls. Because of their relative ease of detection, Common Nighthawks may have been present in low numbers or absent from blocks where they were not recorded. The confirmation of breeding by Common Nighthawks proved difficult during the Atlas Project, due to the inaccessibility of their rooftop nest sites. This species likely nested in most blocks in which it was observed, especially those near towns and cities.

Distribution

Common Nighthawks were distributed statewide. Blocks where the species was recorded correspond to locations of cities. Some cities, of course, were missed due to random sampling. The density of blocks recording nighthawks generally increased toward the western Missouri border. There were many counties—in eastern, central and at the northern edge of Missouri—where nighthawks were not recorded.

Phenology

Compared to other goatsuckers, Common Nighthawks arrive rather late on their breeding grounds in Missouri. While the earliest individuals arrive in late April, peak migration occurs in mid-May (Robbins and Easterla 1992). Most nesting documented during the Atlas Project apparently commenced from late May through early June. Common Nighthawks raise only one brood per season (Harrison 1975).

Breeding Phenology

EVIDENCE (# of Records)	MAR	APR	MAY	JUN	JUL	AUG	SEP
NE (1)			6/09	6/09			
NY (1)					8/05	8/05	

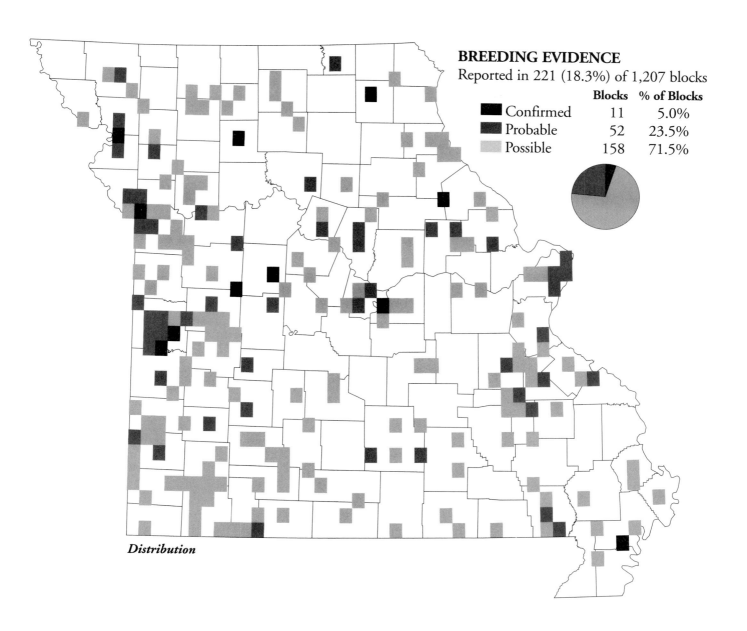

BREEDING EVIDENCE
Reported in 221 (18.3%) of 1,207 blocks

	Blocks	% of Blocks
Confirmed	11	5.0%
Probable	52	23.5%
Possible	158	71.5%

Distribution

Chuck-will's-widow
Caprimulgus carolinensis

> **Rangewide Distribution:** Southeastern United States
> & Gulf Coast
> **Abundance:** Locally common
> **Breeding Habitat:** Open pine-oak woods & forest
> edges
> **Nest:** Lays in approximately same dead-leaf area yearly
> **Eggs:** 2(?) cream, pink, or white; usually with brown,
> purple or gray marks
> **Incubation:** 20+ days
> **Fledging:** 17 days

Due to their nocturnal behavior, Chuck-will's-widows, like Whip-poor-wills, are most often detected by their calls. The largest of our nightjars, Chuck-will's-widows emit a low, whistled vocalization that begins with an explosive "chuck." They are commonly heard in the rural forests of southern Missouri from mid-spring through September. Chuck-will's-widows nest on the forest floor, usually in fairly dense, brushy cover. They have expanded their range northward during this century (Peterjohn and Rice 1991).

Code Frequency

Chuck-will's-widows were easily detected and identified by sound during night surveying. Thus, in most blocks they were recorded as possible breeders. Territoriality, also documented by vocalizations, accounted for most of the probable breeding evidence. As is true for most goatsuckers, their nests were difficult to find. Chuck-will's-widows may have bred in the majority of the blocks where they were recorded.

Distribution

Chuck-will's-widows ranged throughout the Ozarks and Ozark Border natural divisions in regions where extensive forest and forest edges provided suitable breeding habitat. They were absent from the largely-deforested Mississippi Lowlands and a large portion of the Springfield Plateau. The northern-most confirmation, a nest with eggs, was in northern Boone County. This species is a very rare breeder in the Glaciated Plains of northern Missouri (Robbins & Easterla 1992). Thus, scattered possible breeding locations across the northern counties were expected. Some of these may have been breeding-range pioneers or vagrant, unmated individuals.

Phenology

Chuck-will's-widows arrive in late April and early May (Robbins and Easterla 1992). The few observations of nests or nesting behavior provided a limited picture of breeding phenology. The late date for a nest with young could be attributed to a renesting attempt as Chuck-will's-widows are reported by Rohwer (1971) to be single-brooded.

Breeding Phenology

EVIDENCE (# of Records)	MAR	APR	MAY	JUN	JUL	AUG	SEP
NB (1)			6/10	6/10		
NE (4)		6/01		6/10		
NY (1)					8/05	8/05

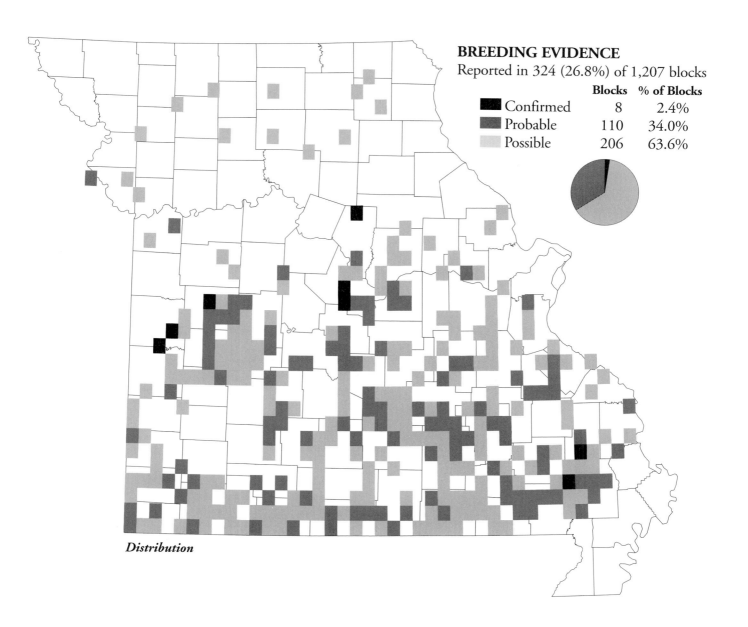

BREEDING EVIDENCE
Reported in 324 (26.8%) of 1,207 blocks

		Blocks	% of Blocks
■	Confirmed	8	2.4%
■	Probable	110	34.0%
▨	Possible	206	63.6%

Distribution

Whip-poor-will
Caprimulgus vociferus

Rangewide Distribution: Southwestern, northeastern & north central United States, most of Mexico

Abundance: Fairly common & widespread

Breeding Habitat: Deciduous & mixed woods with good litter cover

Nest: None, on well-drained ground near edge of woods

Eggs: 2 white, dotted with brown, olive, or lavender

Incubation: 19–20 days

Fledging: 20 days

Listening to the oft-repeated "whip-poor-will" song in the evening has become a pastime for some admirers of this species. Calls of 50–100 repetitions by one individual are common. John Burroughs, a well-known turn-of-the-century naturalist and namesake for the Burroughs Audubon Society in Kansas City, counted 1,088 repetitions (Terres 1987).

Code Frequency

Ninety-six percent of all records were possible and probable observations which were likely based on voice only. Breeding was confirmed in only 4 percent of the records, attesting to the difficulty of locating nest sites. Whip-poor-wills likely nested in most blocks where they were found.

Distribution

Whip-poor-wills were located statewide with the exception of the Mississippi Lowlands. Gaps in distribution in the Mississippi Lowlands and Glaciated Plains natural divisions may have resulted from limited nesting habitat for this ground-nesting, forest-dwelling species. The few records in the agricultural parts of the Mississippi Lowlands occurred June 2 through June 15 and likely represented late migrants or unmated singing males.

Phenology

This species sometimes raises two broods (Ehrlich et al. 1988). Reproduction in this species is apparently synchronized with the lunar cycle to facilitate foraging to feed nestlings on moonlit nights (Ehrlich et al. 1988).

Breeding Phenology

EVIDENCE (# of Records)	MAR	APR	MAY	JUN	JUL	AUG	SEP
NE (12)		5/02				7/25	
NY (8)			5/16			8/12	

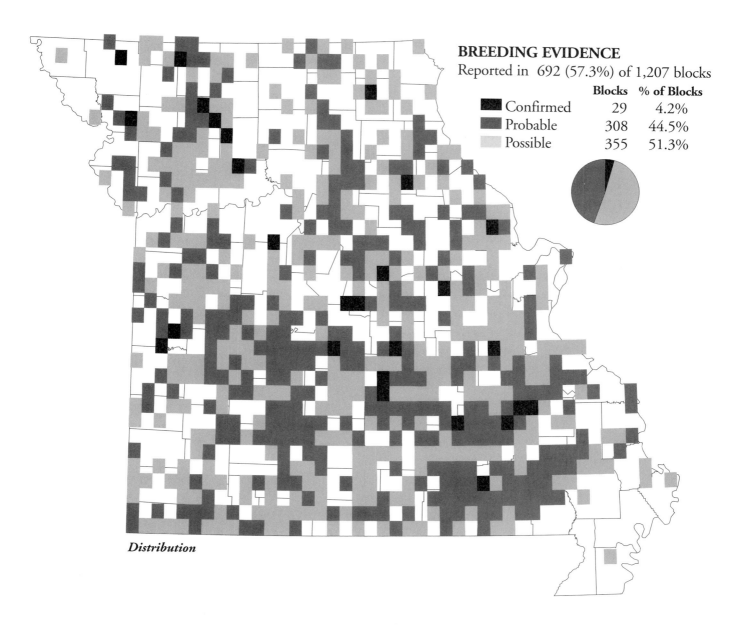

BREEDING EVIDENCE
Reported in 692 (57.3%) of 1,207 blocks

	Blocks	% of Blocks
Confirmed	29	4.2%
Probable	308	44.5%
Possible	355	51.3%

Distribution

Chimney Swift
Chaetura pelagica

Rangewide Distribution: Southeastern Canada & eastern United States
Abundance: Common
Breeding Habitat: Open areas near human habitat, silos & air shafts
Nest: Half saucer of twigs & saliva attached to chimney wall
Eggs: 4–5 white & unmarked
Incubation: 19–21 days
Fledging: 28–30 days

Small flocks of these stiff-winged aerial insectivores are familiar sights in cities, suburbs and small towns. Although Chimney Swifts usually place their nests in chimneys of residences, schools and industrial buildings, they occasionally select barns, silos, wells and hollow trees for nest sites (Fischer 1958). Before settlement, hollow trees were apparently the most common nest sites and Chimney Swifts were presumably less common than today.

Code Frequency

Because of the ease of detecting Chimney Swifts, they were likely absent from blocks where they were not recorded. Breeding confirmations were rare, perhaps because nesting localities such as chimneys are difficult to inspect. Nests were actually observed in only 27 blocks. Additionally, Chimney Swifts often range several kilometers while foraging. Most were sighted as flyovers, perhaps distant from nest sites. In almost all blocks in which probable breeding was recorded, birds were observed entering a nest site (usually a chimney) or occurred in pairs.

Distribution

The Atlas Project documented Chimney Swifts statewide. There were, however, small groups of blocks where they were not reported, primarily in north central and southern central counties. Perhaps the low number of records in south central Missouri relates to a shortage of nest sites in more heavily-forested regions. In all other areas, nesting habitat was likely available. There was some variation in code frequency across the state, with more probable and confirmed records being reported in the western regions of the state. This may be related to Atlaser effort (fig. 5).

Abundance

As expected, because of their association with chimneys, Chimney Swifts were abundant in and around the state's largest cities. This produced high counts within the Big Rivers Natural Division. The Chimney Swift's avoidance of heavily-forested and intensively farmed regions may have resulted in fewer being reported within the Ozark and Mississippi Lowlands natural divisions, respectively.

Abundance by Natural Division
Average Number of Birds / 100 Stops

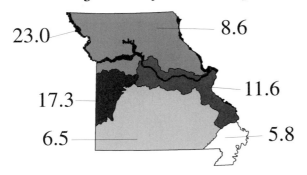

23.0
8.6
17.3
11.6
6.5
5.8

Breeding Phenology

EVIDENCE (# of Records)	MAR	APR	MAY	JUN	JUL	AUG	SEP
NB (2)			5/16	6/04			
NE (2)			6/06	6/24			
NY (25)			5/15				9/05
FY (4)			5/24	6/06			

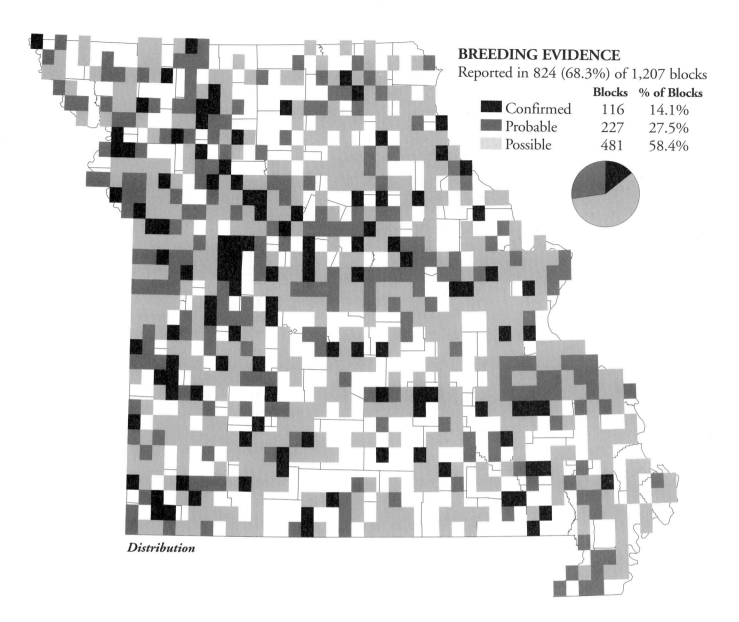

BREEDING EVIDENCE

Reported in 824 (68.3%) of 1,207 blocks

	Blocks	% of Blocks
Confirmed	116	14.1%
Probable	227	27.5%
Possible	481	58.4%

Distribution

Phenology

Although Chimney Swifts are single-brooded (Harrison 1975), the nest event dates were extremely variable, perhaps due to renesting attempts following nesting failure. Especially remarkable was a nest with young discovered on September 5.

Ruby-throated Hummingbird
Archilochus colubris

Rangewide Distribution: Southern Canada, eastern United States & Gulf Coast
Abundance: Fairly common
Breeding Habitat: Areas with scattered trees, gardens & flowers
Nest: Bud scales with lichen exterior & spider silk, lined with plant down, in trees
Eggs: 2 white & unmarked
Incubation: 11–14 days
Fledging: 14–28 days

Among Missouri's best known birds, Ruby-throated Hummingbirds have a special place in the hearts of many. Although Ruby-throated Hummingbirds are traditionally residents of rich bottomland forests, the abundance of hummingbird feeders has attracted this species into open areas where it can be observed and enjoyed. Hummingbirds frequent parks and towns and sometimes nest near feeders and flower gardens. However, finding a nest is still a difficult task. Feeders containing sugar water easily attract Ruby-throated Hummingbirds, which can become quite tame. Several hummingbirds at one locality can consume a gallon or more of sugar water a day.

Code Frequency

Although breeding was confirmed in only 11 percent of the blocks, this species was observed in appropriate habitat in 65 percent of blocks. Most confirmations were based on observations of fledglings, but remarkably, 26 nests were discovered. Atlasers frequently checked for feeders outside each house to document this species. In some instances, Atlasers relied on landowners to document the occurrence of hummingbirds. Reports from the feeder operator about nests they had seen was accepted as confirmation of breeding. Because nests were hard to locate, hummingbirds may have bred in most blocks where they were found. However, lingering migrants may have been included in these records and the distribution may not accurately reflect the state's breeding population.

Distribution

Ruby-throated Hummingbirds were found statewide, with fewer recorded in the Mississippi Lowlands and northern half of the Glaciated Plains natural divisions. These regions, with their extensive cropland, do not contain adequate habitat for hummingbirds.

Abundance

Results from Miniroutes and Breeding Bird Surveys suggest that the greatest relative abundance was in the Ozark and Osage Plains natural divisions, with slightly fewer in the Ozark Border Natural Division. Widmann (1907) similarly postulated that this species was most numerous in the Ozarks and the areas around bluffs along large rivers. Because areas in the Osage Plains are now much more wooded than in Widmann's time, this natural division may presently be more attractive to Ruby-throated Hummingbirds than in the past.

Abundance by Natural Division
Average Number of Birds / 100 Stops

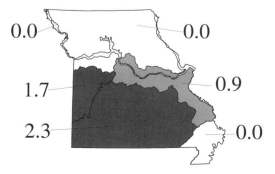

Breeding Phenology

EVIDENCE (# of Records)	MAR	APR	MAY	JUN	JUL	AUG	SEP
NB (7)			5/13	6/16			
NE (5)			5/18	6/21			
NY (5)				6/24	7/26		
FY (4)				6/24	7/30		

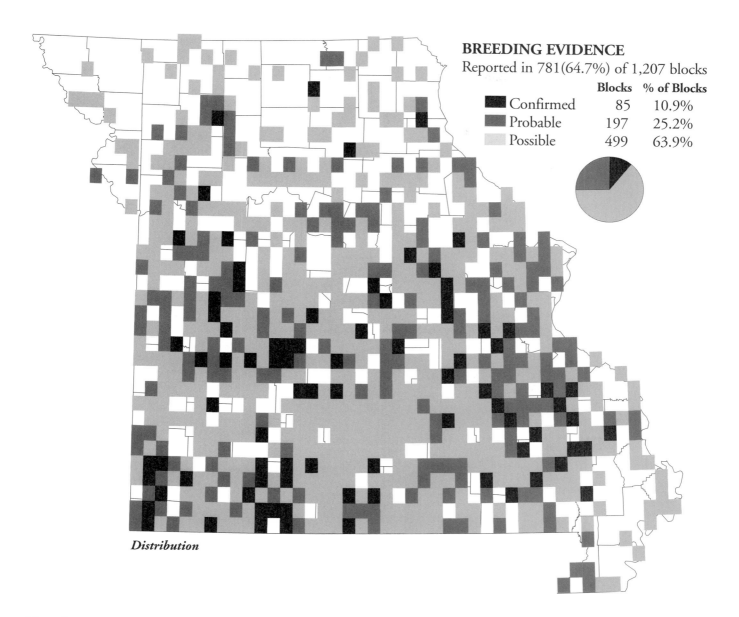

BREEDING EVIDENCE

Reported in 781(64.7%) of 1,207 blocks

	Blocks	% of Blocks
Confirmed	85	10.9%
Probable	197	25.2%
Possible	499	63.9%

Distribution

Phenology

Ruby-throated Hummingbirds typically rear two, sometimes three broods (Ehrlich et al. 1988). Nests found May 8 through July 8 may provide evidence of multiple nesting in Missouri. Nest building observed between May 13 and June 16 is consistent with two broods. Incubation and fledging require 25–42 days (Ehrlich et al. 1988).

Belted Kingfisher
Ceryle alcyon

Rangewide Distribution: Canada, Alaska & northern to southeastern United States
Abundance: Common & widespread
Breeding Habitat: Banks of rivers, streams & fresh & marine swamps
Nest: Saucer of grass & leaves in burrow in vertical bank near water
Eggs: 6–7 white
Incubation: 21–24 days
Fledging: 23+ days

This species requires both permanent water to support accessible prey (primarily small fish and crayfish) and nearby earthen banks in which to excavate nesting burrows (Hamas 1994). Belted Kingfishers often select areas with stream-side snags that serve as hunting and roosting perches. Most birders quickly learn the kingfisher's rattling call, and these birds are easily sighted as they fly along streams or hover, then plunge-dive for fish.

Code Frequency

Because they are conspicuous and well-known, Belted Kingfishers are likely to be found and identified wherever they occur. Thus, gaps on the map likely accurately reflect where the species was absent. Although nesting burrows are easily recognized, Atlasers may have had difficulty finding them because Belted Kingfishers often forage a few hundred meters up or down streams from nest sites. Additionally, burrow entrances usually face the stream and may be difficult to see unless searching from a boat. Thus, many of the blocks where possible or probable breeding evidence was recorded were likely breeding areas.

Distribution

Although essentially breeding statewide, kingfishers had a few intriguing gaps in distribution. There was a scarcity of blocks in which Kingfishers were recorded in north central Missouri, primarily in Chariton, Macon and Putnam counties. There were also fewer blocks where kingfishers were recorded in portions of the Osage Plains and the Mississippi Lowlands. The latter region apparently lacks appropriate nesting habitat.

Abundance

Belted Kingfishers were most abundant in the Ozark Border Natural Division. None were recorded on survey routes in the Big Rivers and Mississippi Lowlands natural divisions.

Phenology

A kingfisher was observed entering a nest site on April 26. The late observation of a nest containing young provides evidence that this species has a protracted breeding season. Belted Kingfishers have been reported to have only one brood per season (Harrison 1975).

Abundance by Natural Division
Average Number of Birds / 100 Stops

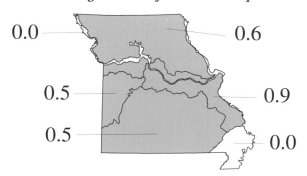

0.0   0.6

0.5   0.9

0.5   0.0

Breeding Phenology

EVIDENCE (# of Records)	MAR	APR	MAY	JUN	JUL	AUG	SEP
NB (3)		5/05		5/12			
NY (1)				6/23	6/23		
FY (34)			5/14				8/09

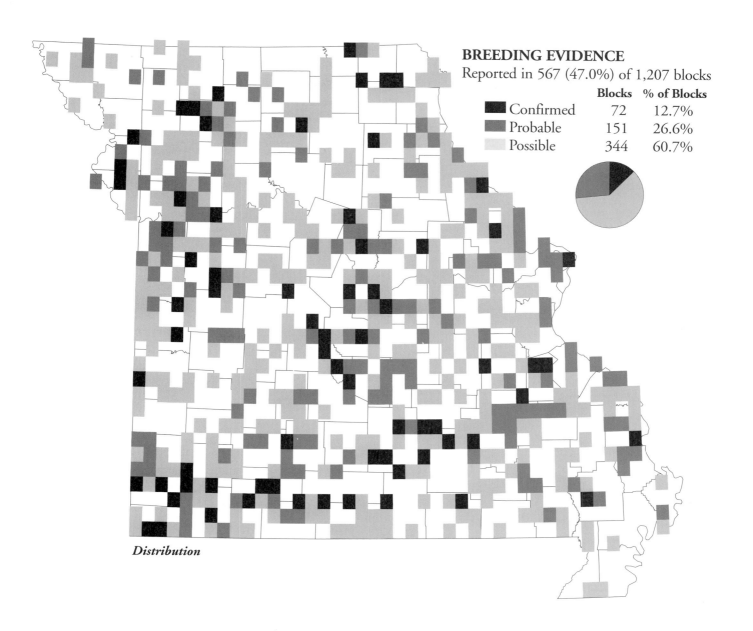

BREEDING EVIDENCE
Reported in 567 (47.0%) of 1,207 blocks

	Blocks	% of Blocks
Confirmed	72	12.7%
Probable	151	26.6%
Possible	344	60.7%

Distribution

Red-headed Woodpecker
Melanerpes erythrocephalus

Rangewide Distribution: Southern Canada, northern to eastern & southeastern United States

Abundance: Common

Breeding Habitat: Woods & suburbs with large deciduous trees

Nest: Deep cup of vines, bark, grass, forbs & spider webs in deciduous tree

Eggs: 4 white with spots of browns, blacks at large end

Incubation: 11–14 days

Fledging: 10–12 days

Red-headed Woodpeckers frequent open forest land, old burns and recent clearings (Ehrlich et al. 1988). Widmann (1907) stated that the Red-headed Woodpecker was the "best known and most familiar, summer resident in all parts of the state ... When most of the states were covered with tree growth the Redhead's home was on the towering giants with which the woods were richly sprinkled." Unlike many forest-dwelling birds, this species has adapted as the landscape has changed. This species now occupies open lands where it uses utility poles, cemetery trees and even cavities in buildings as nest sites. In the winter, Red-headed

Woodpeckers throughout the United States, including Missouri, migrate southward to areas where there is an abundance of acorns (Bent 1939). The spring and summer diet is from 34 percent to 37 percent acorns, fruit and other plant material with the remainder consisting of animal material. In contrast, the winter diet is comprised of 91 percent acorns and other plant material (Martin et al. 1951).

Code Frequency

These noisy, strikingly-colored birds were easy for observers to locate and watch for long periods. Because nests were usually too high to be accessible, 92 percent of breeding confirmations involved activities outside the nest that related to fledglings and feeding young.

Distribution

Although found statewide, this species was most frequently recorded in the less-forested, northern half of the state. Robbins and Easterla (1992) suggested that the spring and summer presence of this species is correlated with the abundance of acorn accumulation on the forest floor the previous winter. Although acorn abundance may influence the fall and winter distribution of this species, it apparently does not significantly affect summer distribution (Graber et al. 1977).

Abundance

Breeding Bird Surveys and Miniroutes may not provide an accurate picture of the relative abundance of Red-headed Woodpeckers. Abundance survey routes likely did not adequately sample river and reservoir habitat.

Abundance by Natural Division
Average Number of Birds / 100 Stops

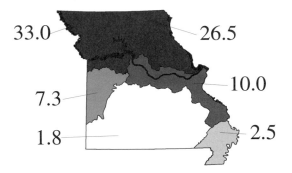

33.0   26.5

7.3   10.0

1.8   2.5

Breeding Phenology

EVIDENCE (# of Records)	MAR	APR	MAY	JUN	JUL	AUG	SEP
NE (2)				6/14 6/17			
NY (33)			5/15				8/18
FY (129)		5/05				8/08	

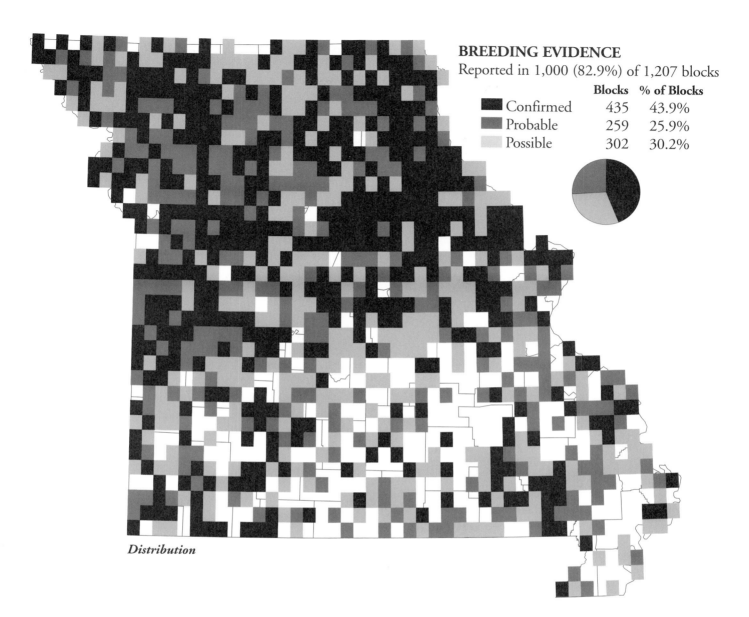

BREEDING EVIDENCE
Reported in 1,000 (82.9%) of 1,207 blocks

	Blocks	% of Blocks
Confirmed	435	43.9%
Probable	259	25.9%
Possible	302	30.2%

Distribution

Phenology

Red-headed Woodpeckers arrive in southern Missouri in early May and reach northern Missouri by the end of the month (Robbins & Easterla 1992). Observations of food being carried to young on May 18 suggest that nesting activity may have initiated as early as late April or early May, with young in nests seen mainly during the last half of May. Double broods or a very delayed renesting likely account for the spread of observations.

Red-bellied Woodpecker
Melanerpes carolinus

Rangewide Distribution: Eastern United States to East Coast & Gulf Coast

Abundance: Common, extending breeding range north

Breeding Habitat: Open woodlands with snags & hollow trees

Nest: Tree hole in deciduous snags, poles & birdhouses

Eggs: 4–5 white & unmarked

Incubation: 12–14 days

Fledging: 24–27 days

Red-bellied Woodpeckers breed in urban areas as well as heavily- forested regions. Like several other woodpecker species, they hoard nuts, fruits and insects and defend them within a territory. They later feed the fruits and insects to their young. They also feed on the ground like Northern Flickers and consume the sap from Yellow-bellied Sapsucker "sapwells" (Terres 1987).

Code Frequency

Their loud "chif-chif" calls give away their presence in backyard trees and forests, making them easy to record. Observations of fledglings and food being fed to young accounted for 71 percent of all confirmed records. Red-bellied Woodpeckers likely nested in most blocks where recorded as possible breeders.

Distribution

Red-bellied Woodpeckers were distributed statewide and were recorded in 94 percent of the state's blocks. However, they were recorded in only 69 percent of blocks in the Mississippi Lowlands Natural Division, reflecting that area's lack of forest cover.

Abundance

The highest relative abundances occurred in the Ozark Borders, Ozark and Osage Plains natural divisions. The lowest relative abundances were recorded in the Big Rivers and Mississippi Lowlands natural divisions. While primarily forest dwellers, Red-bellied Woodpeckers can persist in open agricultural and urban areas where sufficient food and nest sites are available (Terres 1987).

Phenology

The three-month spread for observations of nests with young and fledglings suggest two broods are likely in Missouri. According to Bent (1939), Red-bellied Woodpeckers are single-brooded at northern latitudes and double- to triple-brooded in southern areas of their range. Reports of activity at nest sites were first recorded in early April. Activities related to brood rearing, food being delivered to young and nests with young were observed in late April and continued through August.

Abundance by Natural Division
Average Number of Birds / 100 Stops

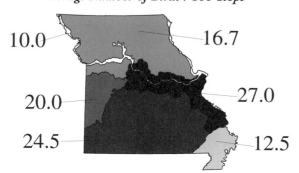

10.0   16.7

20.0   27.0

24.5   12.5

Breeding Phenology

EVIDENCE (# of Records)	MAR	APR	MAY	JUN	JUL	AUG	SEP
NB (2)		5/18		6/07			
NY (35)		4/22				8/12	
FY (122)		4/27				8/12	

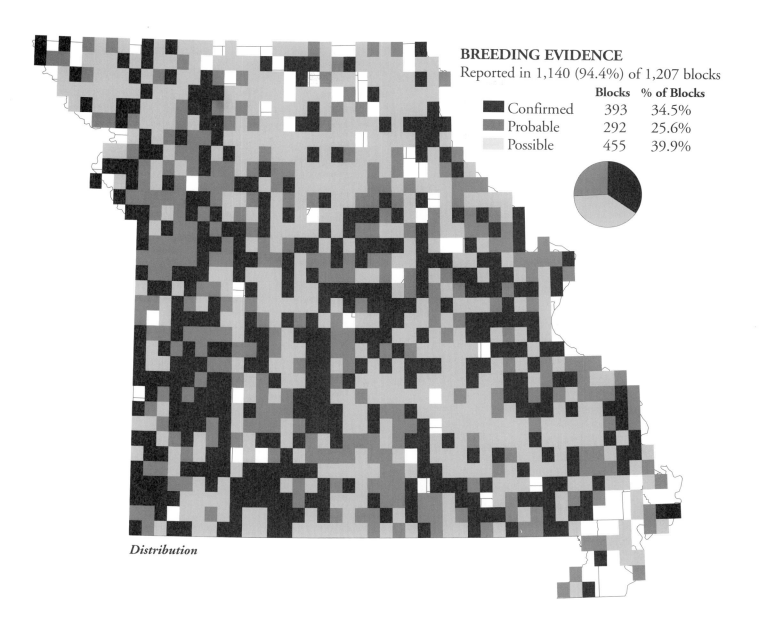

BREEDING EVIDENCE

Reported in 1,140 (94.4%) of 1,207 blocks

	Blocks	% of Blocks
Confirmed	393	34.5%
Probable	292	25.6%
Possible	455	39.9%

Distribution

Downy Woodpecker
Picoides pubescens

> **Rangewide Distribution:** Southern Canada, central &
> eastern Alaska & most of United States
> **Abundance:** Common
> **Breeding Habitat:** Most any area with trees
> **Nest:** Cavity in snag or tree cavity lined yearly with
> chips, entrance hidden by moss, lichen & fungus
> **Eggs:** 4–5 white
> **Incubation:** 12 days
> **Fledging:** 20–25 days

The smallest of our resident woodpeckers occupies all types of woodlands from extensive mature forests to small, isolated second-growth woodlots. Downy Woodpeckers breed wherever there are trees with decayed branches suitable for nesting and foraging. Woodland edge and wooded riparian corridors seem especially attractive to these birds. James and Neal (1986) determined that Downy Woodpeckers were more than twice as numerous in bottomland than in upland woods in Arkansas during the summer.

Code Frequency

Downy Woodpeckers are well known to most bird watchers and are easily located by their frequent "pik" call and rattle. They can usually be sighted as they move over the bark of trees in search of insects. Nests can be difficult to find, especially when located high above the ground. Downy Woodpeckers likely nested in the majority of blocks in which they were observed.

Distribution

As expected, Downy Woodpeckers were distributed throughout Missouri. This species was not recorded in 12 blocks in the lower Ozarks, perhaps because dense regional forests provide inappropriate habitat or caused difficulty in detection. Breeding confirmations were less frequently reported in northern counties. This was unexpected considering this woodpecker's wide range, and the ease of observing nesting evidence in fragmented woodlands. The level of breeding confirmation for this species seemed highly correlated with Atlasing effort (fig. 5).

Abundance

Downy Woodpecker relative abundance generally corresponded to the amount of forest cover in the natural divisions.

Phenology

Downy Woodpeckers are permanent residents and they initiate breeding early in the season. They were first observed entering cavities under circumstances indicating nesting on March 31. Two broods per season are possible in the southern United States (Harrison 1975) and Atlas observations indicate this may have occurred in Missouri.

Abundance by Natural Division
Average Number of Birds / 100 Stops

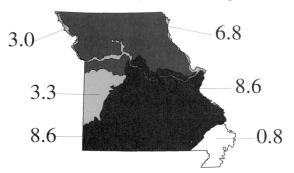

3.0   6.8

3.3   8.6

8.6   0.8

Breeding Phenology

EVIDENCE (# of Records)	MAR	APR	MAY	JUN	JUL	AUG	SEP
NB (1)					8/01	8/01	
NY (24)		4/13		6/15			
FY (46)		5/07				8/11	

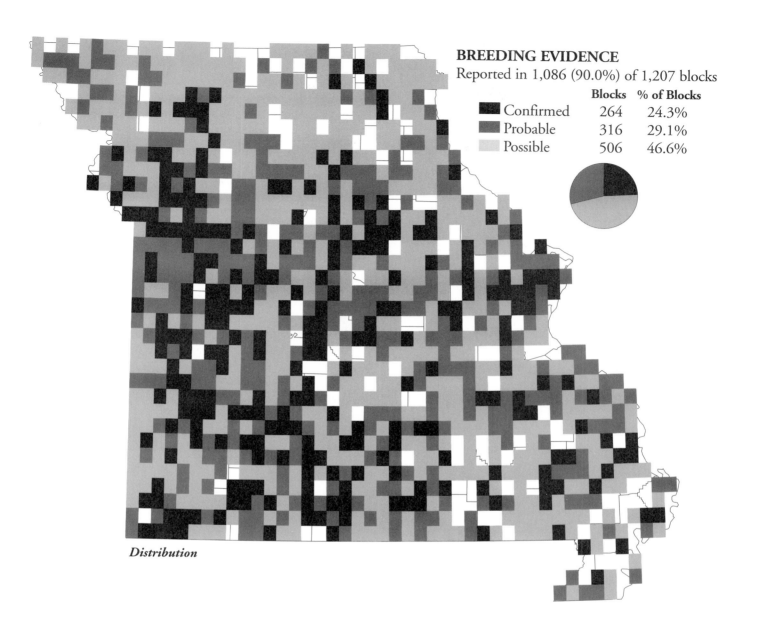

BREEDING EVIDENCE
Reported in 1,086 (90.0%) of 1,207 blocks

	Blocks	% of Blocks
Confirmed	264	24.3%
Probable	316	29.1%
Possible	506	46.6%

Distribution

Hairy Woodpecker
Picoides villosus

> **Rangewide Distribution:** Southern & northwestern Canada, southeastern Alaska, most of United States
> **Abundance:** Fairly common
> **Breeding Habitat:** Open & dense woods with snags & cavity trees
> **Nest:** Cavity in snag excavated in 20 days average & lined with chips
> **Eggs:** 4 white
> **Incubation:** 11–15 days
> **Fledging:** (24?-)28–30 days

Slightly larger than similarly-marked Downy Woodpeckers, Hairy Woodpeckers apparently have more restrictive habitat requirements than their smaller cousins. Hairies are most often associated with mature and over-mature forests with a prevalence of mid-story and under-story vegetation (Kilhan 1983). Otherwise, they are flexible regarding tree species or woodlot size, and would be expected in every township of Missouri although not as numerously as Downy Woodpeckers.

Code Frequency

Hairy Woodpeckers are relatively easy to detect and identify, however, their vocalization can be confused with that of Downy Woodpeckers. Evidence of probable and confirmed breeding was apparently difficult to obtain for this species. Because they often forage far from nest sites, they are difficult to confirm as breeders. This species likely bred in most blocks where it was recorded.

Distribution

As expected, Hairy Woodpeckers ranged throughout Missouri. However, breeding was unexpectedly infrequent in several regions, most notably in the northeastern quarter of the state, even though appropriate habitat would seem to be available in that region. The reduced Atlasing effort in the northeast (fig. 5), may have resulted in fewer detections as compared to the more common and more expected Downy Woodpecker. The same puzzle applies in western Bates and Vernon counties where essentially no Hairy Woodpeckers were found. It was expected that fewer Hairy Woodpeckers would be recorded in the Mississippi Lowlands because of that natural division's shortage of forest cover.

Abundance

Hairy Woodpecker relative abundance was greatest in the Big Rivers and Ozark natural divisions. They were approximately one-fifth as abundant in the Ozarks as were Downy Woodpeckers.

Phenology

Atlas Project data indicate these permanent residents do not initiate breeding until April. Birds were on nests by May 1. Hairy Woodpeckers raise only one brood a year (Ehrlich et al. 1988) presumably in late Spring through early summer. Cavity excavation recorded on August 1 and food for young recorded on July 30 were probably erroneously recorded as breeding evidence. The latter may have actually been food caching.

Abundance by Natural Division
Average Number of Birds / 100 Stops

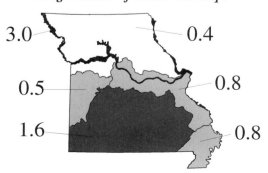

3.0 0.4
0.5 0.8
1.6 0.8

Breeding Phenology

EVIDENCE (# of Records)	MAR	APR	MAY	JUN	JUL	AUG	SEP
NB (1)					8/01	8/01	
NY (7)			5/19	6/14			
FY (15)			5/20			7/30	

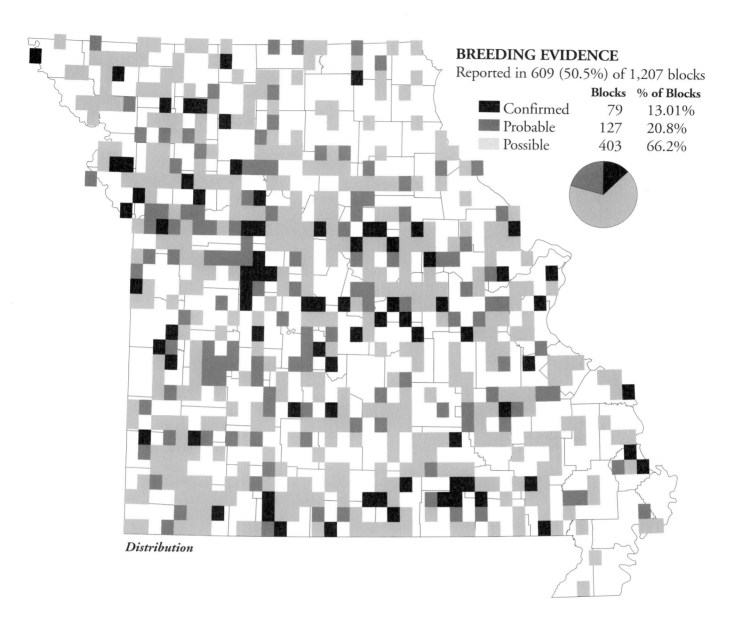

BREEDING EVIDENCE
Reported in 609 (50.5%) of 1,207 blocks

	Blocks	% of Blocks
Confirmed	79	13.01%
Probable	127	20.8%
Possible	403	66.2%

Distribution

Northern Flicker
Colaptes auratus

Rangewide Distribution: Most of Canada & United States	
Abundance: Widespread & common	
Breeding Habitat: Scattered trees with open areas	
Nest: Cavity in snags, poles, posts, houses, banks & haystacks	
Eggs: 5–8 white	
Incubation: 11–14 days	
Fledging: 25–28 days	

Northern Flickers are ubiquitous where nest sites and open ground for feeding occur together. They consume more ants than any other bird species (Ehrlich et al. 1988). As with most woodpeckers, Northern Flickers excavate nesting cavities in dead trees or in branches of live trees. Cavities are often 5–8 meters above the ground (Moore 1995). Old nest cavities are often used by other species such as squirrels, Eastern Screech-Owls and American Kestrels.

Code Frequency

Although Northern Flickers were relatively easy to record, more observation time was required to determine the status of nest sites. Breeding was confirmed in only 19 percent of the records, mostly by observation of noisy young or parents feeding young. Only 9 percent of all records were associated with an actual nest cavity or cavity construction.

Distribution

The distribution map suggests that in southern Missouri this species is associated with larger rivers and reservoirs where there is probably an abundance of large, dead or dying trees essential for nesting. Northern Flickers were recorded in 68 percent of the blocks south of Rolla (the 38th parallel) compared to 89 percent of blocks to the north of this line. The lower number in the forested Ozark section is similar to a study conducted in Arkansas where higher numbers were reported in the open forest of the Gulf Coastal Plain and the Mississippi Alluvial Plain (James and Neal 1986).

Abundance

The greater relative abundance in the Ozark Border and Glaciated Plains natural divisions may have been due to the greater interspersion of woodlands and open lands, and the abundance of foraging and nesting sites. Most of the Mississippi Lowlands lacked potential nest cavity trees except for forested regions along the Mississippi and St. Francis rivers, which had extensive forested land adjacent to the river levees.

Phenology

The first migrants arrived about a month before the first nesting activity was observed. Most observations of fledglings occurred May 20–26. By July 23 most nest-associated confirmations ceased, but fledglings were observed through

Abundance by Natural Division
Average Number of Birds / 100 Stops

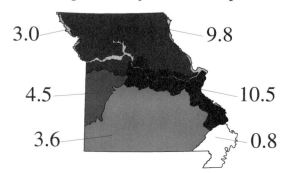

3.0   9.8

4.5   10.5

3.6   0.8

Breeding Phenology

EVIDENCE (# of Records)	MAR	APR	MAY	JUN	JUL	AUG	SEP
NY (26)			5/20		7/05		
FY (32)			5/26		7/23		

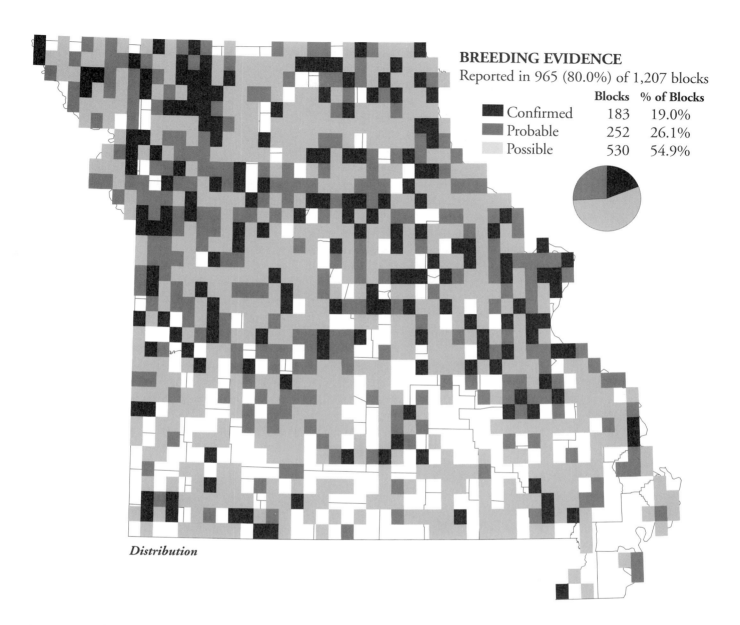

BREEDING EVIDENCE
Reported in 965 (80.0%) of 1,207 blocks

	Blocks	% of Blocks
Confirmed	183	19.0%
Probable	252	26.1%
Possible	530	54.9%

Distribution

August 20 and activity around the nest hole through August 31. Although Northern Flickers may produce a second brood in the southern portion of their range (Ehrlich et al. 1988), Atlas Project observations suggest that only single broods are produced in Missouri with the peak for observations of nest and young occurring around the second week in June.

Notes

Two forms of the Northern Flicker, the Red-shafted Flickers are usually absent from Missouri May through September (Robbins and Easterla 1992). Three forms of this species were formerly treated as separate species (AOU 1983). Atlasers were not asked to separate Red- from Yellow-shafted Flickers, and it is assumed that only Yellow-shafted forms were observed during the Atlas Project. Of these, the Yellow-shafted Flicker and Red-shafted Flicker, are reported in Missouri, the latter as rare primarily in winter (Robbins and Easterla 1992).

Pileated Woodpecker
Dryocopus pileatus

Rangewide Distribution: Southern Canada, eastern United States & West Coast
Abundance: Uncommon & localized in its range
Breeding Habitat: Deciduous & coniferous forests, woods, parks & suburbs
Nest: Cavity in often barkless tree facing east or south & lined with chips
Eggs: 4 white
Incubation: 15–18 days
Fledging: 26–28 days

Pileated Woodpeckers are large, spectacular birds that make even the most avid birdwatcher stop and look. They occasionally visit suet feeders in the winter where they are a welcome sight. These black, crow-sized birds are most commonly seen flying across a road or across a pasture into an adjacent forest. The undulating flight pattern easily distinguishes this species from crows. Nest trees are usually barkless and located in bottomlands or ravines (Renken 1988). Crest raising, wing spreading and head swinging displays, along with noisy vocalizations, are occasionally observed in early spring when pair bonding occurs (Bull and Jackson 1995).

Code Frequency

Fifty-nine percent of records for this secretive species were based on auditory and visual observations. Atlasers discovered 58 nest trees, or about 9 percent of all records. Because the average territory of Pileated Woodpeckers is 53–160 hectares (Renken, 1988), considerable searching was required to locate each nest tree. Presumably Pileated Woodpeckers nested in most of the blocks where they were recorded.

Distribution

The distribution displayed on the map relates to Pileated Woodpeckers being birds of forested landscapes, especially mature, forested floodplains with adjacent wooded ravines. In the Ozark and Ozark Border natural divisions, Pileated Woodpeckers were recorded in 83 percent and 74 percent of blocks respectively, as compared with 18 percent to 42 percent of blocks in other natural divisions. In the Glaciated Plains the population is apparently concentrated in riparian forests along the larger rivers and in the Lincoln Hills Natural Section.

Abundance

Although this species was uncommonly found in the Glaciated Plains and Mississippi Lowlands natural divisions during Atlas Project surveys, none were recorded there on Miniroutes and Breeding Bird Surveys. However, they were recorded on counts conducted in the Ozark Border, Osage Plains, and Ozark natural divisions.

Phenology

Pileated Woodpeckers establish territories during late February and March and lay eggs from late April through mid-May, as Renken noted in a personal communication. Atlas observations of nests and young of this single-brooded species approximate the expected dates of nesting and renesting attempts.

Abundance by Natural Division
Average Number of Birds / 100 Stops

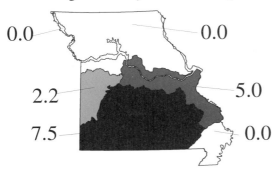

0.0   0.0

2.2   5.0

7.5   0.0

Breeding Phenology

EVIDENCE (# of Records)	MAR	APR	MAY	JUN	JUL	AUG	SEP
NY (10)			5/12		7/17		
FY (2)				6/01	7/05		

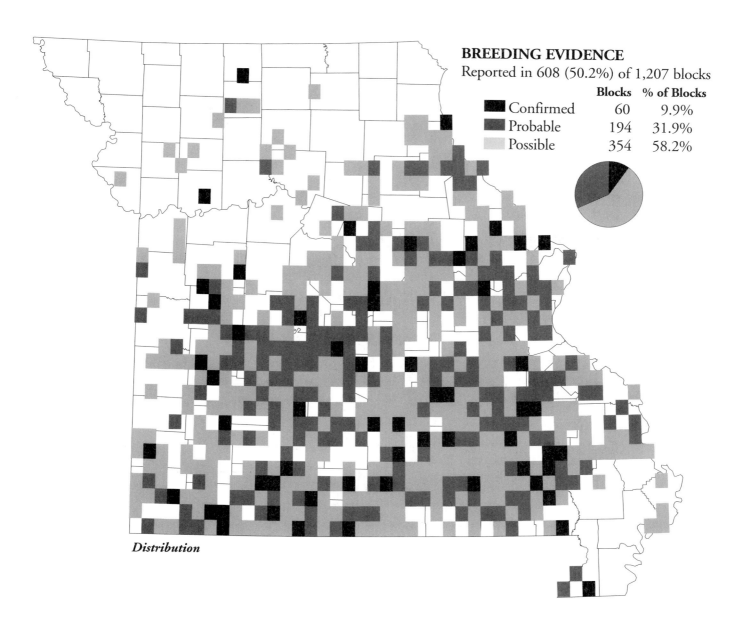

BREEDING EVIDENCE
Reported in 608 (50.2%) of 1,207 blocks

	Blocks	% of Blocks
Confirmed	60	9.9%
Probable	194	31.9%
Possible	354	58.2%

Distribution

Eastern Wood-Pewee
Contopus virens

Rangewide Distribution: Southeastern Canada & eastern
  United States
Abundance: Common & widespread
Breeding Habitat: Openings & edges of deciduous & mixed
  forests & woods
Nest: Lichen exterior on cup of grass, weeds, bark, lichens &
  cocoons, lined with fine material, in tree away from trunk
Eggs: 3 white to cream, with brown or purple marks, often
  wreathed
Incubation: 12–13(14) days
Fledging: 14–18 days

As their name implies, Eastern Wood-Pewees are normally associated with wooded habitats. While most numerous in mature deciduous woods with well-developed canopies and understories, they regularly occupy mixed woodlands (Peterjohn and Rice 1991). They breed from the forest interior to forest edge and, occasionally, even in small woodlots. Although most numerous in bottomland forests, they also breed in dry upland sites (Hicks 1935). Since 1965 Eastern Wood-Pewee populations have exhibited a slight decline across North America (Robbins et al. 1986).

Code Frequency

Eastern Wood-Pewees are easily detected by their plaintive "pewee" call which is emitted at the first hint of dawn and often throughout the heat of the day. Because they are easily detected by these vocalizations, the map is believed to accurately portray their breeding distribution in Missouri. It was fairly easy to classify Eastern Wood-Pewees as territorial due to their vociferous nature. Breeding confirmations were apparently difficult to obtain because foliage can often conceal nests and nesting activity.

Distribution

Eastern Wood-Pewees were found to be one of the most widely distributed breeding birds in Missouri. Only in the Mississippi Lowlands were breeding birds recorded in fewer blocks and they ranged throughout that region, presumably in large woodlots. In both the Mississippi Lowlands and northeastern Missouri fewer breeding confirmations were recorded, perhaps indicating these regions are sinks for non-productive populations.

Abundance

Eastern Wood-Pewees were most abundant in the Ozarks where forests are extensive.

Phenology

Eastern Wood-Pewees normally arrive during the last week of April (Robbins and Easterla 1992) and they may initiate breeding upon arrival. The earliest date that was reported for eggs, April 26, is likely erroneous as egg dates earlier than May 6 have not been recorded even in the southern part of the Eastern Wood-Pewee's range (McCarty 1996). Eastern Wood-Pewees are double brooded in the southern part of their breeding range (Ehrlich et al. 1988). The late dates for nestlings and fledglings likely represent second broods.

Abundance by Natural Division
Average Number of Birds / 100 Stops

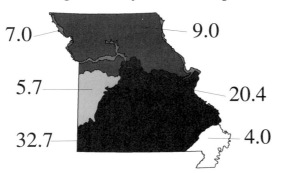

7.0   9.0
5.7   20.4
32.7   4.0

Breeding Phenology

EVIDENCE (# of Records)	MAR	APR	MAY	JUN	JUL	AUG	SEP
NB (15)			5/13		6/30		
NE (11)		4/26			7/01		
NY (19)			5/08				8/22
FY (53)			5/25				8/16

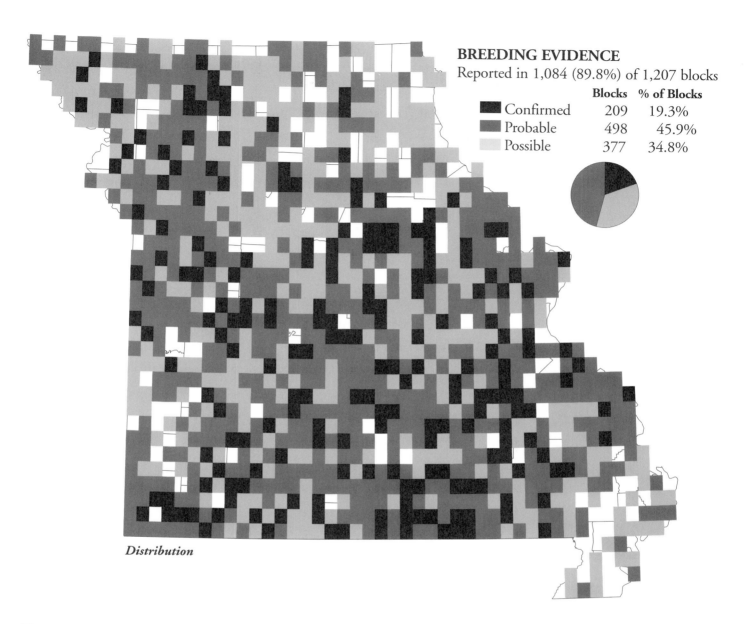

BREEDING EVIDENCE
Reported in 1,084 (89.8%) of 1,207 blocks

	Blocks	% of Blocks
Confirmed	209	19.3%
Probable	498	45.9%
Possible	377	34.8%

Distribution

Notes

Eastern Wood-Pewees are rare hosts to Brown-headed Cowbirds (Terres 1987). Although there were several opportunities to observe cowbird parasitism during the Atlas Project, no records were obtained.

Acadian Flycatcher
Empidonax virescens

Rangewide Distribution: Eastern United States from southeastern South Dakota to southern Texas
Abundance: Common with range expanding northeast
Breeding Habitat: Heavily wooded bottomland, swamps & thicket
Nest: Hammock-like of bark, twigs, stems & grass with hanging streamers of grass, lined with grass, hair & plant down, on trees
Eggs: 3 creamy white spotted with browns
Incubation: 14 days
Fledging: 13–15 days

Visually inconspicuous, these flycatchers are found in rich, multi-layered forest interiors where they feed and nest in the understory. They are especially associated with bottomland forests, swamps and damp ravines, and are found on rare occasions in relatively dry, upland forests (Peterjohn and Rice 1991).

Code Frequency

Acadian Flycatchers are best detected by their explosive, often repeated "peet-suh" call, which can carry for some distance through the forest. Their call is also their best identifi-

er, as Acadian Flycatchers can be difficult to distinguish visually from other summer resident flycatchers. Evidence of breeding beyond simple detection in breeding habitat during the breeding season was apparently difficult to obtain for this forest-dwelling species. They likely nested in many blocks where they were recorded as possible breeders. However, in blocks with more fragmented habitat, Atlasers may have recorded unmated, singing males.

Distribution

Acadian Flycatchers were generally distributed statewide although they were recorded in relatively few blocks in the northern one-third of the state. Blocks where this species was recorded seemed strongly related to forest size. Their scattered, sparse occurrence in northern Missouri likely resulted from the fragmented character of forests in that region. Most occurrences in the north were probably associated with bottomland forests such as along the Grand and Chariton rivers. A nest with eggs located in Daviess County indicates that breeding might have been confirmed in many northern Missouri blocks with additional Atlaser effort. The small forest tracts of the Mississippi Lowlands apparently provided at least some habitat for the species, especially along the Mississippi River. Probable breeding was recorded in the larger forests of that natural division, at Donaldson Point Conservation Area and Big Oak Tree State Park.

Abundance

Acadian Flycatcher abundance was greatest in the Ozark and Ozark Border natural divisions. The diminishing abundance of Acadians northward across Missouri was expected based on habitat availability and the fact that Iowa marks the northern edge of their range (Dinsmore et al. 1984).

Phenology

The early observation of a nest containing young indicated an early breeding season. In the latter half of the nesting

Abundance by Natural Division
Average Number of Birds / 100 Stops

0.0 — 0.6
0.5 — 3.4
2.9 — 0.8

Breeding Phenology

EVIDENCE (# of Records)	MAR	APR	MAY	JUN	JUL	AUG	SEP
NB (3)			6/01		6/20		
NE (1)			6/03	6/03			
NY (4)			6/07		7/11		
FY (4)			6/01		6/25		

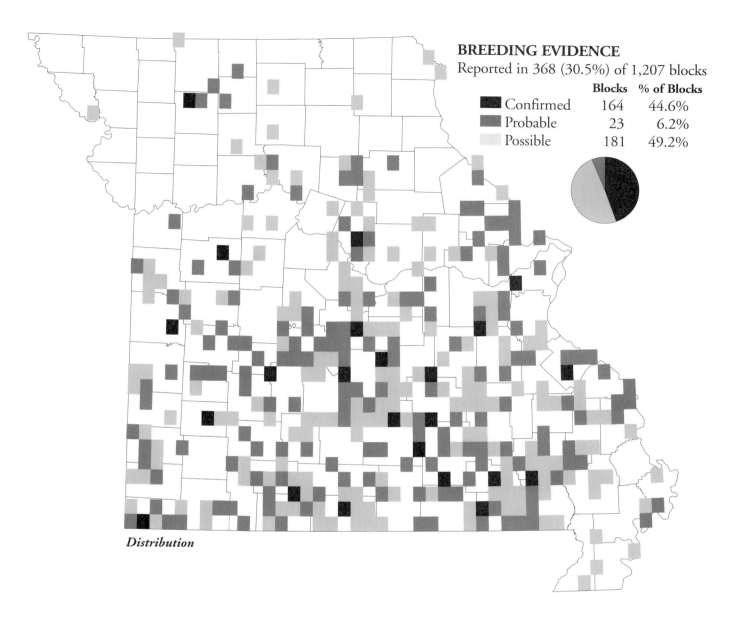

BREEDING EVIDENCE

Reported in 368 (30.5%) of 1,207 blocks

		Blocks	% of Blocks
■	Confirmed	164	44.6%
■	Probable	23	6.2%
■	Possible	181	49.2%

Distribution

season, nest construction was observed on June 20 and a bird on a nest July 19. Ehrlich et al. (1988) suggested that Acadian Flycatchers have two broods per season whereas Harrison (1975) suggested only one. Considering the length of the breeding season detected during the Atlas Project, two broods are likely in Missouri.

Notes

Although Acadian Flycatchers are occasionally victimized by Brown-headed Cowbirds (Ehrlich et al.1988), no such evidence was recorded during the Atlas Project.

Willow Flycatcher

Empidonax traillii

> **Rangewide Distribution:** Southwestern Canada, northern through central United States
> **Abundance:** Common
> **Breeding Habitat:** Swamps, scrubby areas & thickets, especially willow
> **Nest:** Compact cup of bark, weeds & grass lined with fine material in tree
> **Eggs:** 3–4 buff, occasionally white, with brown spots at large end
> **Incubation:** 12–13 days
> **Fledging:** 12–14 days

Willow Flycatchers frequent willows in wetlands and saplings in ungrazed pastures near streams. The "fitz-bew" calls are usually the first indication of their presence, easily distinguishing them from the "we-beo" songs of the look-alike migrant Alder Flycatchers. Willow and Alder Flycatchers were formerly considered the same species, the Trail's Flycatcher. This species has apparently benefited by the reduction of grazing and the retention of willows along watercourses (Ehrlich et al. 1988.)

Code Frequency

This species was recorded in only 5.9 percent of blocks and breeding was confirmed in only 11.3 percent of those where it was observed. The restricted patchy nature of its habitat probably contributes to its sparse distribution and limited detection.

Distribution

Willow Flycatcher records were rather sparsely distributed statewide and were likely associated with wetland areas containing young willows. Widmann (1907) indicated that they were found in the Glaciated and Osage Plains natural divisions and south along the Big Rivers Natural Division to Ste. Genevieve County. This was a much more restricted range than Atlasers detected.

Phenology

With only eight confirmed records, there was insufficient information to assess the phenology of this species. Sight records began on May 16, but were outside safe dates.

Notes

One instance of Brown-headed Cowbird brood parasitism was observed. The Willow Flycatcher is one of the more heavily parasitized bird species (Ehrlich et al. 1988).

Breeding Phenology

EVIDENCE (# of Records)	MAR	APR	MAY	JUN	JUL	AUG	SEP
NB (1)			5/31	5/31			
NE (3)			6/11		6/23		
NY (1)			6/10	6/10			
FY (2)			6/16			8/05	

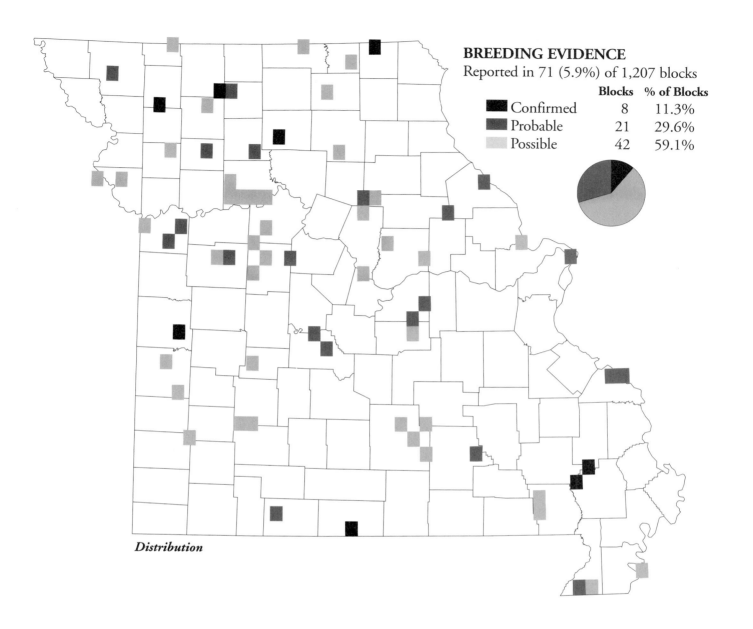

BREEDING EVIDENCE

Reported in 71 (5.9%) of 1,207 blocks

	Blocks	% of Blocks
Confirmed	8	11.3%
Probable	21	29.6%
Possible	42	59.1%

Distribution

Least Flycatcher
Empidonax minimus

Rangewide Distribution: North central Canada &
north central United States to East Coast
Abundance: Common in eastern & rare in western
United States
Breeding Habitat: Brushy areas with scattered trees
Nest: Compact bark, weeds, grass & feathers in trees
Eggs: 4 white & unmarked
Incubation: 13–14 days
Fledging: 12–16 days

The Least Flycatcher is the rarest of three *Empidonax*
flycatchers that breed in Missouri, the other two being the
Willow Flycatcher and the Acadian Flycatcher. Although an
extremely rare breeder, Least Flycatchers are relatively common during migration. The persistent "chebec - chebec -
chebec" call is somewhat ventriloquistic and difficult to
locate. Least Flycatchers frequent mature forests with some
open areas, and a variety of second growth habitats. They
are associated with dryer habitats than those of the Acadian
Flycatcher (Walkinshaw 1966) or Willow Flycatcher.

Code Frequency

Atlasers documented only four records. Confirmed
breeding was documented by a nesting attempt in
Montgomery County in June and July 1992 where a nest
was blown from a tree in the Americus block (McKenzie
and Jacobs 1992). It was confirmed by Lloyd Kiff, Western
Foundation of Vertebrate Zoology, to be that of a Least
Flycatcher. This was the first documented record since an
1891 record in Jackson County (Robbins and Easterla
1992).

Distribution

According to Briskie's (1994) breeding range maps, this
species should extend south to Missouri along the Missouri
and Mississippi rivers, yet its detection rate on Breeding Bird
Survey routes is very low south of mid-Minnesota and
Wisconsin. Based on Atlas Project records, and June and
July records discussed by Robbins and Easterla (1992), this
species may breed occasionally and irregularly throughout
the state. Northern and eastern Missouri are the most likely
areas to search for evidence of breeding pairs.

Phenology

This species is a common spring migrant in late April
and early May with fall migrants arriving in early August
(Robbins and Easterla 1992). The Atlas Project documented
territoriality in late June and a nest in early July. No evidence of eggs or young was found near the nest (McKenzie
and Jacobs 1992).

Notes

Least Flycatchers are listed as an uncommon Brown-
headed Cowbird host (Ehrlich et al. 1988).

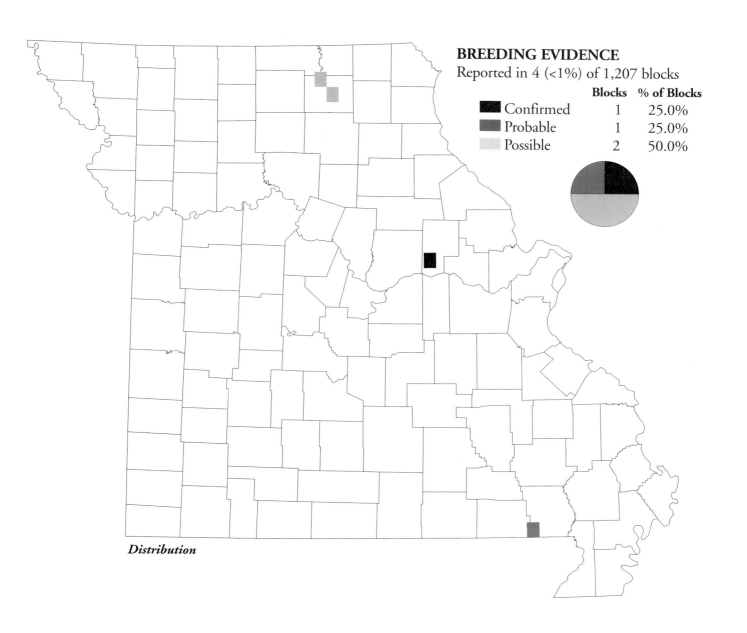

Eastern Phoebe
Sayornis phoebe

> **Rangewide Distribution:** Central & south central Canada & eastern United States
> **Abundance:** Common
> **Breeding Habitat:** Open riparian woods, ravines & farmland
> **Nest:** Mud pellets, plant fiber & moss, lined with hair, feathers & grass, on banks, cliffs, bridges & man-made structures
> **Eggs:** 4–5 white, mostly unmarked, some small brown spots
> **Incubation:** 16 days
> **Fledging:** 15–16 days

These adaptable flycatchers have thrived as a result of their association with humans. Buildings and bridges that provide shelter for their nests have enabled them to expand beyond their traditional nesting distribution on cliffs and rock outcrops. Despite their association with human habitation, Eastern Phoebes remain primarily rural birds. They are especially associated with woodland edges and streams, but can nest several hundred meters from water and woodland openings (Weeks 1994).

Code Frequency

Because this well-known species is easily detected and identified, it would only rarely have been missed in blocks where it occurred. Breeding individuals remain close to nest sites and confirmed breeding evidence can usually be obtained even during brief observation periods. The Eastern Phoebe was the species most commonly confirmed by the observation of used nests.

Distribution

Eastern Phoebes were densely distributed throughout Missouri with the exception of the Mississippi Lowlands. Their absence from most blocks in the Mississippi Lowlands may have been related to the scarcity of trees and woodland edge, or perhaps to the use of agricultural pesticides. For unknown reasons, they were relatively scarce in the Springfield vicinity and in the northeastern and northwestern corners of the state. Eastern Phoebes presumably bred in most blocks in which they were detected.

Abundance

Eastern Phoebes were most abundant in the Ozark and Ozark Border natural divisions, where forest cover and natural nest sites are apparently abundant. They were less than half as plentiful on the Glaciated and Osage Plains and least abundant in the Mississippi Lowlands and Big Rivers natural divisions.

Phenology

Eastern Phoebes arrive in early to mid-March (Robbins and Easterla 1992) and apparently initiate nesting soon afterwards. Nesting chronology was well documented by Atlas observations. The unusual observations of a bird on the nest on March 10 and a nest with young on March 21 provide evidence that Eastern Phoebes can nest very early. Weeks' (1994) summary of Eastern Phoebe breeding phe-

Abundance by Natural Division
Average Number of Birds / 100 Stops

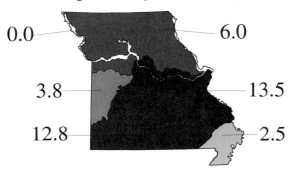

0.0 — 6.0
3.8 — 13.5
12.8 — 2.5

Breeding Phenology

EVIDENCE (# of Records)	MAR	APR	MAY	JUN	JUL	AUG	SEP
NB (33)	4/01					7/21	
NE (86)		4/13				8/04	
NY (127)	3/21					7/27	
FY (60)			5/04				8/26

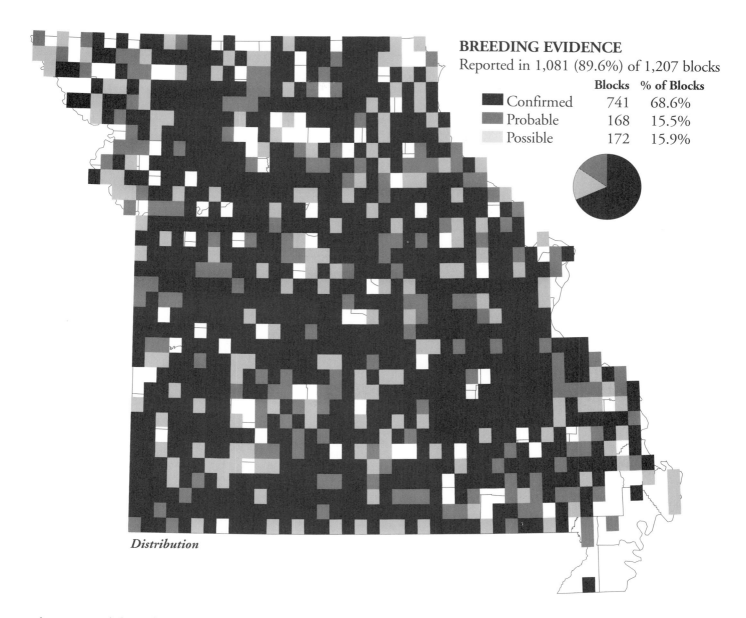

BREEDING EVIDENCE
Reported in 1,081 (89.6%) of 1,207 blocks

		Blocks	% of Blocks
■	Confirmed	741	68.6%
▨	Probable	168	15.5%
▨	Possible	172	15.9%

Distribution

nology reported the earliest eggs in late March in the southern United States. He suggests that breeding season events vary with weather conditions, however, the two Missouri Atlas Project observations were so exceptionally early they would have merited study had they been reported immediately. Eastern Phoebes typically produce two broods (Weeks 1994).

Notes

Eastern Phoebes are classified as an "acceptor" of Brown-headed Cowbird eggs and are heavily parasitized (Weeks 1994). However, parasitism was recorded in only 2.8 percent (6/213) of Atlas blocks in which there were opportunities to record this activity.

Great Crested Flycatcher
Myiarchus crinitus

Rangewide Distribution: Eastern Canada & eastern United States
Abundance: Common
Breeding Habitat: Open woods, forest edges, orchards & parks
Nest: Leaves, fur, feathers & snake skin within one foot of opening in snags
Eggs: 5 creamy white, buff, with brown, olive or lavender marks
Incubation: 13–15 days
Fledging: 12–21 days

Great Crested Flycatchers occupy canopies of mature and second-growth forests as well as isolated woodlots. They are especially associated with wooded riparian corridors over 70–100 meters wide (Peterjohn and Rice 1991). They nest in cavities in dead snags and also reside in urban areas with large shade trees in yards, parks and cemeteries.

Code Frequency

Great Crested Flycatchers are easily detected by their recognizable calls and were frequently categorized as territorial by the repeat observation of a calling bird. Pairs were also observed frequently. Breeding confirmations were difficult, as is expected for a cavity nester. Most breeding confirmations were based on observations of fledged young. It is likely that this species bred in the majority of blocks where it was detected.

Distribution

Great Crested Flycatchers are able to inhabit a wide variety of woodland habitats and are widely distributed in Missouri. Only in the less-forested parts of the Mississippi Lowlands were they not found in several adjacent blocks.

Abundance

Great Crested Flycatchers were somewhat more abundant in east central and south central Missouri. They were much less abundant in the Mississippi Lowlands and Big Rivers natural divisions where large agricultural fields provide less nesting habitat.

Phenology

Great Crested Flycatchers begin arriving in Missouri in late April (Robbins and Easterla 1992) with the bulk of individuals appearing in early to mid-May. Although Ehrlich et al. (1988) suggested that Great Crested Flycatchers have only one brood, the late dates recorded during the Atlas Project seem to indicate a second brood.

Notes

Great Crested Flycatchers are rare hosts to Brown-headed Cowbirds (Ehrlich et al. 1988) and there were no records of parasitism recorded during the Atlas Project.

Abundance by Natural Division
Average Number of Birds / 100 Stops

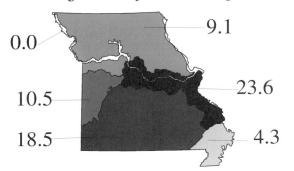

0.0
9.1
10.5
23.6
18.5
4.3

Breeding Phenology

EVIDENCE (# of Records)	MAR	APR	MAY	JUN	JUL	AUG	SEP
NB (23)		5/04				7/19	
NY (6)				6/13		7/30	
FY (46)			5/25			7/30	

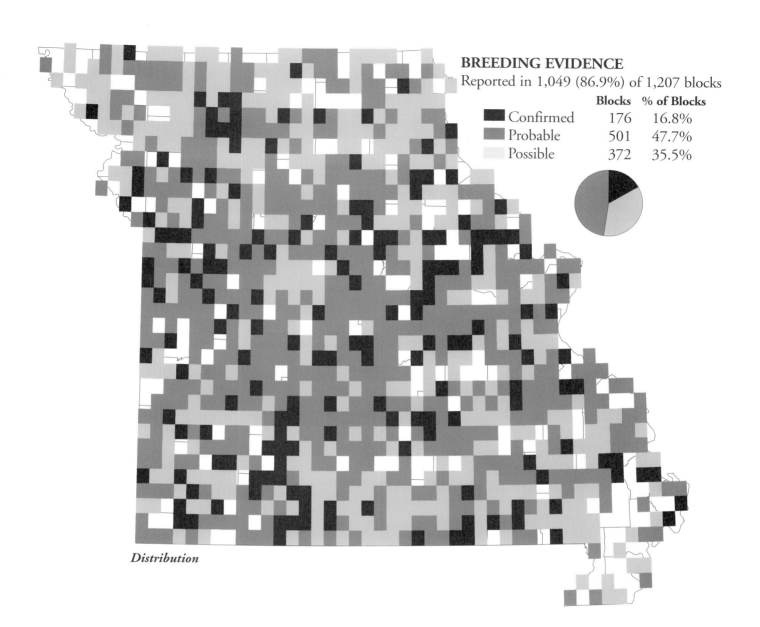

	Blocks	% of Blocks
Confirmed	176	16.8%
Probable	501	47.7%
Possible	372	35.5%

Distribution

Western Kingbird

Tyrannus verticalis

Rangewide Distribution: Southern Canada, western & west central United States, northern Mexico
Abundance: Common
Breeding Habitat: Most habitats, especially agricultural regions
Nest: Cup of variable material, thickly lined, near tree trunk
Eggs: 3–4 white, cream or pinkish, mottled with brown, gray or lavender
Incubation: 18–19 days
Fledging: 16–17 days

Yellow breasts and gray upper parts easily distinguish Western Kingbirds from Eastern Kingbirds. Their voice is distinctive, with its loud "whit" call and its forceful song. The song is especially useful in distinguishing this species from several look-alike species that may stray into the state. Western Kingbirds typically frequent dry riparian woodlands and savannas, but they have expanded into habitats that include agricultural lands, trees around farm complexes and urban areas.

Widmann (1907) noted that little was known about this species in the late 1800s. He mentioned a single sighting in Vernon County and noted that in the 1880s H. Nehrling called it fairly common in Lawrence County.

Code Frequency

Because they are noisy and conspicuous, Western Kingbirds are relatively easy to locate in appropriate habitats. Due to their tendency to perch conspicuously in the open, they were likely to be observed wherever they occurred.

Distribution

Western Kingbird records were located primarily in the Western Glaciated Plains Natural Section. Most confirmed breeding records were located from Kansas City north to Iowa. The one or two pairs located in the St. Louis area since 1986 (Wilson 1988) were documented by Atlasers. Other observations were scattered mainly in the Osage Plains Natural Division. At least one breeding pair was documented in Kennett, Dunklin County in 1991 (Wilson 1991b), but it was located outside an Atlas Project block.

Phenology

Western Kingbirds arrive late and likely rear a single brood. The first bird was recorded by the Atlas Project on May 16.

Breeding Phenology

EVIDENCE (# of Records)	MAR	APR	MAY	JUN	JUL	AUG	SEP
NB (1)			5/27	5/27			
NY (1)					7/15	7/15	
FY (1)				7/06	7/06		

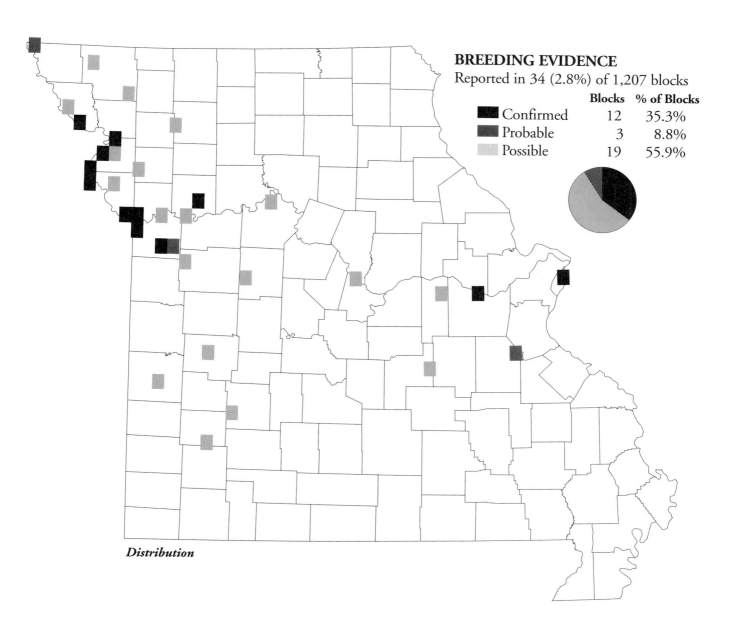

BREEDING EVIDENCE
Reported in 34 (2.8%) of 1,207 blocks

		Blocks	% of Blocks
■	Confirmed	12	35.3%
■	Probable	3	8.8%
■	Possible	19	55.9%

Distribution

Eastern Kingbird

Tyrannus tyrannus

> **Rangewide Distribution:** Southern Canada & most of United States
> **Abundance:** Common & widespread
> **Breeding Habitat:** Open woods, farmlands & forest edges
> **Nest:** Lined cup of weeds, grass & plant down in tree or fence post
> **Eggs:** 3–4 white, creamy, pinkish, with brown, olive or lavender
> **Incubation:** 16–18 days
> **Fledging:** 16–18 days

These distinctively-marked flycatchers occupy fairly open habitats, including roadsides, pastures with scattered large trees and woodland edges bordering open fields. They characteristically perch in full view on branches or wires from which they make short sallying flights to pursue flying insects. Known for their aggressiveness, Eastern Kingbirds are often seen chasing other birds.

Code Frequency

Because of the open habitat in which they reside, these well-known birds are easily detected. Their aggressiveness and other evidence of territorial behavior enabled many Atlasers to classify them as probable breeders. Despite nests being bulky and easy to locate, breeding was confirmed in fewer than half of the blocks. Eastern Kingbirds likely nested in most blocks in which they were recorded as possible breeders.

Distribution

Eastern Kingbirds were distributed statewide. Breeding was apparently less common in the eastern Mississippi Lowlands, perhaps due to that region's intensive rowcropping. There were fewer confirmed and probable records in the eastern half of Missouri. In the Ozarks, this may have resulted from extensive forest which is inappropriate for this species. Diminished Atlasing effort may account for fewer records in northeastern Missouri.

Abundance

As expected, Eastern Kingbirds were most abundant in western Missouri, primarily in the Osage Plains. They were least abundant in the Mississippi Lowlands and Big River natural divisions.

Phenology

Eastern Kingbirds arrive in Missouri during April (Robbins and Easterla 1992). Nesting was apparently initiated during April, because the earliest bird on a nest was detected May 1 and the first young in the nest, as evidenced by fecal sac removal, was recorded May 7. Despite the contention that Eastern Kingbirds produce only one brood per season, birds were observed nesting well into late summer. A bird was on the nest on July 19 and a nest with young was observed on August 2.

Abundance by Natural Division
Average Number of Birds / 100 Stops

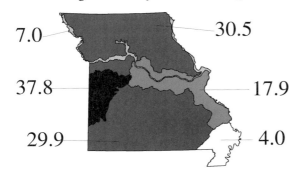

7.0   30.5
37.8   17.9
29.9   4.0

Breeding Phenology

EVIDENCE (# of Records)	MAR	APR	MAY	JUN	JUL	AUG	SEP
NB (76)			5/08		7/01		
NE (10)			5/18	6/23			
NY (34)			5/23			8/02	
FY (111)			5/16			8/18	

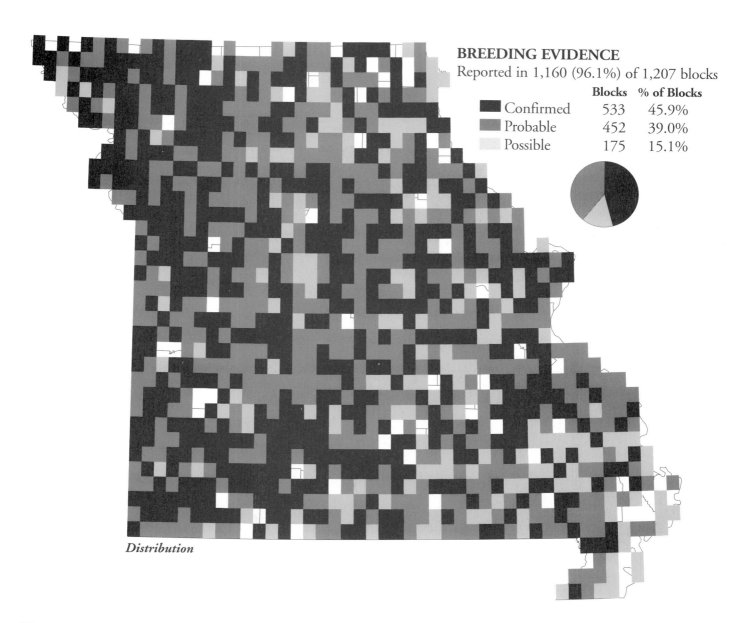

	Blocks	% of Blocks
Confirmed	533	45.9%
Probable	452	39.0%
Possible	175	15.1%

Distribution

Notes

Brown-headed Cowbird parasitism of Eastern Kingbirds
occurs occasionally (Ehrlich et al. 1988), and Atlas surveys
revealed one parasitized nest.

Scissor-tailed Flycatcher
Tyrannus forficatus

Rangewide Distribution: South central United
   States to Texas & Mexico border
Abundance: Common through its breeding range
Breeding Habitat: Open country with scattered trees
   & pastures
Nest: Cup of twigs, roots, weeds, moss, plant down &
   occasionally feathers or hair, low in trees & shrubs
Eggs: 3–5 white, cream or pink, mottled with red,
   brown, gray or olive
Incubation: 14–17 days
Fledging: 14–16 days

Scissor-tailed Flycatchers are found in open grasslands with widely-scattered trees. It is a spectacular species, with long scissor-shaped tail feathers streaming behind or flared out into a forked tail shape resembling a Barn Swallow. Widmann (1907) provided no breeding records for Missouri, but suggested that several contemporary books mentioned southwestern Missouri as within the breeding range of this species. The range of Scissor-tailed Flycatchers has expanded well into northern and eastern Missouri since the turn of the century (Robbins and Easterla 1992). The conversion of forest to grassland in south central Missouri may have contributed to this species' expansion eastward into that area where it has become a regular nester in Howell and Oregon counties.

Code Frequency

More confirmed breeding records were reported for this species than possible or probable records. Because this is a conspicuous species that nests in open situations, and nests are built in small trees that are relatively easy to find, Scissor-tailed Flycatchers were probably absent from blocks where they were not recorded. Additionally, post-breeding dispersal may account for some of the outlying records. Activity around the nest accounted for most of the confirmed records rather than nests with eggs or young. This may have been due to Atlaser reluctance to disturb the nests or because nests were too high above the ground for surveyors to observe nest contents.

Distribution

The percentage of blocks where this species was reported was highest (80 percent) in the Springfield Plateau Natural Section, followed by the Elk River Natural Section (69 percent), and the Osage Natural Division (66 percent). Although scattered observations were reported from northern Missouri, most nests were located south of the Missouri River. This species had one of the most restricted breeding ranges of any species that was recorded in at least 200 blocks.

Abundance

As expected, the highest relative abundance was recorded in the Osage Plains Natural Division. The only other area where this species was observed was in the Ozark Natural Division, with most records originating from the western edge of that division in southwestern Missouri.

Abundance by Natural Division
Average Number of Birds / 100 Stops

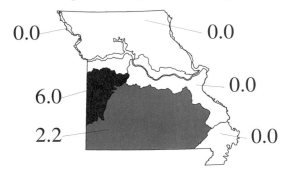

0.0   0.0

6.0   0.0

2.2   0.0

Breeding Phenology

EVIDENCE (# of Records)	MAR	APR	MAY	JUN	JUL	AUG	SEP
NB (13)			5/08		6/25		
NE (1)				6/14 6/14			
NY (6)			5/12			7/31	
FY (8)			5/29		7/15		

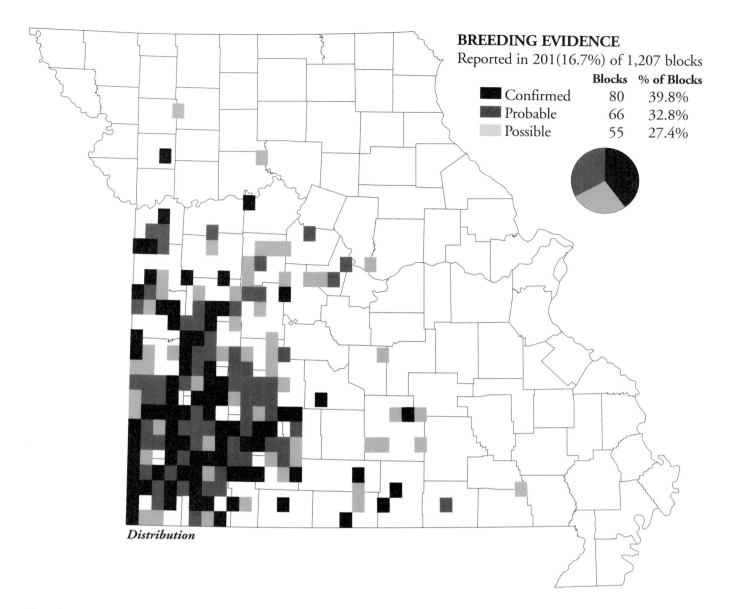

BREEDING EVIDENCE

Reported in 201(16.7%) of 1,207 blocks

		Blocks	% of Blocks
■	Confirmed	80	39.8%
■	Probable	66	32.8%
■	Possible	55	27.4%

Distribution

Phenology

Atlasers noted the first individual on April 15, but most nest building occurred May 8–June 25. Atlas observations correspond to an April–July egg-laying period described by Terres (1987). Because this species is considered single-brooded (Ehrlich et al. 1988), late nesting dates such as August 3 may have involved a renesting attempt rather than a second brood.

Horned Lark
Eremophila alpestris

Rangewide Distribution: Canada, most of United
   States, Mexico, Europe, Asia & Africa
Abundance: Widespread & common
Breeding Habitat: Open areas with sparse vegetation
   & low shrubs
Nest: Depression on ground, rimmed with pebbles &
   dirt clods, lined with roots, grass & hair
Eggs: 3–4 variable, gray, greenish, heavily spotted with
   brown
Incubation: 11–12 days
Fledging: 9–12 days

Horned Larks are well adapted to Missouri's farmlands.
They are especially prevalent on well-drained, flat, treeless
terrain within habitats that range from short grass to
exposed soil (Peterjohn and Rice 1991). They place their
nests in closely-grazed pastures, crop stubble and even
plowed fields. Occasionally, they occupy more rolling terrain
and short-grass habitats in cities such as airfields, golf courses, parks and large disturbed areas (Peterjohn and Rice
1991).

Code Frequency

Horned Larks are easily detected by their jingling songs
and, in the proper season, by their impressive courtship
flights. They are our earliest nesting passerine. Because most
of the Atlasing field effort was initiated after the onset of the
breeding season, potential breeding evidences may have been
missed. Additionally, although nests are often easy to find
because of the open, sparse cover in which they are built,
Atlasers may have been reluctant to enter pastureland and
other farm fields. Therefore, Horned Larks likely nested in
most blocks in which breeding evidence was found.

Distribution

The distribution of Horned Larks mirrors the distribution of open grasslands of Missouri. They were solidly distributed throughout the Osage Plains, Glaciated Plains and
Mississippi Lowlands. These natural divisions contain considerable acreage of the flat, short-grass and open cropland
terrain most often utilized by Horned Larks. Horned Larks
were sparsely distributed through the Ozarks and Ozark
Border despite the apparent suitability of habitat in those
natural divisions.

Abundance

Abundance information available from Miniroutes and
Breeding Bird Surveys confirmed that greater numbers of
Horned Larks occurred within the Mississippi Lowlands,
Big Rivers and Glaciated Plains natural divisions than other
natural divisions. Horned Larks were essentially absent
from the eastern Ozarks and along the southern border of
the state, except for the Mississippi Lowlands.

Phenology

Horned Larks are found in appropriate breeding habitats
throughout the year, facilitating the early initiation of breed-

Abundance by Natural Division
Average Number of Birds / 100 Stops

33.0   19.6

6.0   4.1

2.9   60.8

Breeding Phenology

EVIDENCE (# of Records)	MAR	APR	MAY	JUN	JUL	AUG	SEP
NB (6)		4/28			7/09		
NE (3)	4/01		5/15				
FY (18)		4/26			7/26		

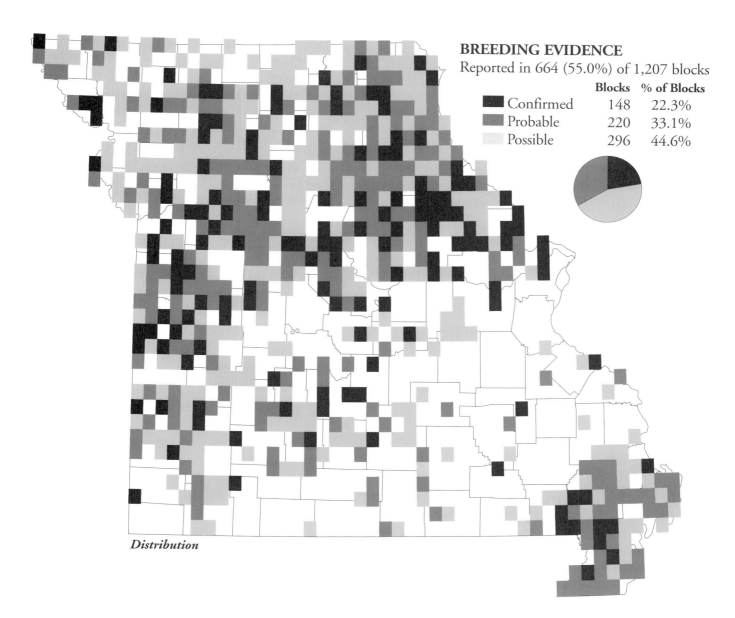

Distribution

ing when conditions become suitable. In Ohio, males will perform their aerial courtship displays during warm days in January and nesting activities may begin during the latter half of February (Peterjohn and Rice 1991). Early Atlas dates for a nest with eggs and a bird seen carrying food for young are likely typical for the first brood. Some pairs may nest twice or even three times a year (Peterjohn and Rice 1991). Breeding evidence from late July and August may have involved a second or third brood.

Notes

Although Ehrlich et al. (1988) described Horned Larks as uncommon Brown-headed Cowbird hosts, no evidence of Brown-headed Cowbird brood parasitism was discovered during this project. However, nest contents were observed in only three blocks.

Purple Martin
Progne subis

Rangewide Distribution: South central Canada, all
   eastern United States & East Coast
Abundance: Locally common, population decreasing
   in west
Breeding Habitat: Open areas, savanna, rural areas
   near water
Nest: Grass, leaves, mud & feathers in tree cavities or
   birdhouses
Eggs: 4–5 white & unmarked
Incubation: 15–18 days
Fledging: 26–31 days

In Missouri, Purple Martins associate with human habitation and appear to be almost totally dependent on nest boxes people provide. As a result, they are found in suburban areas, towns of all sizes and near rural residences. Elaborate martin houses can be found in nearly every town statewide.

Code Frequency

Because martins usually nest near human habitation, they were probably detected if present. Surprisingly, this species was recorded in only 61 percent of all blocks. Forty-five percent of all records were observations of individuals entering or leaving a potential nest site. Because this species nests in artificial structures, 76 percent of all records involved breeding confirmations.

Distribution

This species was distributed statewide, except for several adjacent blocks in the heavily-wooded regions of the Lower Ozark Natural Section. In the northern one-fifth of the state, distribution was more patchy, perhaps because few nest boxes are available in areas with lower human population (Campbell 1991). Atlas Project observations support the findings of Price et al. (1995) who reported lower abundance in the Glaciated Plains.

Abundance

About four times as many Purple Martins were reported in the Ozark and Mississippi Lowlands natural divisions than in the Glaciated Plains and the Osage Plains. The highest relative abundances were associated with the more populated areas that likely have more nest boxes.

Phenology

Martins began to occupy nest sites in early April shortly after the first migrants returned. Most nest-related observations began in early May. An individual carrying a fecal sac was observed on June 11. A nest with young observed on August 9 may represent a second or even a third brood.

Abundance by Natural Division
Average Number of Birds / 100 Stops

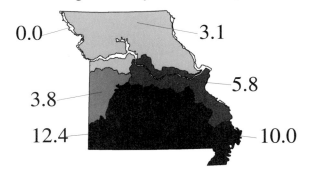

0.0   3.1

3.8   5.8

12.4   10.0

Breeding Phenology

EVIDENCE (# of Records)	MAR	APR	MAY	JUN	JUL	AUG	SEP
NB (15)		4/21		6/18			
NE (7)			5/09	6/16			
NY (70)			5/13			8/09	
FY (31)			5/18		7/20		

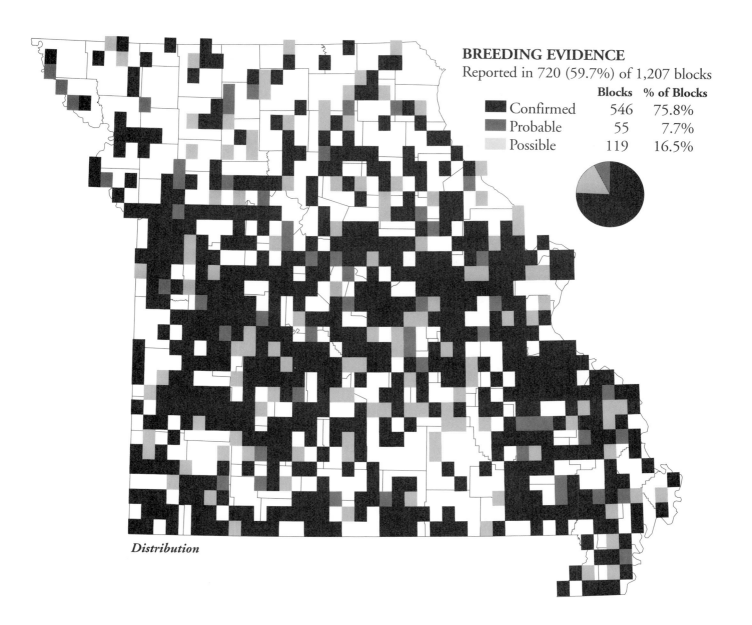

BREEDING EVIDENCE
Reported in 720 (59.7%) of 1,207 blocks

	Blocks	% of Blocks
Confirmed	546	75.8%
Probable	55	7.7%
Possible	119	16.5%

Distribution

Tree Swallow
Tachycineta bicolor

> **Rangewide Distribution:** Southern & northwestern Canada, Alaska & southern to west central and east central United States
> **Abundance:** Common
> **Breeding Habitat:** Near water, abundant dead trees & open woods
> **Nest:** Grass lined with feathers in tree hole or cavity
> **Eggs:** 4–6 white & unmarked
> **Incubation:** 13–16 days
> **Fledging:** 16–24 days

Like many cavity nesters, Tree Swallows return early to Missouri. They usually select cavities produced by woodpeckers, but will use other cavities including bird boxes and mail boxes. In large reservoirs where trees standing in water have died, large colonies of this species can form. As the trees succumb to the elements and fall into the water, the colony moves on. On natural sites, such as beaver ponds or flooded riparian areas, abundant dead trees may supply ample nest cavities (Robertson et al. 1992). Tree Swallows frequently nest in bird boxes intended for bluebirds even though this isolates them from a colony.

Code Frequency

Most breeding confirmations resulted from activities around the nest site, with only a few observations of actual nests. Nests are usually difficult to inspect because they are located in trees standing in several feet of water. Although Tree Swallows farther north in the United States frequently nest in bird boxes, pairs nesting in bird boxes seemed scarce in this survey.

Distribution

The Atlas Project found most sites in Missouri around lakes, ponds and reservoirs with standing dead trees in the water. Most blocks encompassing parts of Truman Reservoir recorded Tree Swallows. That reservoir was filled in 1979 and still contained numerous standing dead trees at the time of the Atlas Project. Tree Swallows were found throughout the state. However, few records were from the central part of the Ozark Natural Division, where few large impoundments were present. The species was noted on several 5–10 acre farm ponds in the Ozark Natural Division, but these sites were outside Atlas blocks.

Abundance

Data available from Breeding Bird Surveys and Atlas Project Miniroutes were insufficient to evaluate regional variations in abundance for this rarely recorded colonial species. Overall, Tree Swallows were most abundant in large stands of dead trees in lakes and reservoirs. Tree Swallows were considered a scarce resident in the forests along the big rivers during Widmann's (1907) time.

Phenology

Early Tree Swallows arrive in late February, but most arrive from late March to early May (Robbins and Easterla 1992). Atlasers observed birds on nests from April 16 to July 17. Because swallows migrate over a period of several

Breeding Phenology

EVIDENCE (# of Records)	MAR	APR	MAY	JUN	JUL	AUG	SEP
NB (1)		5/07	5/07				
NE (1)			5/28	5/28			
NY (1)				6/14	6/14		
FY (5)				6/08	6/23		

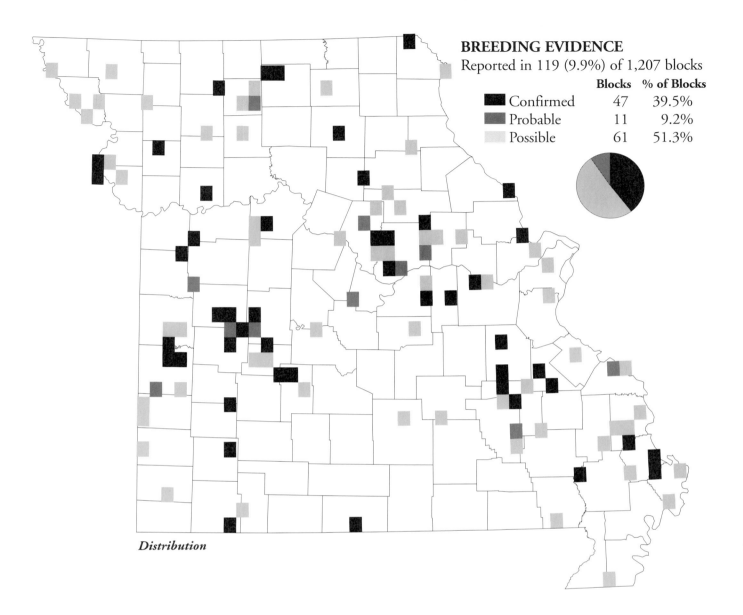

BREEDING EVIDENCE
Reported in 119 (9.9%) of 1,207 blocks

		Blocks	% of Blocks
■	Confirmed	47	39.5%
■	Probable	11	9.2%
░	Possible	61	51.3%

Distribution

months, confirmation codes are especially necessary to evaluate breeding phenology for species in this family.

Notes

As with most cavity nesters, brood parasitism was difficult to detect and less likely than with open-cup nesters.

None was reported during the Atlas Project. Tree Swallows aggressively defend their nest sites, even risking their lives, which explains many dead birds with head injuries found in nest boxes (Robertson et al. 1992).

Northern Rough-winged Swallow
Stelgidopteryx serripennis

> **Rangewide Distribution:** Southern Canada throughout United States & Central America
> **Abundance:** Widespread & fairly common
> **Breeding Habitat:** Open country & road cuts near running water
> **Nest:** Grass, leaves & weeds, unlined in cliff, crevice or niche
> **Eggs:** 5–6 white & unmarked
> **Incubation:** 12 days
> **Fledging:** 19–21 days

A cavity nester found statewide, Northern Rough-winged Swallows are normally solitary nesters in cut banks along bends in rivers and streams. Sometimes their burrows are among those of bank swallows, a colonial nester. With channelization and bank stabilization, many river bank nest sites have been eliminated. Northern Rough-winged Swallows have, however, expanded their nesting habitat to include road-cut cliffs along highways where they become somewhat colonial.

Code Frequency

Most of the breeding confirmations documented during the Atlas Project were obtained by observing fledglings or parent birds at nest holes. Because of the difficulty of observing a nest, nests with eggs (3) or young (15) were recorded in only 3 percent of blocks. Some fully-flighted birds with immature plumage may have been incorrectly classified as fledglings.

Distribution

Atlaser observations documented this species in 50 percent to 61 percent of blocks in all natural divisions except the Mississippi Lowlands, where it was recorded in only 20 percent of blocks. The channelized watercourses and relatively level terrain of the Mississippi Lowlands apparently do not support many nest sites.

Abundance

About six times as many individuals were recorded in the Big Rivers Natural Division as in the other natural divisions. No individuals were recorded for the Mississippi Lowlands. Roadside abundance counts are not an effective method for determining the status of a species associated with river ways. Riparian surveys would more accurately count this species.

Phenology

The first birds appear in late March (Robbins and Easterla 1992). Late dates for nests with young may indicate renesting or a rare second brood (Ehrlich et al. 1988). The late dates for fledglings observed during the Atlas Project may represent a very late nesting attempt or an immature bird misidentified as a fledgling. Most birds have left the state by mid-October (Robbins and Easterla 1992).

Abundance by Natural Division
Average Number of Birds / 100 Stops

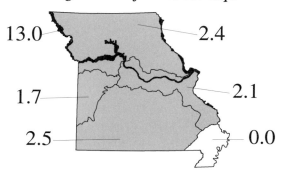

13.0   2.4

1.7   2.1

2.5   0.0

Breeding Phenology

EVIDENCE (# of Records)	MAR	APR	MAY	JUN	JUL	AUG	SEP
NB (19)			4/19	6/19			
NE (3)			5/28	6/03			
NY (15)			5/26		7/19		
FY (19)				6/01		8/06	

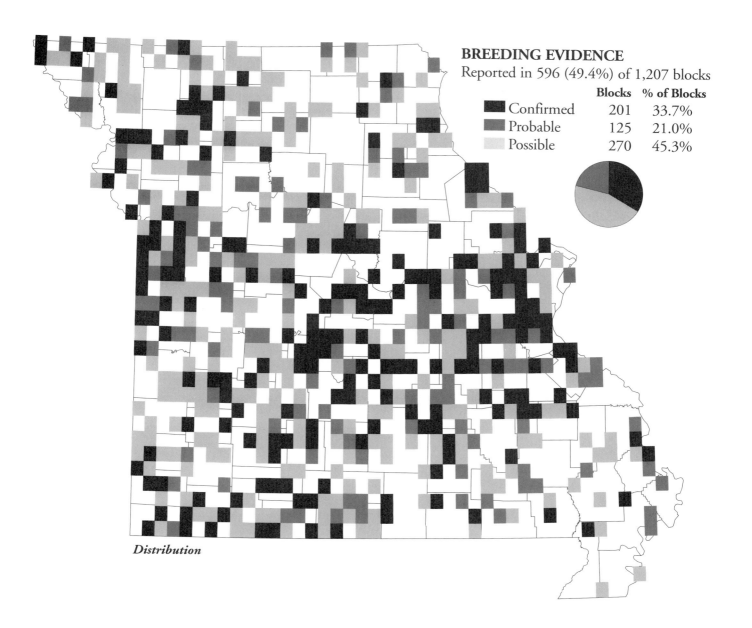

BREEDING EVIDENCE

Reported in 596 (49.4%) of 1,207 blocks

	Blocks	% of Blocks
Confirmed	201	33.7%
Probable	125	21.0%
Possible	270	45.3%

Distribution

Bank Swallow
Riparia riparia

> **Rangewide Distribution:** Canada, Alaska, northern & central United States & northern Eurasia
> **Abundance:** Locally common
> **Breeding Habitat:** Open area, road cuts near running water & stream banks
> **Nest:** Grass, roots, stems & feathers at end of 2–3 foot burrow
> **Eggs:** 4–5 white & unmarked
> **Incubation:** 14–16 days
> **Fledging:** 18–24 days

These easily-identified brown swallows prefer open areas where they can fly low to forage. Their typical flight pattern alternates between brief climbs and low swoops over fields of grass or water surfaces. Bank Swallows typically excavate nesting burrows in river banks and other vertical faces of earth. They often nest singly or in small colonies but sometimes form dense colonies numbering as many as several hundred burrows (Harrison 1975).

Code Frequency

These aerial insectivores are relatively easy to sight. Therefore, where not recorded by the Atlas Project, they may not have occurred or were present in low numbers. Even when sighted, they may not have nested precisely in that block, as they will forage as far as 1.5 kilometers from nest sites (Gross 1942). Although Bank Swallow nest excavations are conspicuous, Atlasers often had difficulty accessing areas such as river banks. Therefore, Bank Swallows likely nested in most of those blocks where Atlasers listed them as possible breeders. Nesting was confirmed in 40 percent of the blocks, most often by the observation of adults entering or leaving nest holes. Atlasers rarely observed nests with eggs or young as would be expected for a cavity nester.

Distribution

The pattern of Bank Swallow distribution appears related to the location of rivers, presumably because cut banks provide a suitable substrate for burrows. No regional difference in distribution was apparent, although Robbins and Easterla (1992) suggested Bank Swallows are more common north of the Missouri River. The Atlas Project did not confirm Robbins and Easterla's (1992) statement that Bank Swallows are rarer in the Ozark or Osage Plains natural divisions.

Abundance

Bank Swallows displayed their greatest abundance in the Ozark Border Natural Division followed by the Glaciated Plains Natural Division. This perhaps relates to the prevalence of appropriate nesting situations. Abundance data, however, may be unreliable due to the colonial nesting behavior of the species. Survey points that happened near a colony would record greater abundance than those distant from a colony.

Abundance by Natural Division
Average Number of Birds / 100 Stops

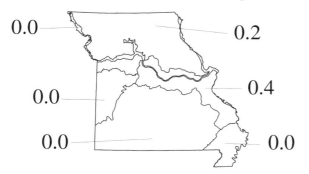

0.0   0.2
0.0   0.4
0.0   0.0

Breeding Phenology

EVIDENCE (# of Records)	MAR	APR	MAY	JUN	JUL	AUG	SEP
NB (4)		5/03		6/09			
NE (2)			6/09	6/17			
NY (2)			6/10				8/11
FY (3)			6/04	6/23			

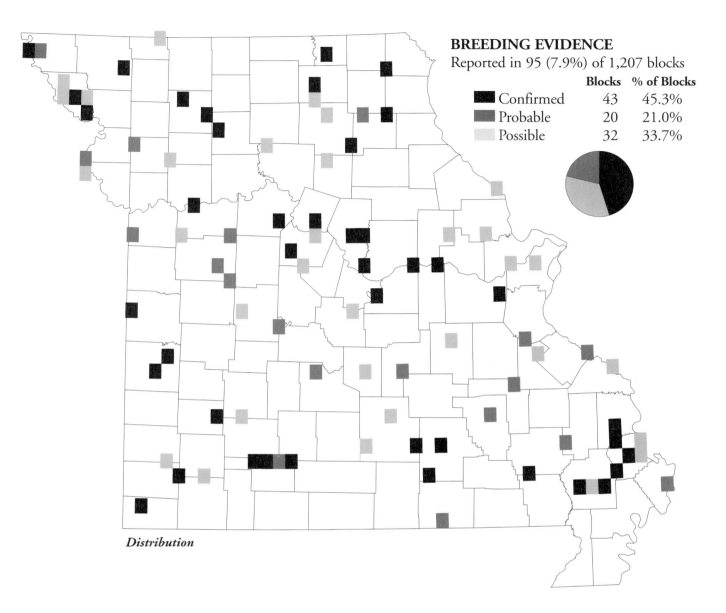

BREEDING EVIDENCE
Reported in 95 (7.9%) of 1,207 blocks

		Blocks	% of Blocks
■	Confirmed	43	45.3%
■	Probable	20	21.0%
■	Possible	32	33.7%

Distribution

Phenology

Nesting season events were well documented by the Atlas Project. The earliest recorded date for birds on the nest was May 25. Graber et al. (1972) reported a great range in egg-laying dates in northern Illinois extending from May 7 to July 9. They also reported young in the nest as late as August 14 in southern Illinois. No evidence of double-brooding for the Bank Swallow was found. Their nesting appears less synchronized than most species, perhaps because of renesting following nest site disturbances.

Cliff Swallow
Hirundo pyrrhonota

> **Rangewide Distribution:** Canada, Alaska & most United States to central Mexico
> **Abundance:** Locally common, with range expanding south in eastern United States
> **Breeding Habitat:** Open country, usually near running water
> **Nest:** Gourd-shaped mud pellet construction, lined with grass & feathers, on bridges, cliffs & buildings
> **Eggs:** 4–5 white, creamy, or pinkish-white with brown spots
> **Incubation:** 14–16 days
> **Fledging:** 21–24 days

These widely-distributed swallows are usually seen in flight, when they are best identified by their tan rump patches. Like other swallows, they venture great distances in pursuit of insect prey. When nesting, however, they typically remain near nest sites. The most social of the swallows, Cliff Swallows nest colonially, plastering their funnel-shaped mud nests on vertical surfaces sheltered from rain. They often build their nests on bridges, barns, large dams and sheltered bluff faces. Originally summer residents of the western mountains, Cliff Swallows have expanded their breeding range during the last 100–150 years in response to highway, bridge and building construction (Brown and Brown 1995).

Code Frequency

Cliff Swallows are easy to see and identify wherever they occur. Therefore, they probably did not reside in blocks where they were not reported. They were obviously easy to confirm perhaps because of their easily-identified nests. As with other colonial species, several of their breeding colonies were no doubt missed because of the random selection of sampling blocks.

Distribution

Cliff Swallows were distributed sparsely statewide. Most nest sites were associated with structures such as bridges and dams as at Truman, Long Branch and Table Rock lakes. Other breeding evidence was found in blocks that included bridges on the Missouri and Mississippi rivers. Four colonies were on natural sites: three on overhanging cliff faces and one in a cave entrance. Two cliffs were along Lake of the Ozarks in Camden County, one was along Auxvasse Creek in Morgan County and the cave site was along Gravois Creek in Morgan County.

Abundance

Areas of high Cliff Swallow abundance were associated with Table Rock, Taneycomo and Truman Lakes and the Missouri River in central Missouri. The largest colonies reported contained hundreds of birds at Lock and Dam 25 in Lincoln County and under a Table Rock Lake bridge in Barry County. Other large colonies were at Stockton and Truman lakes and on bridges over the Missouri River. Half of the colonies reported contained more than 50 nests. The abundance map was not included because the colonial nesting habits of the species biased the results.

Phenology

Cliff Swallows normally arrive during the second week of April (Robbins and Easterla 1992). Birds were on nests by

Breeding Phenology

EVIDENCE (# of Records)	MAR	APR	MAY	JUN	JUL	AUG	SEP
NB (6)		-					-
NY (10)			5/23		7/17		
FY (3)			5/27	6/21			

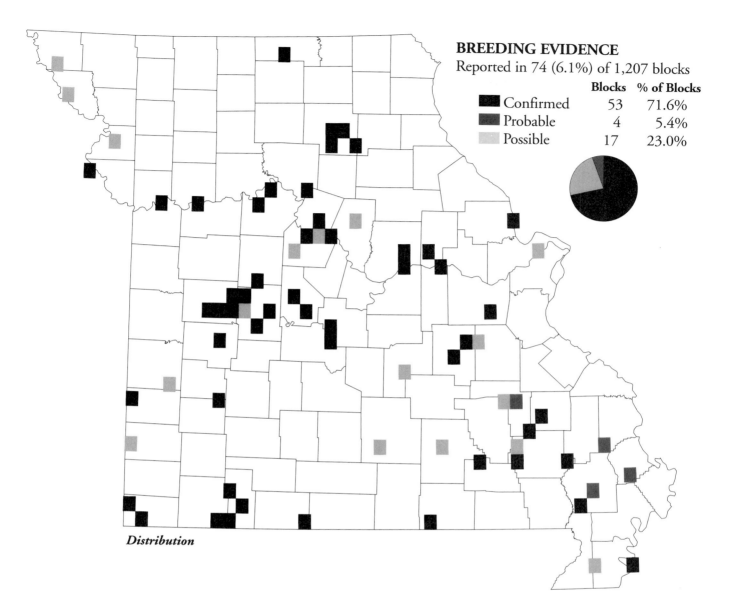

BREEDING EVIDENCE
Reported in 74 (6.1%) of 1,207 blocks

		Blocks	% of Blocks
◼	Confirmed	53	71.6%
◼	Probable	4	5.4%
◻	Possible	17	23.0%

Distribution

May 15. Cliff Swallows can produce two broods (Harrison 1975). Second broods or renesting may account for later nesting event dates.

Notes

Brown-headed Cowbird parasitism is unknown for Cliff Swallows, however, interspecific parasitism by House Sparrows and intraspecific parasitism among Cliff Swallows has been documented (Brown and Brown 1989). House Sparrows also compete with Cliff Swallows by occasionally usurping nests (Brown and Brown 1989). However, of 28 colonies for which Atlasers could recall House Sparrow presence or absence, only four were shared by House Sparrows.

Barn Swallow
Hirundo rustica

Rangewide Distribution: Northwestern & southern Canada, United States & worldwide
Abundance: Common
Breeding Habitat: Open country, savannas, near water & agricultural areas
Nest: Mud pellets & straw, lined with feathers, plastered on ledges, walls & buildings
Eggs: 4–5 white with brown spots
Incubation: 13–17 days
Fledging: 18–23 days

Perhaps the most familiar of the swallows, Barn Swallows are commonly seen flying low in pursuit of insects over fields, ponds and farm lots. Barn Swallows construct mud nests where sheltered from above, in barns and other buildings, under bridges and, rarely, under overhanging cliffs (Campbell 1968). Although often nesting singly, Barn Swallows will nest colonially where conditions are favorable.

Code Frequency

Barn Swallows are an easily-seen species and their nests were easily found. Forty-five percent of confirmed breeding evidence was obtained by observing adults tending nests, eggs or young. Some Atlasers were reluctant to ask farmers if they could check outbuildings for evidences of nesting. Therefore, Barn Swallows likely also bred in the 105 blocks where Atlasers simply observed them as possible breeders. Although Barn Swallows can forage a distance from nesting locales, their nests were likely located within the block where they were observed.

Distribution

The only region of the state that showed any scarcity of Barn Swallows was the central Ozarks. This region's forested conditions apparently provide insufficient open habitat for this species. Additionally, a portion of northwestern Missouri contained fewer blocks where Barn Swallows were recorded.

Abundance

Barn Swallows were least abundant in the Mississippi Lowlands Natural Division. They were most abundant in the Big Rivers, Osage Plains and Glaciated Plains natural divisions. Overall, more Barn Swallows were recorded in northern than in southern Missouri.

Phenology

The sequence of nesting events was predictable and observation dates conform to the Barn Swallow's 15-day incubation and 18-day rearing periods (Harrison 1975). Atlas Project findings support the two broods documented by Ehrlich et al. (1988).

Notes

Brown-headed Cowbird brood parasitism of Barn Swallows has not been reported (Ehrlich et al. 1988) but

Abundance by Natural Division
Average Number of Birds / 100 Stops

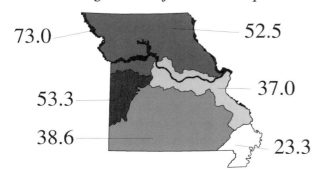

73.0   52.5
53.3   37.0
38.6   23.3

Breeding Phenology

EVIDENCE (# of Records)	MAR	APR	MAY	JUN	JUL	AUG	SEP
NB (65)	4/11				7/19		
NE (58)	4/13					8/11	
NY (141)		5/06				8/12	
FY (38)		5/09				8/14	

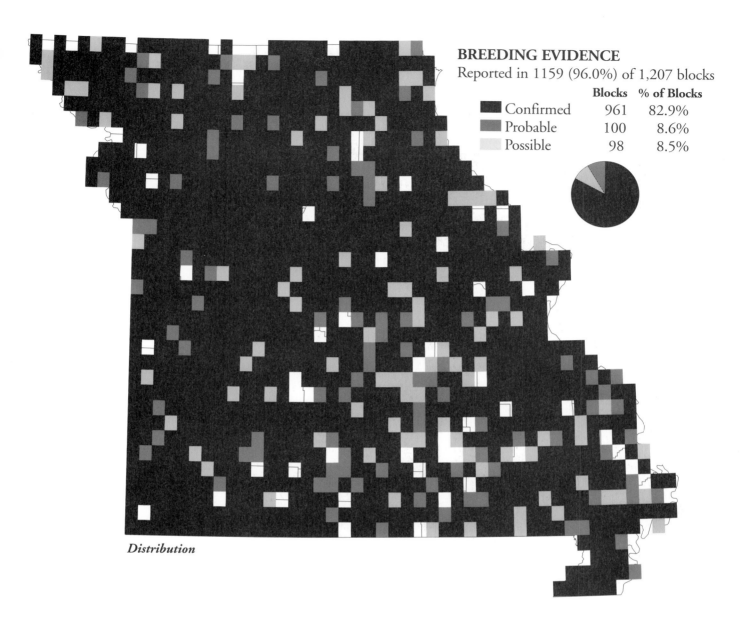

BREEDING EVIDENCE

Reported in 1159 (96.0%) of 1,207 blocks

	Blocks	% of Blocks
Confirmed	961	82.9%
Probable	100	8.6%
Possible	98	8.5%

Distribution

was suggested in one block by the presence of an egg. However, this may have actually been a House Sparrow egg, as House Sparrows are known usurpers of Barn Swallow nests (Ehrlich et al. 1988).

Blue Jay
Cyanocitta cristata

Rangewide Distribution: Central to southeastern Canada, central United States to East Coast
Abundance: Common & widespread
Breeding Habitat: Open deciduous & mixed forests & woods, parks & residential areas
Nest: Bulky, compact construction of twigs, bark, moss, lichens, paper, rags, string, grass & mud, lined with roots, in trees, shrubs & vine
Eggs: 4–5 variable colored of green, buff or blue, spotted with browns
Incubation: 16–18 days
Fledging: 17– 21 days

Blue Jays are familiar inhabitants of parks, cemeteries and shaded residential areas in cities, towns and the Missouri countryside. They also reside and breed in upland forest edges, small, isolated woodlots and narrow riparian corridors (Terres 1987). Breeding Blue Jays prefer second growth forests with dense understories.

Code Frequency

These common birds are usually extremely easy to detect, however, they may become secretive around nest sites. Therefore, they likely bred in many blocks where they were only recorded as possible or probable breeders. Most of the blocks in which they were recorded as possible breeders were in heavily-forested regions where observations of breeding behavior would be more difficult.

Distribution

The Atlas Project confirmed that Blue Jays are distributed statewide. They were not recorded, however, in several adjacent blocks in various parts of the state. Because Blue Jays are usually easy to detect, they were likely absent or present in low numbers in those blocks during the Atlas Project.

Abundance

Survey results suggest a fairly uniform abundance statewide except slightly more are found in the north and fewer in the Mississippi Lowlands and along the western border.

Phenology

Confirmed breeding observations accurately depict Blue Jay nesting phenology. Harrison (1975) reported two or even three broods. The late dates when nesting was observed may be attributable to second or third broods.

Notes

Blue Jays are described as rare Brown-headed Cowbird hosts (Terres 1987) and Alasers observed no evidence of parasitism during this project.

Abundance by Natural Division
Average Number of Birds / 100 Stops

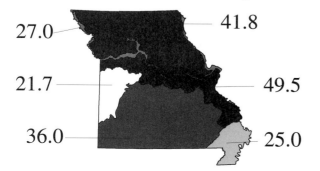

27.0
41.8
21.7
49.5
36.0
25.0

EVIDENCE (# of Records)	MAR	APR	MAY	JUN	JUL	AUG	SEP
NB (50)	4/06				7/21		
NE (7)		4/28		6/28			
NY (36)		5/01			7/26		
FY (133)		5/04				8/16	

Breeding Phenology

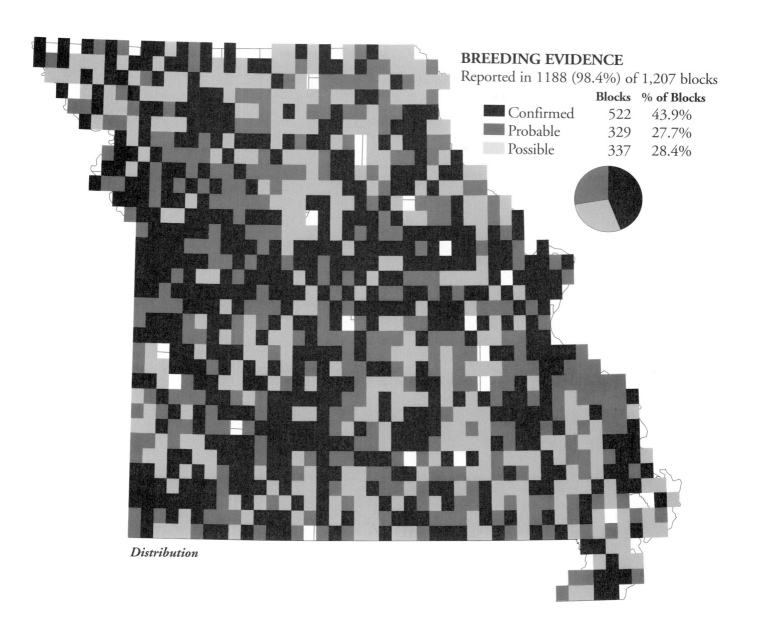

BREEDING EVIDENCE
Reported in 1188 (98.4%) of 1,207 blocks

	Blocks	% of Blocks
Confirmed	522	43.9%
Probable	329	27.7%
Possible	337	28.4%

Distribution

American Crow
Corvus brachyrhynchos

Rangewide Distribution: All southern Canada & entire United States
Abundance: Common
Breeding Habitat: Open woods & forest, farmland & orchards
Nest: Branches, twigs & bark, lined with bark, moss, grass, feathers, hair & leaves in tree or shrub
Eggs: 4–6 blue- or olive-green, with brown, gray marks
Incubation: 18 days
Fledging: 28–35 days

Missouri's largest passerine is one of the most easily seen and recognized birds in the state. Crows nest wherever large trees are found, in forests, rural woodlots, parks and even suburban yards (Harrison 1975).

Code Frequency

Despite the American Crow's conspicuousness and their presumed prevalence in Missouri, Atlasers found it difficult to obtain breeding evidence. This may be because they for-age distantly from nest sites and adopt a secretive behavior when near nests (Peterjohn and Rice 1991). They no doubt initiated breeding before most Atlasing efforts began, and some of the most obvious nesting behaviors might have been detected earlier in the season. As a result, breeding evidence beyond simple occurrence was observed in half the blocks, and actual confirmation of breeding in only 29 percent of the blocks. Most surprising, no evidence of breeding was recorded in 11.5 percent of the blocks.

Distribution

American Crows were recorded and confirmed to breed in more blocks in southern than in northern Missouri. Surprisingly, Atlasers did not record them in western Missouri in several blocks near the Missouri River. Also, this species was not reported in southern Pike County or the Mississippi Lowlands. The reasons for the American Crow's absence are uncertain except in the Mississippi Lowlands, where there may be fewer nest trees.

Abundance

American Crow abundance was unexpectedly variable. Generally, they appeared more abundant in the more forested regions of the state, including the Ozark Border and Ozark natural divisions. There were relatively fewer reported in the Mississippi Lowlands, perhaps due to a lack of suitable nest sites and surprisingly, none were recorded on abundance surveys in the Big Rivers Natural Division.

Phenology

American Crow nesting activity at this latitude commences as early as March (Peterjohn and Rice 1991). Migrating individuals are expected through March (Robbins and Easterla 1992), which may have complicated the efforts of early-season Atlasers to determine which crows were breeders. Data suggest that fledging began in May and

Abundance by Natural Division
Average Number of Birds / 100 Stops

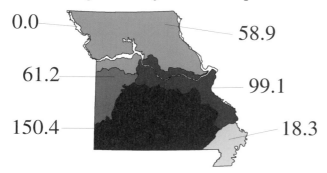

0.0
58.9
61.2
99.1
150.4
18.3

EVIDENCE (# of Records)	MAR	APR	MAY	JUN	JUL	AUG	SEP
NB (13)	4/01			6/01			
NE (1)			5/30	5/30			
NY (30)			5/13		7/12		
FY (41)			5/12				8/15

Breeding Phenology

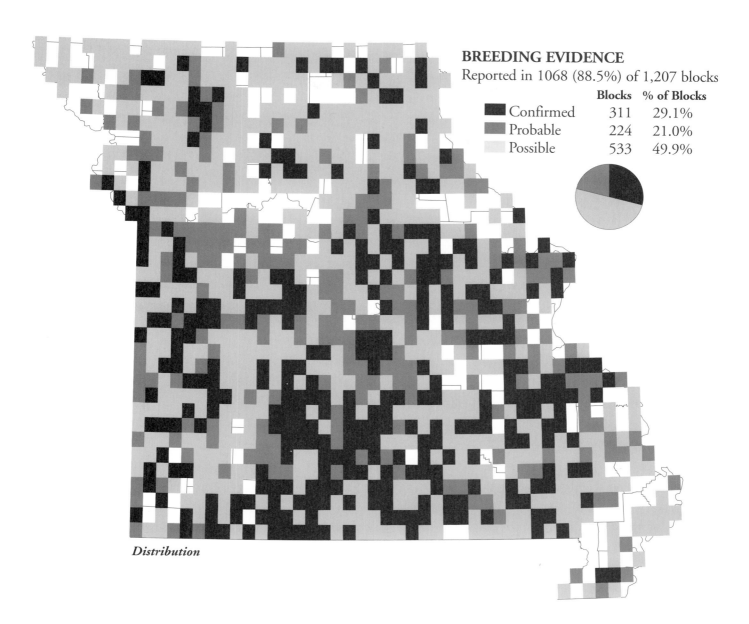

BREEDING EVIDENCE
Reported in 1068 (88.5%) of 1,207 blocks

	Blocks	% of Blocks
Confirmed	311	29.1%
Probable	224	21.0%
Possible	533	49.9%

Distribution

peaked in June. The late Atlas Project records of nest build-ing suggest crows can have a protracted breeding season and may also be evidence of double-brooding, a characteristic of American Crows in the southern United States (Ehrlich et al. 1988).

Notes

Atlasers obtained no evidence of Brown-headed Cowbird parasitism, a rare phenomenon with crows (Harrison 1975).

Fish Crow
Corvus ossifragus

Rangewide Distribution: Eastern coastal United States & along major rivers inland

Abundance: Common southeastern coast; uncommon to rare inland

Breeding Habitat: Beaches, bays, swamps & major inland waters

Nest: Sticks & twigs lined with fine material & dung, high in trees or shrubs

Eggs: 4–5 blue-green or gray-green with brown or gray marks

Incubation: 16–18 days

Fledging: 21 + days

Missouri's rare, smaller crow species is associated primarily with swamps, wooded watercourses and backwaters. Robbins and Easterla (1992) describe it as a permanent Mississippi Lowlands resident and a rare Mississippi River summer resident north to Pike County.

Code Frequency

Few Fish Crows were observed during the Atlas Project. Because the expected range of the Fish Crow is primarily in the Mississippi Lowlands Natural Division, and because much of the Atlasing effort in that division was accomplished by block-busting during June, Fish Crows may have been under-represented. It is also possible, especially outside their expected range, that Atlasers misidentified a few Fish Crows as American Crows.

Distribution

Fish Crows were found in only three blocks in Dunklin, Perry and Pemiscot counties. The Dunklin County site was at Hornersville Swamp Conservation Area; the others were along the Mississippi River. No Fish Crows were confirmed to breed by the Atlas Project, and to our knowledge, they have never been confirmed as Missouri breeders, though they likely breed regularly in southeast Missouri. Breeding at sites such as the Gayosa Bend and Wolf Bayou conservation areas and at Big Oak Tree State Park is likely as well as within the woodlands near the Mississippi and St. Francis rivers in Dunklin, Mississippi, New Madrid and Pemiscot counties.

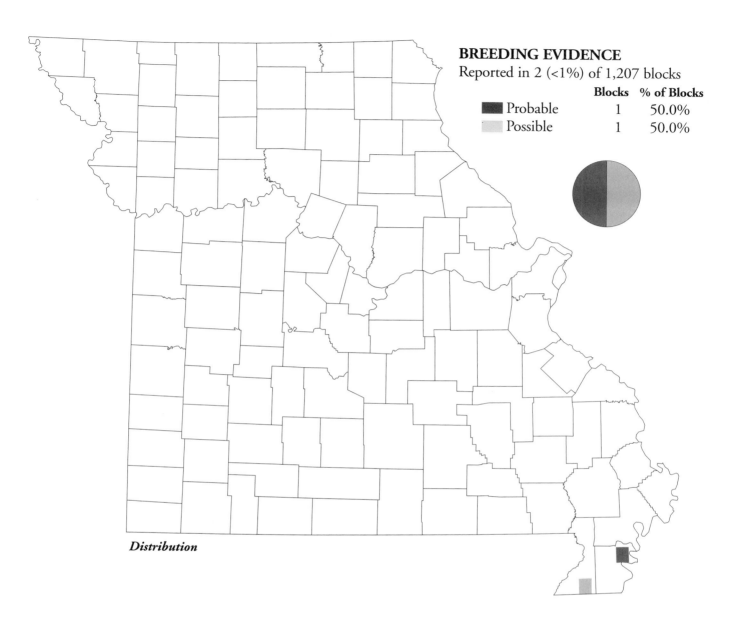

Reported in 2 (<1%) of 1,207 blocks

	Blocks	% of Blocks
Probable	1	50.0%
Possible	1	50.0%

Distribution

Black-capped Chickadee
Parus atricapillus

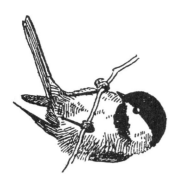

Rangewide Distribution: Southwestern to southeastern Canada, south central Alaska & northern to central United States
Abundance: Common & widespread
Breeding Habitat: Deciduous & mixed forests, thickets, parks & riparian areas
Nest: Excavated cavity lined with plant down, moss, feathers, hair & cocoons
Eggs: 6–8 white with reddish-brown fine marks
Incubation: 11–13 days
Fledging: 14–18 days

Black-capped Chickadees breed primarily in deciduous, open woodlands. They are generally more common near edges of wooded areas but can be found in the middle of large wooded tracts (Smith 1993). They occur in upland and riparian habitats, and where large trees are available in well-shaded residential areas, parks and small woodlots (Terres 1987). They nest in dead snags, rotten branches, old woodpecker holes, bird boxes and even in the ground (Smith 1991).

Code Frequency

Black-capped Chickadees are one of the easiest birds to detect. Early in the season when they are calling most, they are easy to separate from Carolina Chickadees by their song. Because of their small size and the forested habitat in which they reside, chickadees were apparently difficult to elevate to the probable or confirmed categories. Accordingly, many probable records, especially those in northern Missouri, likely indicated breeding areas.

The Atlas Project required that documentation forms be completed in an attempt to insure the proper identification of Black-capped versus Carolina chickadees, especially where their ranges meet or where they were reported outside their expected range. Atlasers recorded both chickadee species in most southern Missouri blocks, suggesting they could identify both. The hybridization known to occur between Black-capped and Carolina chickadees (Braun and Robbins 1986) adds a source of error Atlas Project methodology did not address.

Distribution

If blocks with possible and probable records are considered actual breeding blocks, as evidence suggests, then Black-capped Chickadees had a nearly solid distribution throughout northern Missouri. A line from St. Louis to Jefferson City to Nevada delineates most of their northern Missouri range. This species was recorded in almost all blocks north of that line compared with few blocks south of that line. The boundary between the two species revealed by the Atlas Project corresponds well with the boundary documented by Robbins et al. (1986) in southwestern Missouri. Of the 580 blocks that recorded Black-caps, only 43 (7.4 percent) also recorded Carolinas. This supports Smith's (1993) contention that there is very little range overlap between the two chickadee species.

Abundance by Natural Division
Average Number of Birds / 100 Stops

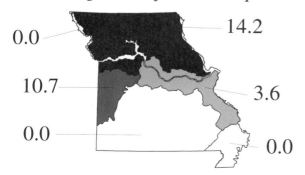

0.0   14.2

10.7   3.6

0.0   0.0

Breeding Phenology

EVIDENCE (# of Records)	MAR	APR	MAY	JUN	JUL	AUG	SEP
NB (15)	3/25				7/07		
NE (3)		5/13			7/15		
NY (15)		5/03				7/21	
FY (57)		5/07				8/08	

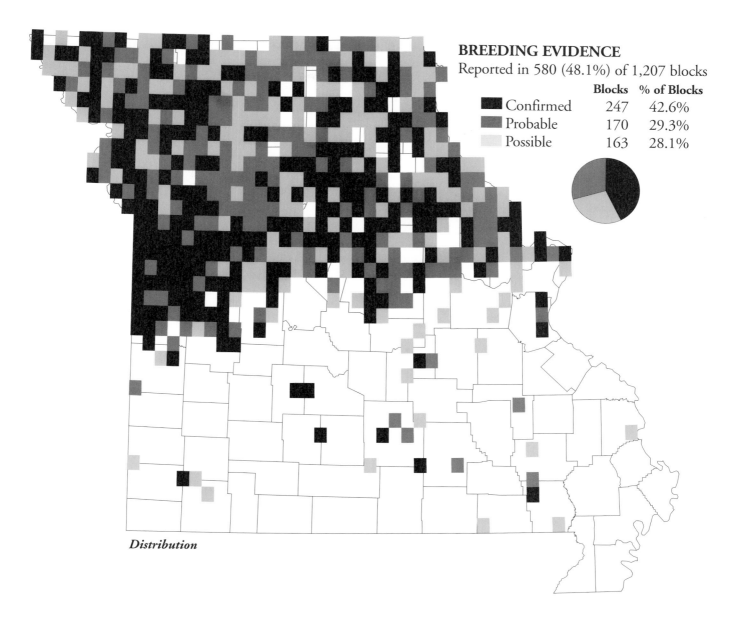

BREEDING EVIDENCE
Reported in 580 (48.1%) of 1,207 blocks

	Blocks	% of Blocks
Confirmed	247	42.6%
Probable	170	29.3%
Possible	163	28.1%

Distribution

Abundance

Black-capped Chickadees were most abundant on surveys in west central and north central Missouri. Their abundance generally diminished toward the southern edge of their range.

Phenology

Black-capped Chickadees initiate nesting early based on the dates that nest building and bird on the nest were observed. Late observations may have been second broods. Smith (1993) reported that although second broods are rare in chickadees, they may be more frequent in low-density populations.

Notes

Atlasers observed no evidence of brood parasitism during this project. Black-capped Chickadees are rarely a host to Brown-headed Cowbirds as their cavity entrance is usually too small (Harrison 1975).

Carolina Chickadee

Parus carolinensis

Carolina and Black-capped chickadees are virtually identical in nesting ecology and habitat. Like its northern counterpart, the Carolina Chickadee is primarily an occupant of mature and second-growth forests and is often associated with woodland edge and riparian habitats. It can breed in shaded urban habitats, including residential areas and parks. Chickadees excavate nest cavities in decaying trees or will nest in existing cavities such as woodpecker holes, broken branches or nest boxes (Terres 1987).

Code Frequency

Chickadees are easy to detect due to their abundance, despite their tendency to become more quiet and reclusive when raising young. Confirmation of breeding was considered difficult for this species because they nested before most of the Atlasing effort took place. Therefore, the majority of the possible and probable breeding blocks shown on the map likely did contain breeding areas. They presumably also bred, undetected, in most of the blocks within their southern Missouri range. Atlasers were usually able to distinguish Carolina from Black-capped chickadees according to the documentation forms that were required for unexpected locations. Although confusion between the two species may have produced a few erroneous records, most observations within the expected range should be considered valid.

Distribution

Carolina Chickadees appear to have a more clearly-defined breeding range than Black-capped Chickadees. A line from St. Louis to Jefferson City to Nevada roughly delineates the northern edge of their range as determined by the Atlas Project. The range edge in western Missouri appeared more feathered. The few records in northern Missouri might have resulted from misidentifications. In addition to subtle differences in appearance between the two species, hybridization and the learning of alternate calls can occur, especially where the two ranges abut (Robbins et al. 1986).

Although the Mississippi Lowlands is within the Carolina Chickadee's southern Missouri range, breeding evidence was scarce in that natural division, perhaps because of a shortage of woodland habitat. They were listed only as possible breeders in a number of adjacent blocks in the central Ozarks. This may have been due to the species' association with open forests and forest edge habitats, which are less available in the more densely-forested regions of the state. However, it may simply have been more difficult to observe evidence of breeding in more-forested regions.

Abundance by Natural Division
Average Number of Birds / 100 Stops

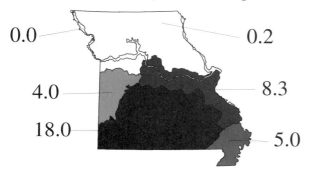

0.0   0.2

4.0   8.3

18.0   5.0

Breeding Phenology

EVIDENCE (# of Records)	MAR	APR	MAY	JUN	JUL	AUG	SEP
NB (7)	4/01			5/26			
NE (6)		4/15		6/01			
NY (17)		4/17			6/25		
FY (67)		4/23					8/26

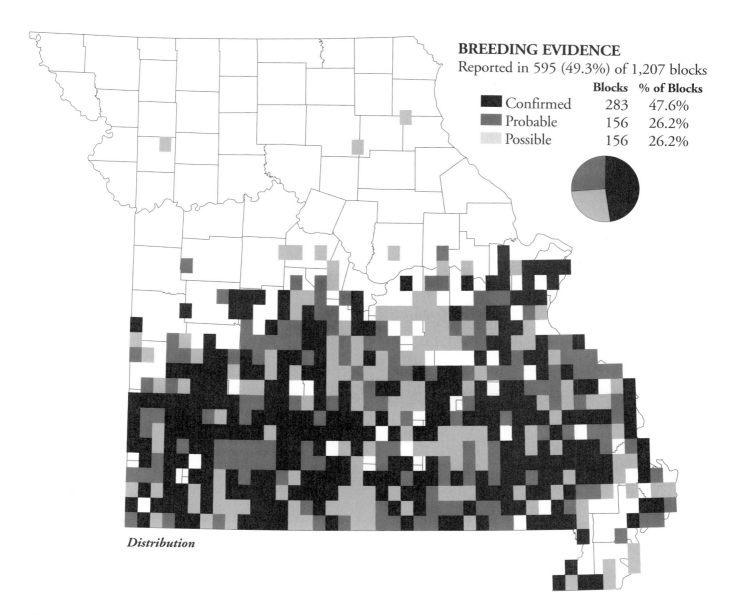

BREEDING EVIDENCE

Reported in 595 (49.3%) of 1,207 blocks

	Blocks	% of Blocks
Confirmed	283	47.6%
Probable	156	26.2%
Possible	156	26.2%

Distribution

Abundance

Carolina Chickadees were most abundant in the Ozarks. Their abundance generally diminished towards the northern edge of their range.

Phenology

Replacement broods begun soon after the loss of a first brood may have accounted for the late observations of nest building, eggs and young. Actual second broods, begun after the first clutch fledges, are rare in chickadees (Smith 1991).

Tufted Titmouse
Parus bicolor

Rangewide Distribution: Eastern United States, Gulf
   Coast & eastern Mexico
Abundance: Common
Breeding Habitat: Thickets, river bottoms & partly
   open areas
Nest: Tree cavity, lined with moss, fur, bark, leaves &
   grass
Eggs: 5–7 white to creamy white, with brown spots &
   occasionally wreathed
Incubation: 13–14 days
Fledging: 15–18 days

The Tufted Titmouse is well-known for its noisy, conspicuous habits. The species frequents bird feeders and is
believed to be expanding its range northward due to the
artificial provision of food (Grubb and Pravosudov 1994).
Although mostly limited to deciduous forests, these birds
also breed in orchards, parks and suburban areas (Terres
1987). Nest sites are usually abandoned woodpecker holes
although these birds will occasionally accept man-made nest
boxes (Elder 1985).

Code Frequency

The conspicuous traits of Tufted Titmice made them
easy to locate and easy to elevate to probable breeding status,
at least in regions where they were most numerous.
Territoriality and the observation of fledglings were most
often used to classify Tufted Titmice as probable and confirmed breeders. A most unusual nest building evidence
resulted when Atlaser Rick Thom observed a titmouse collecting hair from the head of fellow Atlaser, author Jim
Wilson.

Distribution

Tufted Titmice were found in nearly every block in every
natural division except the Mississippi Lowlands. Fewer
reports from that natural division indicated they were comparatively scarce there, perhaps because of fewer forest tracts
of adequate size. The northern edge of the state is near the
northern edge of the species' breeding range. This may
account for the preponderance of possible records in this
area, which also has less woodland acreage.

Abundance

Tufted Titmice were most abundant in the Ozark Border
Natural Division followed by the Ozark Natural Division.
Nearly twice as many recorded in those natural divisions as
in the Glaciated Plains and Osage Plains.

Phenology

As permanent residents, Tufted Titmice are able to initiate nesting relatively early in the season. The earliest report
of fledglings seems unreasonably early and may have been
courtship feeding instead, which is common for Tufted
Titmice (Ehrlich et al. 1988). Late observations of nests
with eggs and nests with young indicate Tufted Titmice may
be double-brooded in Missouri.

Abundance by Natural Division
Average Number of Birds / 100 Stops

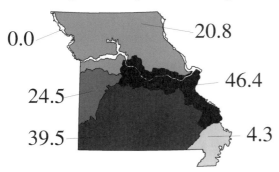

0.0   20.8

24.5   46.4

39.5   4.3

Breeding Phenology

EVIDENCE (# of Records)	MAR	APR	MAY	JUN	JUL	AUG	SEP
NB (22)	4/11				7/28		
NE (5)		5/03		6/15			
NY (12)		5/12			7/25		
FY (137)		4/27				8/03	

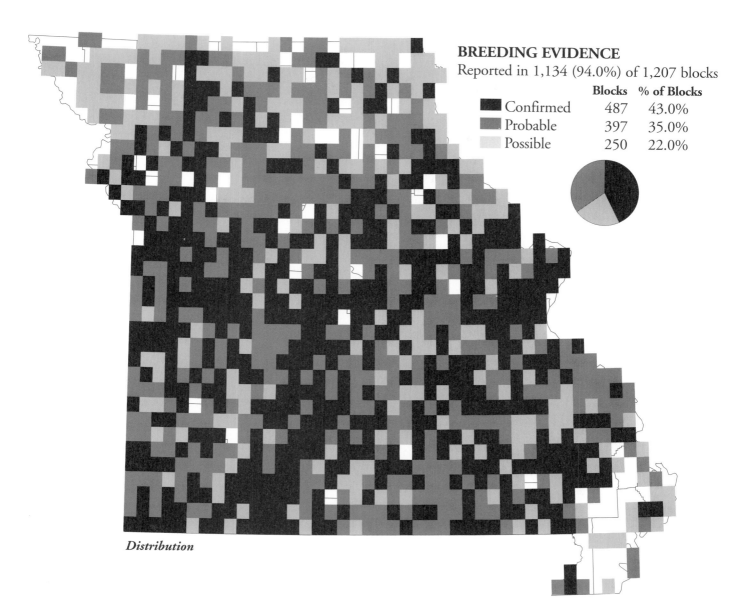

BREEDING EVIDENCE
Reported in 1,134 (94.0%) of 1,207 blocks

	Blocks	% of Blocks
Confirmed	487	43.0%
Probable	397	35.0%
Possible	250	22.0%

Distribution

Notes

Tufted Titmice are a rare Brown-headed Cowbird host (Ehrlich et al. 1988). No observations of brood parasitism were obtained during the Atlas Project.

White-breasted Nuthatch
Sitta carolinensis

> **Rangewide Distribution:** Southern Canada, most of United States & southern Mexico
> **Abundance:** Common & widespread
> **Breeding Habitat:** Mature woods with decaying trees
> **Nest:** Cavity in tree with bed of bark shreds, hair & feathers
> **Eggs:** 5–8 white to pinkish-white, usually with red or brown marks
> **Incubation:** 12 days
> **Fledging:** 14 days

White-breasted Nuthatches are familiar, permanent residents in Missouri. One study reported an individual spent an entire year within its territory of 10–20 hectares (Terres 1987). Their "yank-yank" call can be heard in most woodlots and large forests. They nest in old woodpecker holes and other cavities (Ehrlich et al. 1988).

Code Frequency

White-breasted Nuthatches are reasonably easy to detect, and Atlasers presumably recorded them in most blocks where they occurred. Most confirmations were based on fledglings and observation of food being carried to young. As finding nest cavities was fairly time consuming and difficult, only three nests with young and no nests with eggs were reported.

Distribution

Except for its relative absence in the Mississippi Lowlands, this was one of the most continuously-distributed breeding birds. It was reported in a few blocks in the Mississippi Lowlands in forests along the St. Francis and Mississippi rivers and in the Ben Cash and Donaldson Point conservation areas. Elsewhere in the state, it was distributed through regions with varied landscapes indicating its tolerance of a broad range of habitat.

Abundance

White-breasted Nuthatches were twice as abundant in the more heavily-forested Ozark and Ozark Border natural divisions as in the open lands of the Osage Plains and Glaciated Plains natural divisions. Abundance surveyors tallied no individuals on the Big Rivers or Mississippi Lowlands natural divisions, where forested habitat is scarce. As the landscape becomes less forested and contains more pastureland and cropland, this species usually disappears.

Phenology

Atlasers observed fledglings mainly from May 15 through August 22. Graphing the sightings of fledglings produced a normal curve from mid-May to mid-August. There were likely several broods involved.

Notes

White-breasted Nuthatches are considered a rare host of Brown-headed Cowbirds by Ehrlich et al. (1988).

Abundance by Natural Division
Average Number of Birds / 100 Stops

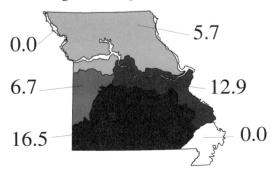

0.0
5.7
6.7
12.9
16.5
0.0

Breeding Phenology

EVIDENCE (# of Records)	MAR	APR	MAY	JUN	JUL	AUG	SEP
NB (4)	3/24			6/16			
NY (3)			5/26	6/19			
FY (39)		5/08				8/04	

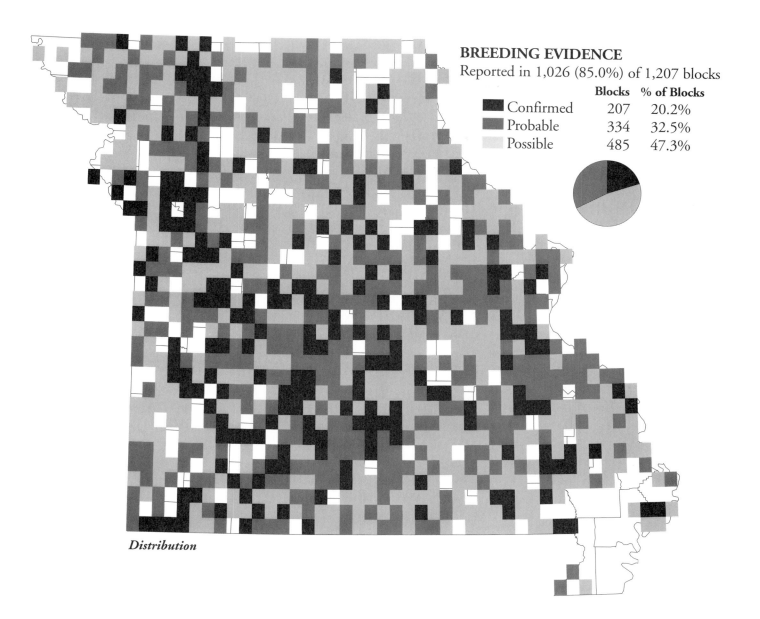

BREEDING EVIDENCE
Reported in 1,026 (85.0%) of 1,207 blocks

	Blocks	% of Blocks
Confirmed	207	20.2%
Probable	334	32.5%
Possible	485	47.3%

Distribution

Brown Creeper
Certhia americana

Rangewide Distribution: Southern & coastal Canada, central & southern Alaska, western to southwestern & northeastern United States & the Appalachians.
Abundance: Fairly common
Breeding Habitat: Pine, mixed & swampy forests
Nest: Hammock-like cup beneath loose bark, made of bark, moss, conifer needles & silk on a base of twigs, & lined with bark & feathers
Eggs: 5–6 white, flecked with red or brown, often wreathed
Incubation: 14–17 days
Fledging: 13–16 days

Brown Creepers are small and inconspicuous birds that peck insects from bark as they spiral up the tree trunk and forage from tree to tree. They typically build nests under the sloughing bark of large trees in bottomland forests (Ehrlich et al. 1988).

Code Frequency

Although primarily a migrant and winter resident in Missouri, summering birds have been recorded. At the turn of the century they were deemed regular inhabitants of the Mississippi Lowlands where they nested under the loose bark of bald cypress trees (Widmann 1907). Since then, virtually the entire swampland habitat of that region has been eliminated. From 1907 until the Atlas Project, only two records of breeding were documented, near St. Louis in 1909 and at Big Oak Tree State Park in 1985 (Robbins and Easterla 1992).

Distribution

Atlasers reported only two records of Brown Creepers during the Atlas Project, both indicating possible breeding. One individual was seen near the Meramec River in the Missouri Botanical Garden Arboretum on June 6, 1987. The other was near Freeburg in Osage County on June 22, 1992. Both individuals were observed feeding on loose bark in wooded areas.

Abundance

Based on the historical records described by Robbins and Easterla (1992) and the locations of Atlas Project records, Brown Creepers could breed anywhere in the state where large, mature or over-mature trees occur in bottomland forests. Because few surveys are conducted in bottomland forests, breeding Brown Creepers may be more numerous than suggested by either the Atlas Project or historical records.

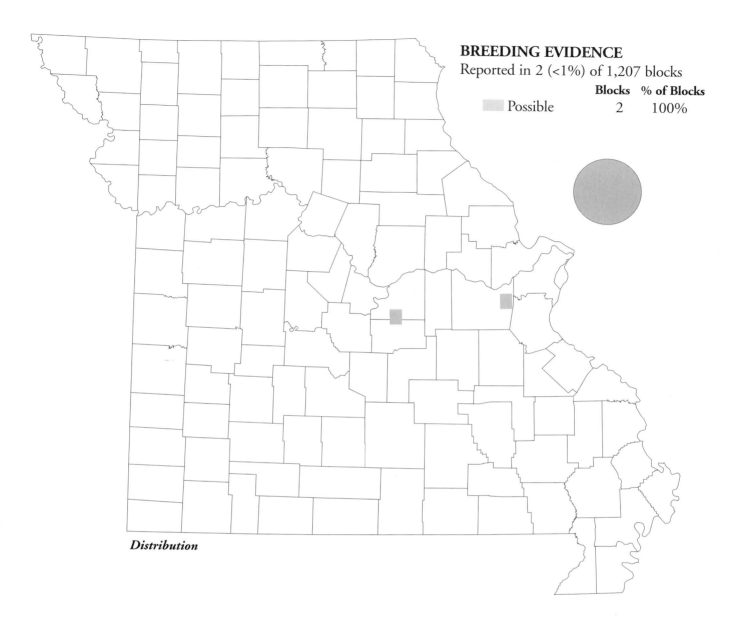

	Blocks	% of Blocks
Possible	2	100%

Distribution

Carolina Wren
Thryothorus ludovicianus

> **Rangewide Distribution:** Southeastern United States through eastern Mexico & Central America
> **Abundance:** Common
> **Breeding Habitat:** Open deciduous woods, farmland, dense undergrowth & thickets
> **Nest:** Tree or other cavity, of twigs, bark, leaves & grass, lined with finer material
> **Eggs:** 5 white, pinkish or creamy, flecked brown or purple & wreathed
> **Incubation:** 12–14 days
> **Fledging:** 12–14 days

Carolina Wrens are distinguished from other Missouri wrens by their larger size, rusty plumage and prominent white eye stripe. They do not migrate and their ringing "teakettle, teakettle, teakettle" song may be heard at any season. Associated mostly with rural underbrush, they typically nest in natural shelters such as woodpecker holes, upturned roots and brush piles. They often reside close to human habitation, and sometimes select unnatural sites for nesting such as rock walls, outbuildings, machinery, mailboxes and hanging flower pots.

Code Frequency

Carolina Wrens are usually first perceived by their loud songs. Most Atlasers probably identified them correctly although their calls are sometimes confused with those of Kentucky Warblers, Ovenbirds and Tufted Titmice. The prevalence of their singing likely increased the chances of detecting them. They were probably scarce or absent in those blocks where they were not recorded. Confirming breeding is difficult for this species unless nests are constructed near human habitation. Carolina Wrens likely nested in many blocks where Atlasers recorded them only as possible or probable breeders.

Distribution

Carolina Wrens were distributed throughout Missouri except the most northern counties where they were sparse and scattered. This species has been extending its range northward during the 20th century (Brewer et al. 1991). In Missouri, it has expanded its range northward after being essentially extirpated by a series of hard winters in the late 1970s. It is likely that north-ranging birds observed during the Atlas Project were pioneers of new breeding range.

Abundance

Carolina Wrens were most abundant at Missouri's south central edge and in the northern part of the Mississippi Lowlands, especially Crowley's Ridge.

Phenology

Carolina Wrens will nest two or even three times a season (Haggerty and Morton 1995). Nesting apparently commenced in March to result in the earliest observation of young in the nest. Late nesting was evident from the observations late in the season including the removal of a fecal sac on August 3.

Abundance by Natural Division
Average Number of Birds / 100 Stops

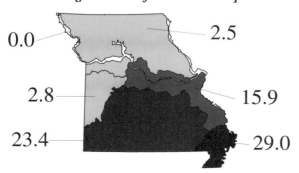

0.0   2.5
2.8   15.9
23.4   29.0

Breeding Phenology

EVIDENCE (# of Records)	MAR	APR	MAY	JUN	JUL	AUG	SEP
NB (7)		4/28			7/24		
NE (7)		4/29		7/02			
NY (27)		4/25				8/01	
FY (40)	4/19					8/19	

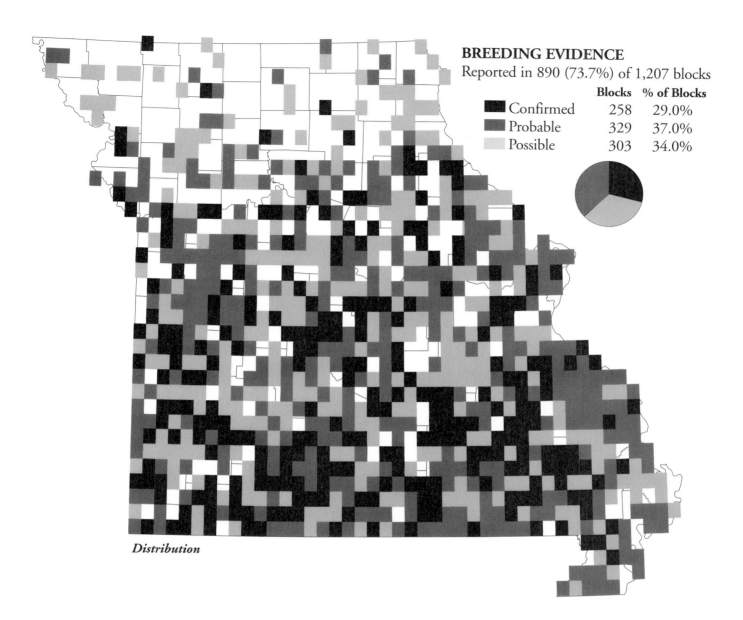

BREEDING EVIDENCE

Reported in 890 (73.7%) of 1,207 blocks

		Blocks	% of Blocks
■	Confirmed	258	29.0%
■	Probable	329	37.0%
■	Possible	303	34.0%

Distribution

Notes

Although Carolina Wrens are known Brown-headed Cowbird hosts (Ehrlich et al. 1988), brood parasitism was recorded in only one of 168 potential blocks during the Atlas Project.

Bewick's Wren
Thryomanes bewickii

Rangewide Distribution: Southwestern edge of Canada, West Coast & southern United States & Mexico
Abundance: Common but declining in eastern United States
Breeding Habitat: Open woods, thickets, brushy areas & gardens
Nest: Twigs or grass lined with feathers & grass in natural cavity or brushpile
Eggs: 5–7 white with brown or purple flecks & occasionally wreathed or unmarked
Incubation: 12–14 days
Fledging: 14 days

Breeding Bewick's Wrens occupy upland shrub habitats and woodland edges which offer cavities for nesting. They are especially detectable around rural houses and brushy fencerows. Most nests are located within two meters of the ground (Peterjohn 1989). The Bewick's Wren population is apparently declining rapidly in Missouri. Missouri Breeding Bird Survey data reveal an average annual decline of 6.1 percent from 1967 through 1989 (Wilson 1990). The reasons for this decline are unclear. Peterjohn (1989) suggests that in Ohio they are retreating after their expansion into the state in the late 19th and early 20th centuries. Perhaps this expansion was in response to the availability of shrubby cover that followed extensive clearing. The Bewick's population may be declining as these shrubby areas succeed to forest.

Code Frequency

Bewick's Wrens are relatively easy to detect when around human habitation. Therefore, the blocks in which they were recorded likely provide a reasonably accurate representation of their true frequency and distribution in the state. Their loud singing and their cavity nesting habits allowed many Atlasers to elevate them to probable or confirmed breeding status. Despite this, in 38 percent of the blocks in which they were found the highest evidence of breeding observed was their occurrence at the proper season in appropriate habitat.

Distribution

The distribution of Bewick's Wrens was primarily confined to the Ozark and Ozark Border natural divisions. The south central part of the state represented the heart of the Missouri range. Inexplicably, Bewick's Wrens were sparse in the eastern Ozarks. They were essentially absent from the Mississippi Lowlands, perhaps due to the scarcity of nesting habitat. Scattered locations in northern Missouri indicate they can also nest there. The numerous possible breeding records in northern Missouri may indicate a higher proportion of unmated Bewick's Wrens in the northern part of their range. In forty-nine percent of the 410 blocks in which Bewick's Wrens were recorded, House Wrens were also seen. This suggests some separation between the two species. Newman (1961) suggests that House Wrens compete more successfully then Bewick's Wrens for limited nest sites. Robbins and Easterla (1992), however, suspect that the decline of Bewick's Wrens and the increase in House Wrens is coincidental.

Abundance by Natural Division
Average Number of Birds / 100 Stops

0.0   0.8

0.0   3.4

5.7   0.0

Breeding Phenology

EVIDENCE (# of Records)	MAR	APR	MAY	JUN	JUL	AUG	SEP
NB (7)		4/16		5/29			
NE (13)		5/02			6/30		
NY (23)		5/10				8/01	
FY (23)		5/15				8/12	

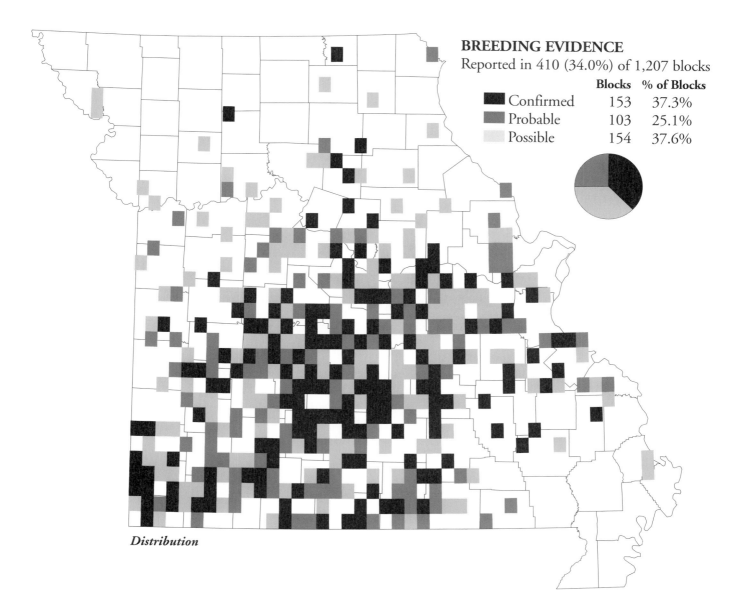

BREEDING EVIDENCE
Reported in 410 (34.0%) of 1,207 blocks

	Blocks	% of Blocks
Confirmed	153	37.3%
Probable	103	25.1%
Possible	154	37.6%

Distribution

Abundance

Bewick's Wrens were most abundant in south central Missouri, presumably because of heterogeneity of forests, forest openings, brushy cover and nesting cavities.

Phenology

Peterjohn (1989) reported Bewick's Wrens have two broods per season, perhaps the reason for the late observation of young in the nest.

Notes

Although Brown-headed Cowbird parasitism of Bewick's Wrens is insignificant, there is evidence of the species' nest being disturbed by both House Wrens and Carolina Wrens (Harrison 1975).

House Wren
Troglodytes aedon

> **Rangewide Distribution:** Southern Canada to Mexico, except southeastern United States from Texas to Florida
> **Abundance:** Common in most areas
> **Breeding Habitat:** Backyards, forest edges & shrubs
> **Nest:** Twigs & grasses in a cavity with finer grass lining
> **Eggs:** 6–8 white with brown markings & occasionally wreathed
> **Incubation:** 13 days
> **Fledging:** 12–18 days

House Wrens are common birds of backyards, parks and brushy landscapes. Their highly-vocal song with cascading, twittering calls is familiar to many who have a bird house in their backyard. An unusual trait, which puzzles many bird house enthusiasts, is this species' habit of building extra nests in their territory. They sometimes fill all the cavities in their territory with twigs, called "dummy nests." This may be an attempt to confuse and frustrate predators trying to locate eggs or to persuade a cavity-nesting competitor to nest elsewhere.

Code Frequency

Reported in 60 percent of all survey blocks, this species is easily detected by its familiar musical song. The most frequently used confirmation code for this species documented its easily-observed visits to its cavity nest. Only physiological evidence and distraction display codes were not used to report this easily-observed backyard bird.

Distribution

House Wrens were widely distributed in the northern half of Missouri south to the northern edge of the forested Ozark Natural Division. The Ozark and Mississippi Lowlands natural divisions include the southern edge of the species' breeding range. In the southern part of the state they are seen primarily in urban areas. However, in northern Missouri settings they are more widespread, except in heavily-forested regions. Since the 1960s, the range has apparently been expanding southward beyond its historic edges in Ste. Genevieve and Jasper counties (Widmann 1907). The distribution map indicates more sites in the Ozarks region and southwestern Missouri than suggested by Robbins and Easterla (1992). James and Neal (1986) reported an increase in the presence and breeding of House Wrens in northeastern and northwestern Arkansas during 1960–1980.

Abundance

This species was about 10 times as abundant in the Glaciated Plains as in the Ozark and Ozark Border natural divisions. The Glaciated Plains' extensive acreage of old fields, hedgerows and scattered wood lots with an abundant edge habitat may explain this great difference. The House Wren's relative abundance was greater in the Big Rivers Natural Division than in the more heavily-forested Ozark and Ozark Border natural divisions.

Abundance by Natural Division
Average Number of Birds / 100 Stops

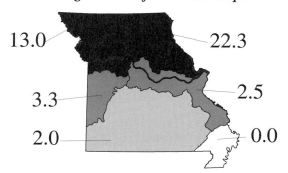

13.0   22.3
3.3   2.5
2.0   0.0

Breeding Phenology

EVIDENCE (# of Records)	MAR	APR	MAY	JUN	JUL	AUG	SEP
NB (10)		5/01			6/30		
NE (14)		5/11			7/03		
NY (44)		5/14				8/06	
FY (54)		5/06				8/03	

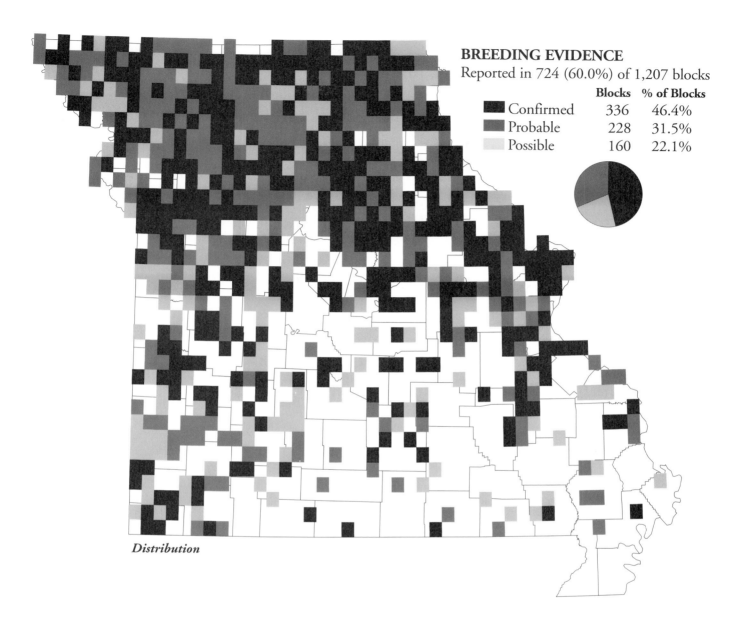

BREEDING EVIDENCE
Reported in 724 (60.0%) of 1,207 blocks

	Blocks	% of Blocks
Confirmed	336	46.4%
Probable	228	31.5%
Possible	160	22.1%

Distribution

Phenology

This species rears 2–3 broods between its arrival in April and May and departure by mid-October (Robbins and Easterla 1992). Summer resident House Wrens begin nesting when they arrive as evidenced by the early observation of fledglings in southwestern Missouri at Bolivar.

Notes

The only report of brood parasitism by Brown-headed Cowbirds came from the Thayer block in southern Oregon County. It is considered a rare cowbird host by Ehrlich et al. (1988).

Sedge Wren
Cistothorus platensis

Rangewide Distribution: South central Canada, southeastern border & northeastern United States

Abundance: Common but local, uncommon to rare in the east

Breeding Habitat: Grassy wet meadows, fields, prairies & marshes

Nest: Ball of dry or green grass woven with growing grass, lined with finer material on or near the ground

Eggs: 7 white & unmarked

Incubation: 12–16 days

Fledging: 12–14 days

Sedge Wrens frequent damp meadows and shorelines where sedges, rushes and other wetland vegetation predominate. Occasionally they are found in upland pastures, hayfields and fallow fields. Their ball-shaped nests, composed of fine green stems, are woven into standing vegetation.

Code Frequency

Atlasers documented Sedge Wrens mainly by their songs and by making repeated visits to determine territoriality. Few Atlasers located their well-concealed nests, although several were found outside of Atlas blocks. Sedge Wrens may be more abundant and widely distributed than the maps illustrate as late summer searches may not have been conducted in many regions.

Distribution

Observations of Sedge Wrens were somewhat grouped in north central counties and sparsely distributed in the southern part of the state, although observations were made in every natural division. This species was once a common breeder in marshes along the Mississippi and Missouri rivers before these floodplains were converted to agriculture. Sedge Wrens also nested in the wet prairies in the northern and western parts of the state (Widmann 1907).

Abundance

Abundance surveyors detected Sedge Wrens in very low numbers in the northwestern Glaciated Plains and Osage Plains natural divisions. This species is better surveyed by walking through appropriate habitat than by roadside counts.

Phenology

In Missouri (Robbins and Easterla 1992) and Kansas (Thompson and Ely 1992) Sedge Wrens are frequently reported to nest during August and September. This is well after the May-to-July nesting period (Terres 1987) in this species' main breeding range to the north. Singing territorial males were observed in the extreme north central part of the Glaciated Plains in June. Presumably, these represented the southern edge of late spring/early summer nesting. Many additional observations were made statewide after mid-July in marshes and wet, grassy lowlands. The only confirmations occurred after the statewide influx of potential "second-nesting" birds in mid-July.

Breeding Phenology

EVIDENCE (# of Records)	MAR	APR	MAY	JUN	JUL	AUG	SEP
FY (1)						8/25	8/25

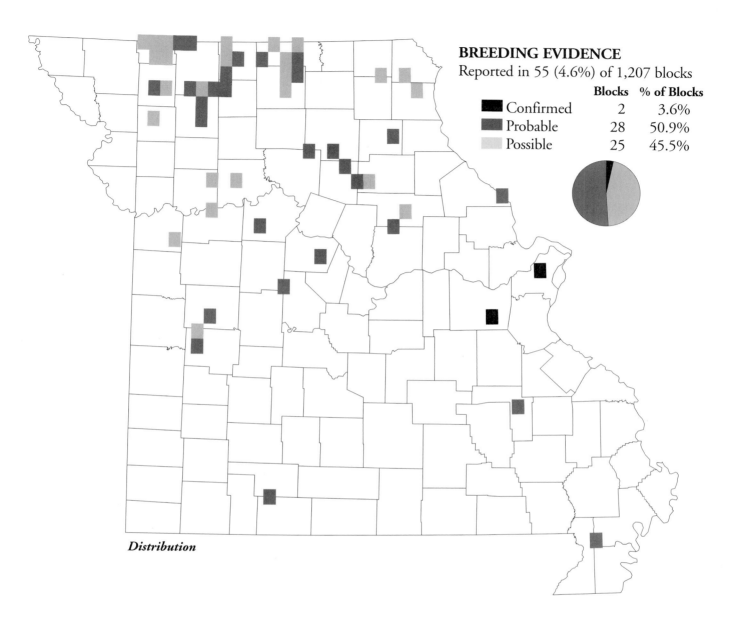

BREEDING EVIDENCE
Reported in 55 (4.6%) of 1,207 blocks

	Blocks	% of Blocks
Confirmed	2	3.6%
Probable	28	50.9%
Possible	25	45.5%

Distribution

Marsh Wren

Cistothorus palustris

> **Rangewide Distribution:** Southern Canada, northern & coastal United States
> **Abundance:** Fairly common but local
> **Breeding Habitat:** Marsh vegetation, reeds & cattails
> **Nest:** Reeds & grass, lined with finer plant material & feathers, attached to reeds above water, with entrance near top
> **Eggs:** 4–6 dull brown, usually with darker brown markings
> **Incubation:** 12–16 days
> **Fledging:** 13–16 days

The discordant clattering song of the Marsh Wren often gives away its presence. These emergent marsh dwellers construct grapefruit-sized nests low in standing vegetation. Although they use only one nest they may build 10 or more false or dummy nests. Most of Missouri's suitable wetland habitat has been altered or drained during this century, reducing habitat for Marsh Wrens, which breed only when conditions are very wet.

Code Frequency

As with other marsh-breeding birds, Marsh Wrens are probably under-represented in the Atlas because of the difficulty in reaching the habitat in which they reside. Atlasers obtained the few confirmed breeding records through intensive searches. Because Marsh Wrens tend to nest late in the season, most of the Atlasing effort may have preceded them. Although obviously quite rare as breeders, Marsh Wrens are likely more abundant and widely distributed as breeders than the Atlas Project indicated.

Distribution

This species is known to breed in several northern Missouri marshes (Robbins and Easterla 1992), but the Atlas Project confirmed breeding in only Vernon and Iron counties. Squaw Creek and Clarence Cannon national wildlife refuges, two of the best-known locations for Marsh Wrens, were not included in Atlas blocks.

Abundance

A scarce breeder in cattail *(Typha* spp.*)* marshes, this species may not breed anywhere statewide in very dry years (Robbins and Easterla 1992). In 1995, assistant refuge manager Beatrix Treiterer and Paul McKenzie, U. S. Fish and Wildlife Service, Wildlife Enhancement Office, Columbia, noted 39 singing males, several nests and some young at Clarence Cannon National Wildlife Refuge.

Phenology

On the average, Marsh Wrens arrive by mid-April with fall migration commencing in mid-September (Robbins and Easterla 1992).

Notes

Because males usually build several "dummy" nests near the brood nest, nest searches are needed to find eggs or young to confirm breeding.

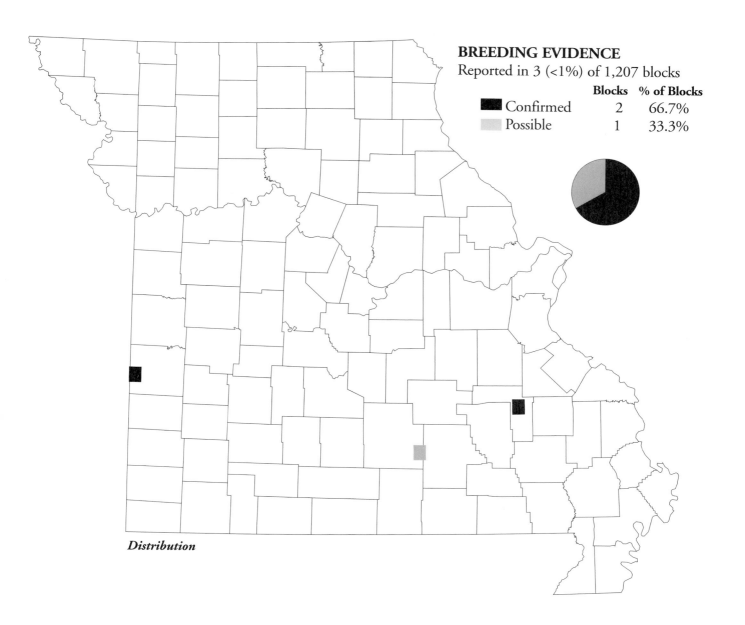

Distribution

Blue-gray Gnatcatcher
Polioptila caerulea

Rangewide Distribution: Central, southern & eastern
  United States to northern Middle America
Abundance: Common
Breeding Habitat: Forests, woods, shrubland, old fields,
  swamp or desert
Nest: Compact plant down held with insect & spider
  silk, covered with lichen, lined with fine material in trees
Eggs: 4–5 pale blue to blue-white with brown flecks,
  occasionally wreathed
Incubation: 13 days
Fledging: 10–12 days

Blue-gray Gnatcatchers inhabit the canopies and under-stories of mature deciduous woods in upland and bottom-land forests. They seem most prevalent in oak-hickory forests in moist situations near habitat edges (Ellison 1992). They have been reported most frequently in woodlands exceeding eight hectares in size or along riparian corridors that are at least 30–60 meters wide (Peterjohn and Rice 1991). They actively glean insects from leaves and small branches and occasionally rush after flushed prey.

Code Frequency

Blue-gray Gnatcatchers are reasonably easy to detect by their distinctive, wheezy calls and their habit of conspicuously flittering about. They are often seen in pairs, which Atlasers recorded as probable breeding. Blue-gray Gnatcatchers superbly camouflage their nests with bits of bark and lichen. Nests can often be found, however, by watching the adults. As with many species, the ability to confirm breeding of Blue-gray Gnatcatchers appeared to relate to their abundance in a particular block. There were likely fewer birds in many of the blocks in which they were found but not confirmed.

Distribution

Blue-gray Gnatcatchers were distributed statewide. However, probable or confirmed breeding observations were mainly associated with the most-forested regions. This association with forests explains their absence from the most intensively-farmed regions of the state: the Mississippi Lowlands and much of northern Missouri. They were solidly distributed throughout the Osage Plains and much of the western Glaciated Plains natural divisions.

Abundance

Blue-gray Gnatcatchers were most abundant along the Arkansas border and through the eastern Ozarks, which matches the distribution of forest cover in the state. Few were detected in northern counties.

Phenology

Blue-gray Gnatcatchers arrive in the south at the beginning of April and about a week later in the north (Robbins and Easterla 1992). Nesting behavior commences soon after arrival. Ellison (1991) reported that in the southeastern United States the earliest-nesting Blue-gray Gnatcatchers had second broods. This may account for the late Atlas observations.

Abundance by Natural Division
Average Number of Birds / 100 Stops

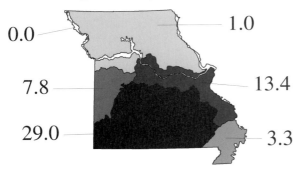

0.0

1.0

7.8

13.4

29.0

3.3

Breeding Phenology

EVIDENCE (# of Records)	MAR	APR	MAY	JUN	JUL	AUG	SEP
NB (34)	4/09				6/27		
NE (5)		5/05		6/15			
NY (21)		5/02				7/27	
FY (67)		5/03				8/06	

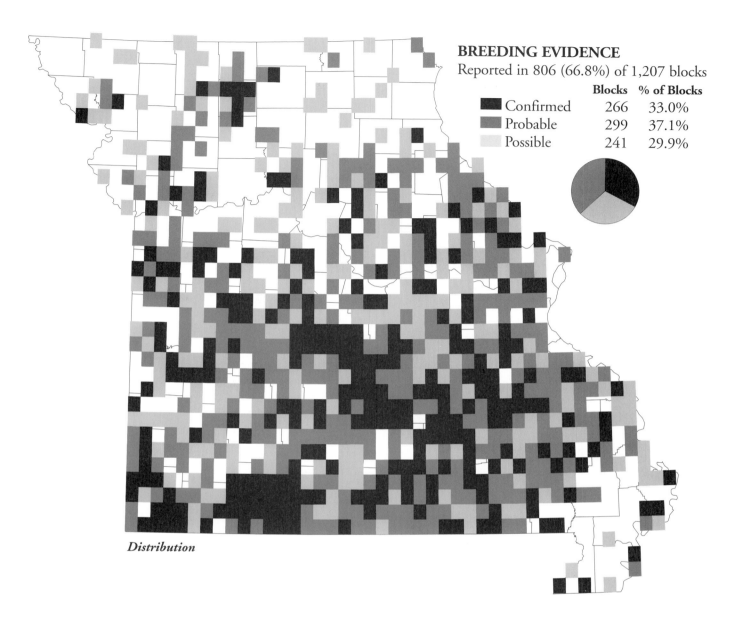

BREEDING EVIDENCE
Reported in 806 (66.8%) of 1,207 blocks

	Blocks	% of Blocks
Confirmed	266	33.0%
Probable	299	37.1%
Possible	241	29.9%

Distribution

Notes

Brown-headed Cowbird parasitism of Blue-gray
Gnatcatchers was reported in only one out of 101 blocks.
Brood parasitism of Blue-gray Gnatcatchers varies regionally
from 82 percent in California to 4 percent in Vermont
(Ellison 1992).

Eastern Bluebird
Sialia sialis

> **Rangewide Distribution:** Southeastern Canada, eastern United States to Central America
> **Abundance:** Common but declining in some areas
> **Breeding Habitat:** Open areas adjacent to woods & forests
> **Nest:** Cup of grass, weeds, twigs & pine needles in tree cavity or nest box
> **Eggs:** 4–5 pale blue, occasionally white & unmarked
> **Incubation:** 12–14 days
> **Fledging:** 15–20 days

Eastern Bluebirds breed primarily in rural areas. They favor rolling grasslands with scattered trees and shrubs that provide tree cavities or other sheltered sites in which they raise their young. Although their numbers cycle in response to the harshness of winter and nesting conditions, the Missouri state bird had a robust population throughout the Atlas Project. The fact that citizens build and place bluebird nest boxes in rural areas is likely responsible in part for the size of the Eastern Bluebird population.

Code Frequency

Where they occur, Eastern Bluebirds are likely the easiest of birds to find, identify and confirm to breed. Even novice birders recognize their field marks and their "cheer-cheery-up" song. Once sighted, adult bluebirds usually quickly reveal the location of their nest sites. The inspection of nest boxes often results in immediate confirmation of breeding. Considering these parameters, Eastern Bluebirds likely did not occur in the white areas on the map.

Distribution

Eastern Bluebirds were distributed statewide except for the Mississippi Lowlands. Scarcity in the Mississippi Lowlands was unexpected but may have been due to pesticides, a scarcity of grasslands and associated insect prey, or a shortage of natural nest sites. The Eastern Bluebird was absent in some blocks in the heavily-forested regions of Shannon and Reynolds counties and scarce in large metropolitan areas, especially St. Louis. These patterns reflect the bluebird's preference for open grasslands. The apparent scarcity of bluebirds in the extreme northwestern corner of the state was unexpected.

Abundance

Eastern Bluebirds were most abundant in the Osage Plains followed by the Ozark and Ozark Border natural divisions, presumably because of the adjunct nature of grasslands and woodlands in those natural divisions. They were less numerous in the Glaciated Plains and essentially absent from the Mississippi Lowlands.

Phenology

Eastern Bluebirds winter as far north as mid-Missouri (Sinnott 1981) insuring an early arrival on the breeding ground. The earliest observation of a nest with young, near Eunice in Texas County, indicates nest initiation in mid-

Abundance by Natural Division
Average Number of Birds / 100 Stops

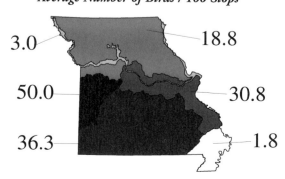

3.0   18.8
50.0   30.8
36.3   1.8

Breeding Phenology

EVIDENCE (# of Records)	MAR	APR	MAY	JUN	JUL	AUG	SEP
NB (23)	4/05				7/31		
NE (48)	4/09				7/09		
NY (113)	4/07					8/12	
FY (144)	4/12					8/25	

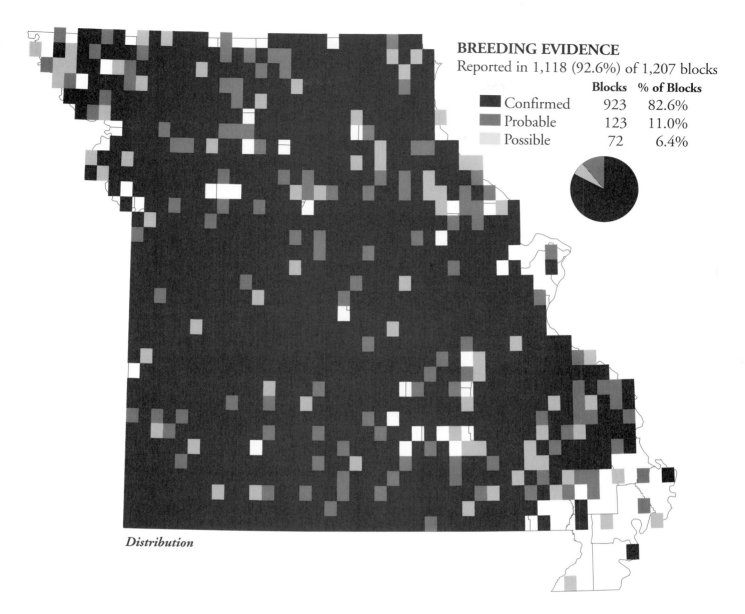

Reported in 1,118 (92.6%) of 1,207 blocks

	Blocks	% of Blocks
Confirmed	923	82.6%
Probable	123	11.0%
Possible	72	6.4%

Distribution

March. Eastern Bluebirds commonly nest two or three times in a season (Harrison 1975). Nest initiation for the first brood occurs in late March, the second in mid- to late May and the third in July (Sinnott 1981). Accordingly, the latest observation of nest building probably documented a third nesting.

Notes

Eastern Bluebirds are extremely rare hosts to Brown-headed Cowbirds (Harrison 1975). The Atlas Project reported no brood parasitism.

Wood Thrush
Hylocichla mustelina

Rangewide Distribution: Southeastern Canada & eastern United States

Abundance: Common, with range expanding northeast

Breeding Habitat: Moist, shady mature woods, occasionally near humans

Nest: Bulky cup of weeds, grass, leaves & mud, lined with rootlets, in trees

Eggs: 3–4 greenish-blue & unmarked

Incubation: 13–14 days

Fledging: 12 days

Of forest-dwelling songbirds, Wood Thrushes may be the best known and most appreciated. Their ringing, bell-like songs resemble the sounds of the flute and the triangle. A recognized "forest interior" species, the Wood Thrush is most often found deep within large forest tracts with high canopies and moderate understories. They occasionally occupy small woodlots.

Code Frequency

Wood Thrushes were located mostly by their song, which carries for great distances. Atlasers may have had difficulty finding individuals heard singing in the distance, thus the low numbers of confirmations. Confirmed breeding evidence was found statewide, clumped in areas where forest cover was more extensive. This may indicate that Wood Thrushes were more likely to mate and breed successfully in large forests. Ragged and bulky Wood Thrush nests were apparently difficult for Atlasers to find.

Distribution

The Wood Thrush was found statewide in forested areas, especially in the Ozark, Big River and Ozark Border natural divisions. In these three areas, Wood Thrushes were recorded in 41 percent to 46 percent of blocks. In the Glaciated Plains, Osage Plains and Mississippi Lowlands natural divisions, which have smaller woodlots, this species was observed in from 22 percent to 33 percent of blocks. In these three natural divisions, clumps of blocks with confirmations may have been the result of exceptional Atlaser effort.

Abundance

The largest relative abundance value for the Wood Thrush was focused mainly in the Ozark Natural Division. The Osage Plains and Ozark Border natural divisions, that surround the Ozark Natural Division, had somewhat lower abundance. Although many Atlasers suggested the Wood Thrush has disappeared from many Glaciated Plains woodlots, nearly two birds per 100 stops were observed there.

Phenology

Nesting activity began soon after Wood Thrushes arrived in mid- to late April (Robbins and Easterla 1992) in order

Abundance by Natural Division
Average Number of Birds / 100 Stops

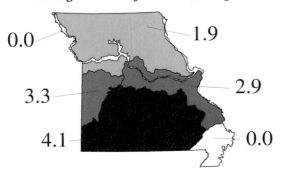

Breeding Phenology

EVIDENCE (# of Records)	MAR	APR	MAY	JUN	JUL	AUG	SEP
NB (2)		5/31			7/11		
NE (4)		5/13		6/03			
NY (4)			6/01		7/06		
FY (5)			6/06		7/27		

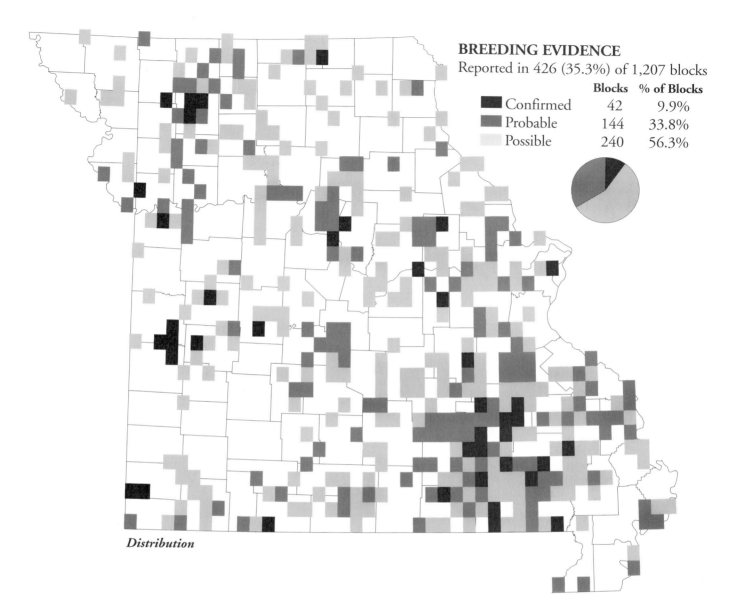

BREEDING EVIDENCE

Reported in 426 (35.3%) of 1,207 blocks

		Blocks	% of Blocks
■	Confirmed	42	9.9%
■	Probable	144	33.8%
▫	Possible	240	56.3%

Distribution

to permit the earliest observations of nests and eggs and fledglings. Second broods likely accounted for the late observations.

Notes

Atlasers documented three records of Brown-headed Cowbird brood parasitism for the Wood Thrush. Ehrlich et al. (1988) lists it as a frequent Brown-headed Cowbird host. The Wood Thrush has become a barometer used by many researchers to determine the reproductive success of forest bird populations in many regions of the Midwest. Studies have shown Ozark Natural Division populations of Wood Thrush are infrequently parasitized and depredated while the opposite is true in the Glaciated Plains (Robinson et al. 1995).

American Robin

Turdus migratorius

Rangewide Distribution: All of Canada, Alaska & United
    States to southern Mexico
Abundance: Common & widespread
Breeding Habitat: Open woods, edges, residential areas,
    parks & fields
Nest: Cup of mud lined with fine grass on a loose founda-
    tion of twigs & grass in any tree, shrub, building or struc-
    ture offering support
Eggs: 4 pale blue, occasionally white, usually unmarked,
    occasionally brown flecked
Incubation: 12–14 days
Fledging: 14–16 days

American Robins thrive in their association with humans perhaps partly because they favor mowed lawns when foraging for invertebrates. They nest in shade trees, shrubs and occasionally on buildings. Residential areas, parks, golf courses and cemeteries as well as farmsteads are frequented by breeding robins. Breeding pairs also frequent open woodlands where understory and ground cover are sparse. These woodland-nesting robins are surprisingly shy and they never congregate as densely as more urban populations (Peterjohn and Rice 1991).

Code Frequency

American Robins are conspicuous in residential settings so Atlasers likely observed them if they were present in a block. In these areas they are easy to confirm by observations of fledglings and food delivery. In forested situations, such as the Lower Ozarks, they may have gone undetected.

Distribution

American Robins were found throughout Missouri. They were found less often in heavily-forested portions of Oregon, Ozark, Reynolds and Shannon counties. Most birds observed in these areas were likely associated with lawns, although the Ozark forests may have supported a few woodland breeders as well.

Abundance

Overall, American Robins were most abundant in northern Missouri, especially near St. Louis, Kansas City and the northeastern corner of the state. They were least abundant through the eastern Ozark and Mississippi Lowlands natural divisions.

Phenology

An exceptionally high number of breeding confirmations (943) depicted a logical sequence of American Robin nesting phenology. American Robins produce two or three broods each year (Ehrlich et al. 1988), with the latest nest building date providing evidence of the lengthy nesting season.

Notes

American Robins are ineffective hosts of Brown-headed Cowbird eggs as they usually recognize and expel them (Harrison 1975). The Atlas Project confirmed parasitism in just three of 153 observations of nest contents.

Abundance by Natural Division
Average Number of Birds / 100 Stops

200.0

108.8

52.2

47.1

38.3

13.3

Breeding Phenology

EVIDENCE (# of Records)	MAR	APR	MAY	JUN	JUL	AUG	SEP
NB (58)	4/05					8/18	
NE (61)	4/06				7/17		
NY (92)	4/15						8/21
FY (274)	4/23					8/11	

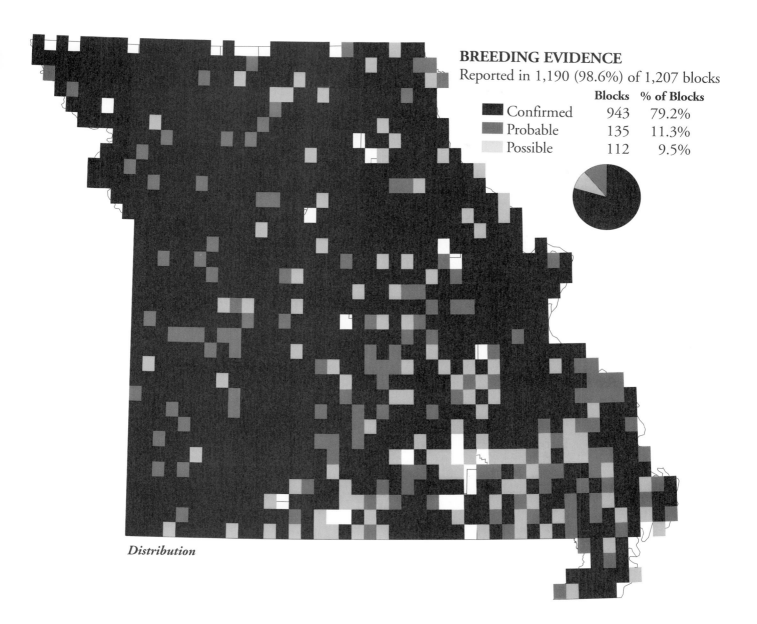

BREEDING EVIDENCE
Reported in 1,190 (98.6%) of 1,207 blocks

	Blocks	% of Blocks
Confirmed	943	79.2%
Probable	135	11.3%
Possible	112	9.5%

Distribution

Gray Catbird
Dumetella carolinensis

Gray Catbirds are best detected by their "mews" and soft "conversations." Vocalization frequency varies seasonally. They are often quiet while raising young, and are most vocal at dawn or dusk. They tend to remain near nest sites enabling Atlasers to observe territorial behavior, mated pairs and other nesting activities. They presumably nested in most blocks in which they were observed.

Distribution

Gray Catbirds were densely distributed across the northern two-thirds of the state becoming progressively scarcer to the south. They were recorded in fewer than half the blocks in the most southern counties. This diminished frequency in southern Missouri was surprising as catbirds are common breeders throughout Arkansas (James and Neal 1986).

Abundance

Gray Catbirds were most numerous on routes in the Big Rivers and Glaciated Plains natural divisions. They were more than 12 times as abundant in the these natural divisions as in the Ozarks.

Phenology

During most springs, Gray Catbirds are not seen until mid-April in the south and about a week later in the north (Robbins and Easterla 1992). Nesting is apparently initiated in late April. Gray Catbirds often raise two or three broods (Cimprich and Moore 1995). The late dates for nest building and a nest with young likely indicated second broods.

Notes

Brown-headed Cowbirds parasitized two of 28 nests in which contents were observed. Cimprich and Moore (1995) reported that Gray Catbirds recognize and remove Brown-headed Cowbird eggs; therefore, this species is parasitized less often than other common birds.

Rangewide Distribution: Southern Canada, northwestern to southeastern United States
Abundance: Generally common, but stays hidden
Breeding Habitat: Dense brush & thickets bordering woods near water
Nest: Grass, forbs, twigs & leaves, lined with fine material in dense thicket
Eggs: 4 blue-green, unmarked, rarely spotted with red
Incubation: 12–13 days
Fledging: 10–11 days

Gray Catbirds are most numerous in shrub-sapling stage successional habitats, shrubby margins of woodlands and shrub-dominated corridors along fence rows and bordering streams (Stauffer and Best 1986), and especially in riparian thickets. They also breed in hedges and ornamental shrubs around homes in cities, towns and farms (Zimmerman 1963).

Code Frequency

Because they tend to remain hidden in shrubby cover,

Abundance by Natural Division
Average Number of Birds / 100 Stops

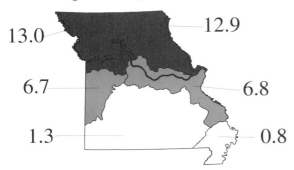

13.0   12.9
6.7   6.8
1.3   0.8

Breeding Phenology

EVIDENCE (# of Records)	MAR	APR	MAY	JUN	JUL	AUG	SEP
NB (20)			5/12		7/15		
NE (17)			5/06	6/28			
NY (13)			5/29		7/22		
FY (49)			5/06			8/17	

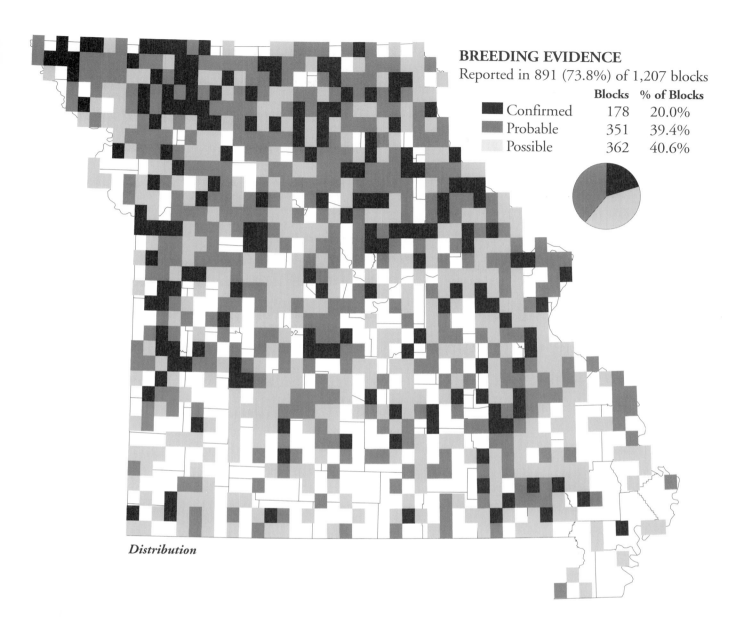

BREEDING EVIDENCE
Reported in 891 (73.8%) of 1,207 blocks

	Blocks	% of Blocks
Confirmed	178	20.0%
Probable	351	39.4%
Possible	362	40.6%

Distribution

Northern Mockingbird

Minus polyglottos

Rangewide Distribution: Southern & northeastern
United States, expanding north
Abundance: Common & widespread
Breeding Habitat: Dense hedgerows, scattered scrubs
& thickets
Nest: Twigs lined with grass & rootlets in shrub or
vines
Eggs: 3–5 blue-green & heavily marked with brown
Incubation: 12–13 days
Fledging: 11–13 days

Northern Mockingbirds closely associate with human habitation in Missouri. Most people in the southern portion of the state recognize this vocal "mimic" in their hedgerows or backyard shrubs. Typical yards provide a combination of shrubs for nesting and open areas and fruits appropriate for the species. Mockingbirds frequent brushy pastures in rural areas but apparently avoid large forests. Derrickson and Breitwisch (1992) suggested that Northern Mockingbirds have expanded their range northward since the 19th century. As permanent residents that rely on fruits and berries, they sometimes suffer a massive reduction in population during severe winter weather.

Code Frequency

Mockingbirds are extremely conspicuous both visually and vocally. Therefore, they were likely observed in all blocks where they occurred. Atlasers revisiting these areas usually documented territoriality and occasionally fledglings. While nests are easy to locate, the presence of thorns, chiggers and ticks may have dissuaded Atlasers from the search. Only 18 percent of the records that confirmed breeding resulted from nest observations.

Distribution

Northern Mockingbirds were distributed statewide. They were detected less frequently near the northern edge of their range, especially in northeastern and north central counties. Atlasers also did not record them in many of the more forested blocks of the state. In the Glaciated Plains the frequency of reports decreases from west to east. This species was present in 93 percent of blocks in the Western Natural Section of the Glaciated Plains as compared with 68 percent in the Central, 57 percent in the Eastern, and 50 percent in the Lincoln Hills natural sections. This decrease from west to east may be associated with a decrease in brush land, hedgerows and thickets and the prevalence of forest cover in the Lincoln Hills.

Abundance

Although this species was distributed throughout Missouri, abundance surveys indicated a sharp reduction in numbers northward through the state.

Phenology

Observations of nest and eggs spanned three months for this two-brooded species. Atlasers recorded initial broods between April 26 and May 16, second broods between May 25 and June 23, and an apparent third brood or renesting

Abundance by Natural Division
Average Number of Birds / 100 Stops

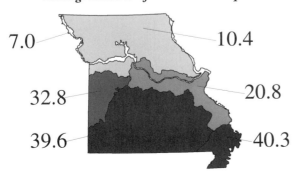

7.0   10.4

32.8   20.8

39.6   40.3

EVIDENCE (# of Records)	MAR	APR	MAY	JUN	JUL	AUG	SEP
NB (21)		4/22			7/22		
NE (27)		4/26			7/27		
NY (27)		5/03				8/08	
FY (79)		5/06				8/05	

Breeding Phenology

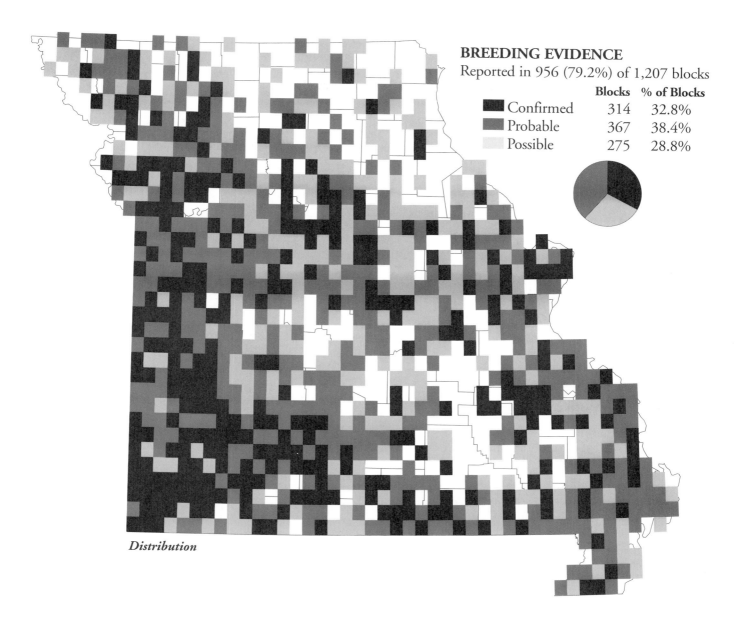

BREEDING EVIDENCE
Reported in 956 (79.2%) of 1,207 blocks

		Blocks	% of Blocks
■	Confirmed	314	32.8%
■	Probable	367	38.4%
■	Possible	275	28.8%

Distribution

between July 23 and 27. Observations of nests and young followed appropriately.

Notes

Atlasers noted only two instances of Brown-headed Cowbird parasitism. Ehrlich et al. (1988) considered the Northern Mockingbird to be a rare host.

Brown Thrasher
Toxostoma rufum

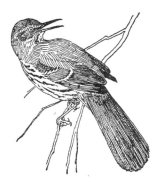

Rangewide Distribution: South central Canada to
   southeastern United States to East & Gulf coasts
Abundance: Common
Breeding Habitat: Thick brush, hedgerows, forest edge
   & clearing
Nest: Twigs, leaves, grass, lined grass & rootlets on
   ground, vine or tree
Eggs: 4–5 pale blue-white, occasionally greenish,
   spotted with red-brown
Incubation: 11–14 days
Fledging: 9–13 days

These rusty-brown members of the mimic-thrush family occur along country roads and brushy landscapes throughout much of Missouri. Associated with thickets, brushy fields and woodland edges, they typically build their bulky nests just above the ground in low vegetation.

Code Frequency

Even novice Atlasers easily recognize this familiar species. Thus, Brown Thrashers may actually have been absent from those few blocks where they were not found. Locating nests and observing nesting behavior were easy for Atlasers due to the Brown Thrasher's commonness and its tendency to nest near the ground.

Distribution

Atlas data support the supposition that Brown Thrashers have a statewide distribution. Records of probable or confirmed breeding were less frequent in the more-forested regions of the state. This may have been due to observations being more difficult in more-forested regions. However, it is more likely that Brown Thrashers, being less abundant in those blocks, were less often seen exhibiting behaviors constituting higher levels of breeding evidence.

Abundance

Except for a surprisingly high number on a route in south central Missouri, Brown Thrashers were most abundant in the Big Rivers, Osage Plains and Glaciated Plains natural divisions. A combination of open fields and brush characterize these natural divisions. This is appropriate habitat for Brown Thrashers (Terres 1987). Sampling error may have accounted for the exceptionally high relative abundance recorded for the Big Rivers Natural Division.

Phenology

Brown Thrashers typically return to Missouri in late March (Robbins and Easterla 1992). The many confirmed breeding evidences provide a rather complete picture of breeding phenology. The surprisingly late nest that contained young on August 8 was likely a second brood, typical for Brown Thrashers (Ehrlich et al. 1988).

Abundance by Natural Division
Average Number of Birds / 100 Stops

43.0   28.9

27.2   15.8

17.7   4.0

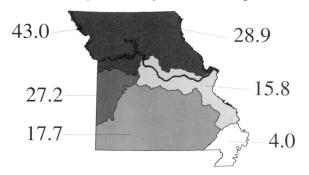

Breeding Phenology

EVIDENCE (# of Records)	MAR	APR	MAY	JUN	JUL	AUG	SEP
NB (39)		4/16			7/10		
NE (66)		4/20		6/21			
NY (34)		4/30				8/08	
FY (195)		5/01				8/08	

BREEDING EVIDENCE
Reported in 1,155 (95.7%) of 1,207 blocks

	Blocks	% of Blocks
Confirmed	581	50.3%
Probable	397	34.4%
Possible	177	15.3%

Distribution

Cedar Waxwing
Bombycilla cedrorum

Rangewide Distribution: Southern Canada & northern United States
Abundance: Fairly common
Breeding Habitat: Forest edge, gardens, parks & residential
Nest: Twigs, grass & moss, lined with fine grass, moss, roots, pine needles & hair, in tree
Eggs: 3–5 pale blue-gray, dotted with black or browns
Incubation: 12 days
Fledging: 16 days

A favorite among bird watchers, Cedar Waxwings are year-round residents of Missouri, although they are less numerous in summer. In fall and winter they feed on berries and in spring they often select buds, flower petals and catkins. In summer they can be seen hawking insects. A gregarious species, they form flocks even in summer and occasionally nest colonially (Harrison 1975).

Code Frequency

Although Cedar Waxwings emit an easily-recognized trill when in a flock, isolated pairs of nesting birds are quiet and secretive. Thus, actual nest locations may have been more widespread and dense than shown on the map because of the difficulty of finding nests and observing breeding behavior. It is suspected, however, that flocks of Cedar Waxwings sometimes summer in Missouri as nonbreeders.

Distribution

Although sparsely distributed, Cedar Waxwings were essentially found statewide. They were less frequently recorded in south central Missouri and more frequently recorded in northern counties.

Abundance

The relative abundance of Cedar Waxwings was surprisingly high in the Ozark Border and Osage Plains natural divisions. The species was not recorded at all on abundance surveys in the Big Rivers and Mississippi Lowlands natural divisions. The flocking nature of the species may have biased results.

Phenology

Most nesting seemed to have been initiated in late May and early June. Second broods are common (Harrison 1975) and the observation of nest building on August 6 indicated that Cedar Waxwings sometimes nest surprisingly late.

Notes

Ehrlich et al. (1988) recorded Brown-headed Cowbird parasitism for Cedar Waxwings, however, none was observed during the Atlas Project.

Abundance by Natural Division
Average Number of Birds / 100 Stops

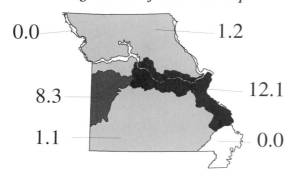

0.0   1.2
8.3   12.1
1.1   0.0

Breeding Phenology

EVIDENCE (# of Records)	MAR	APR	MAY	JUN	JUL	AUG	SEP
NB (10)			5/25			8/06	
NE (1)			6/01	6/01			
NY (4)				6/20	7/14		
FY (3)			6/02			8/05	

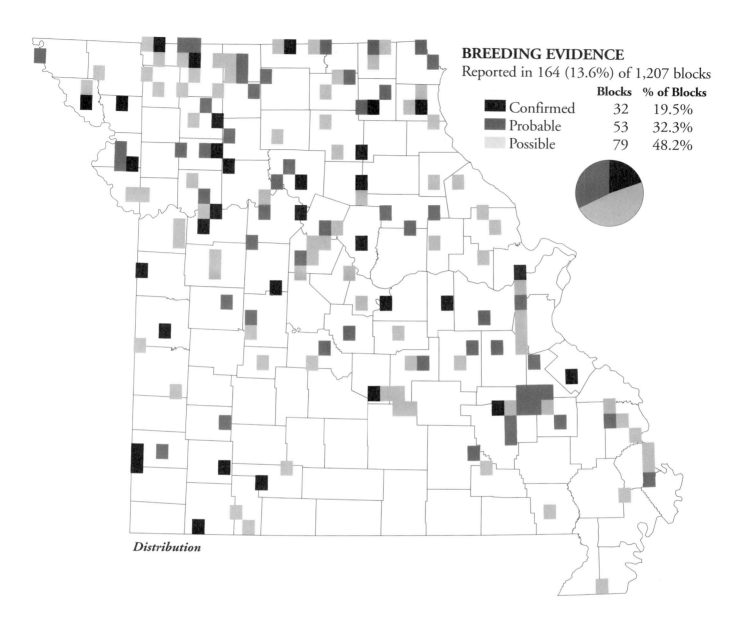

BREEDING EVIDENCE

Reported in 164 (13.6%) of 1,207 blocks

	Blocks	% of Blocks
Confirmed	32	19.5%
Probable	53	32.3%
Possible	79	48.2%

Distribution

Loggerhead Shrike
Lanius ludovicianus

Rangewide Distribution: Central Canada & entire United States
Abundance: Fairly common over much of range
Breeding Habitat: Fields, woodlands, scattered trees & shrubs
Nest: Woven twigs, forbs & bark, lined with fine material in tree or shrub
Eggs: 5–6 grayish-buff, marked with gray, browns or blacks
Incubation: 16–17 days
Fledging: 17–21 days

A passerine "bird of prey," the Loggerhead Shrike feeds on mice, large insects, frogs and even small birds. The Loggerhead Shrike is essentially a bird of open country. It frequents overgrown fields with small trees as perches for territorial defense and hunting. The presence of thorn trees is not vital, but is characteristic of most areas they frequent (Kridlebaugh 1982). Ehrlich et al. (1988) believed that pesticides and habitat loss have contributed to the large decline in this species' numbers. Breeding Bird Survey data indicate an average annual reduction in numbers of 6.7 percent from 1967 through 1989 (Wilson 1990).

Code Frequency

Loggerhead Shrikes select open perches and are easy to detect wherever they occur. Surprisingly, in 56 percent of the blocks where they were found, Loggerhead Shrikes were confirmed as breeders. Two easily-observed codes, fledglings and adults carrying food, comprised 76 percent of all confirmation codes. Only 11 percent of confirmed records were actual nest observations.

Distribution

Loggerhead Shrikes occurred statewide with most of the observations occurring in the Western Glaciated Plains, Springfield Plateau, Elk River and White River natural sections, and Osage Plains and Mississippi Lowlands natural divisions. Seventy-six percent of all records occurred west of the 92nd meridian, a north-south line through Fulton and West Plains. Fifty-six percent lie west of the 93rd meridian, a line through Springfield and Marshall. Shrike records were less frequent throughout the more forested Ozark and Ozark Border natural divisions. Loggerhead Shrikes may be one species that is more common in the Mississippi Lowlands than it was historically. Widmann (1907) described the Mississippi Lowlands shrike populations as scarce. However, at that time the area contained extensive forests and swamps. The row crop fields and pastures that predominate in that natural division today are more appropriate for shrikes.

Abundance

The Osage Plains Natural Division and the Springfield Plateau and Upper Ozark natural sections supported the greatest relative abundance of shrikes. The grasslands, hedge rows, native prairies and old fields provided excellent habitat for this species. Although much of the Mississippi Lowlands was in row crops, Atlasers observed shrikes at nests in farmyard trees, cemetery shrubs and trees that occasionally line drainage ditches. In the Glaciated Plains, shrike relative abundance declined from west to east by 40 percent.

Abundance by Natural Division
Average Number of Birds / 100 Stops

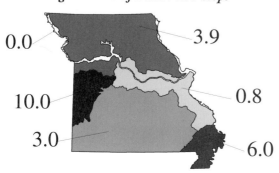

0.0   3.9

10.0   0.8

3.0   6.0

Breeding Phenology

EVIDENCE (# of Records)	MAR	APR	MAY	JUN	JUL	AUG	SEP
NB (5)	4/03			6/18			
NE (10)		4/23			7/23		
NY (20)		4/23			7/27		
FY (57)			5/14		7/10		

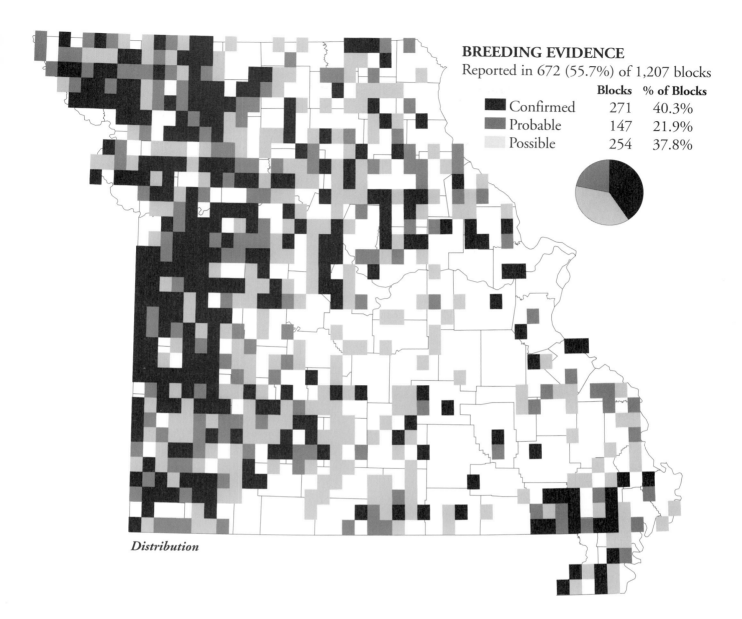

BREEDING EVIDENCE

Reported in 672 (55.7%) of 1,207 blocks

	Blocks	% of Blocks
Confirmed	271	40.3%
Probable	147	21.9%
Possible	254	37.8%

Distribution

Phenology

Loggerhead Shrikes usually establish territories by late March (Robbins and Easterla 1992). For this double-, sometimes triple-brooded species, fledglings were reported until August 19. Later nesting events remained undetected as most Atlasers had ceased surveying for the year.

Notes

Due to its extensively-reduced range, this species was formerly a potential candidate (C2) for listing by the United States Fish and Wildlife Service under the Endangered Species Act. Breeding pairs still occur across much of the state, but many Atlasers found only one pair in a block.

European Starling

Sturnus vulgaris

Rangewide Distribution: Southern Canada, southeastern Alaska, all United States & northern Mexico
Abundance: Abundant
Breeding Habitat: Open fields, woods, suburbia & cities
Nest: Slovenly construction of grass, roots, straw, twigs & forbs in any cavity
Eggs: 4–6 pale blue or green-white, with brown marks
Incubation: 12–14 days
Fledging: 18–21 days

European Starlings were introduced into eastern North America during the late 1880s and spread rapidly across the continent (Cabe 1993). They were first recorded in Missouri in 1928 (Cooke 1928) and by 1930 were established statewide (Bennitt 1932). Starlings select sheltered sites for nests, breeding in cities, suburbs, towns and on farmsteads. They also breed in association with agricultural fields, wooded edges and wetlands, especially in proximity to farms or urban areas (Dunnet 1955).

Code Frequency

Starlings are easy to detect due to their noisy habits and flocking nature. They are also easy to confirm by observing parents delivering food or entering cavities. The young often noisily solicit food from inside and outside nesting cavities. Where not confirmed to breed on the map, they likely were absent or extremely rare breeders.

Distribution

Starlings were found essentially statewide except in the most heavily-forested regions of the state. Peterjohn and Rice (1991) stated that starlings ordinarily avoid large tracts of undisturbed natural habitats such as interiors of mature forests. For unknown reasons, fewer starlings were confirmed to breed across the northern counties than in the western portion of the state.

Abundance

As expected, the highest starling numbers were reported in cities, including St. Louis, Cape Girardeau and Kansas City. The natural divisions in which the greatest numbers were recorded were the Glaciated Plains and Big Rivers, with the fewest recorded in the Mississippi Lowlands.

Phenology

Starlings are permanent residents that initiate nesting as conditions become favorable in spring. Starlings typically raise two broods. Egg laying is associated with the first brood in mid-April with the second around mid-May (Cabe 1993). The latest young typically fledge in early July.

Notes

An Atlaser who recorded one event of Brown-headed Cowbird parasitism may have been mistaken because brood parasitism has not previously been recorded for the European Starling (Cabe 1993).

Abundance by Natural Division
Average Number of Birds / 100 Stops

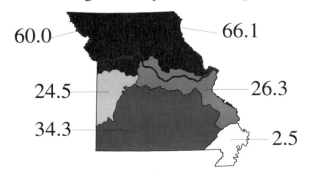

60.0   66.1
24.5   26.3
34.3   2.5

Breeding Phenology

EVIDENCE (# of Records)	MAR	APR	MAY	JUN	JUL	AUG	SEP
NB (54)	4/04				7/05		
NE (7)	4/10			6/12			
NY (62)		4/20			7/15		
FY (177)	4/18					8/11	

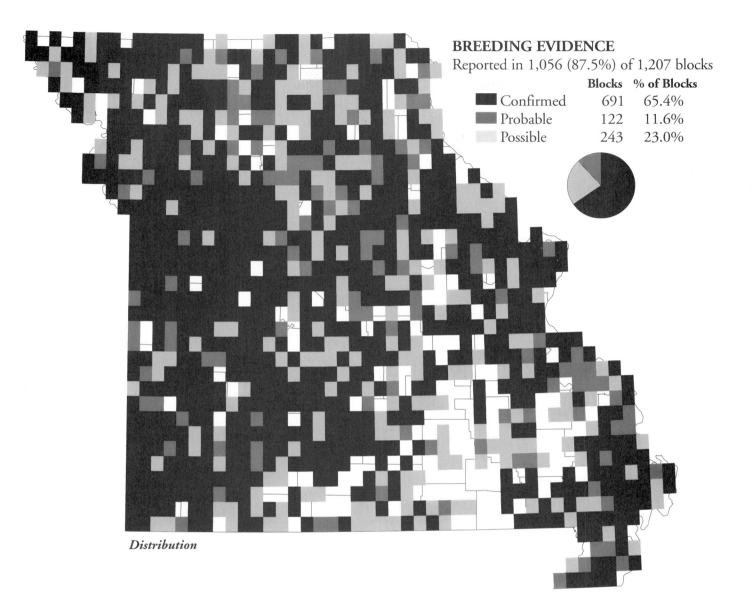

BREEDING EVIDENCE
Reported in 1,056 (87.5%) of 1,207 blocks

	Blocks	% of Blocks
Confirmed	691	65.4%
Probable	122	11.6%
Possible	243	23.0%

Distribution

White-eyed Vireo
Vireo griseus

Rangewide Distribution: Great Lakes though southeastern United States, northeastern Mexico
Abundance: Common
Breeding Habitat: Dense, moist, brushy thickets & tangles
Nest: Roots, twigs & grass, bound with silk & vegetation fiber, lined with fine grass & fibers in shrub or sapling
Eggs: 4 white with brown or black spots
Incubation: 12–16 days
Fledging: 9–11 days

White-eyed Vireos conceal themselves in thickets and tangles along rivers, streams and in pastures overgrown with brush. They are more readily detected by their loud songs than by sight. Because the scrub and tangles this species inhabits are frequently considered to be land of low economic value the areas are often converted to other land uses (Hopp et al. 1995).

Code Frequency

Atlasers easily detected and identified White-eyed Vireos by sight and sound. Most nest site records were of food being carried to young and fledglings. The difficulty of locating nests of this species, which were usually hidden within tangles and briars, was the most likely explanation for the preponderance of other confirmation codes, such as food for young and fledglings. White-eyed Vireos likely nested in most blocks where recorded as possible and probable breeders.

Distribution

Most White-eyed Vireo records were distributed from just north of the Missouri River southward, with only scattered locations to the north in the Glaciated Plains Natural Division. Brushy growth on restored strip-mined land has created some habitat in northern Missouri. The Springfield Plateau and Lowlands natural sections contained fewer records, probably because appropriate habitat has been converted to crop fields and pasture. Forest regeneration in many blocks in the Ozark and Ozark Border natural divisions created extensive areas of second growth and shrub thickets, which provide appropriate habitat for this species.

Abundance

The Ozark Natural Division was the center of abundance for this species. The Lower Ozark and White River natural sections had the largest relative abundance values. More than three times as many birds were estimated for the Ozark Natural Division as for the adjacent Ozark Border. White-eyed Vireos were surprisingly abundant in the Mississippi Lowlands.

Phenology

Atlasers documented most nest site activity in the first week in May. However, a May 5 nest-with-young record in

Abundance by Natural Division
Average Number of Birds / 100 Stops

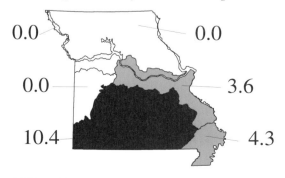

0.0   0.0

0.0   3.6

10.4   4.3

Breeding Phenology

EVIDENCE (# of Records)	MAR	APR	MAY	JUN	JUL	AUG	SEP
NB (3)		5/05		6/15			
NE (4)			6/08		7/02		
NY (4)		5/22			7/15		
FY (17)			6/06			8/03	

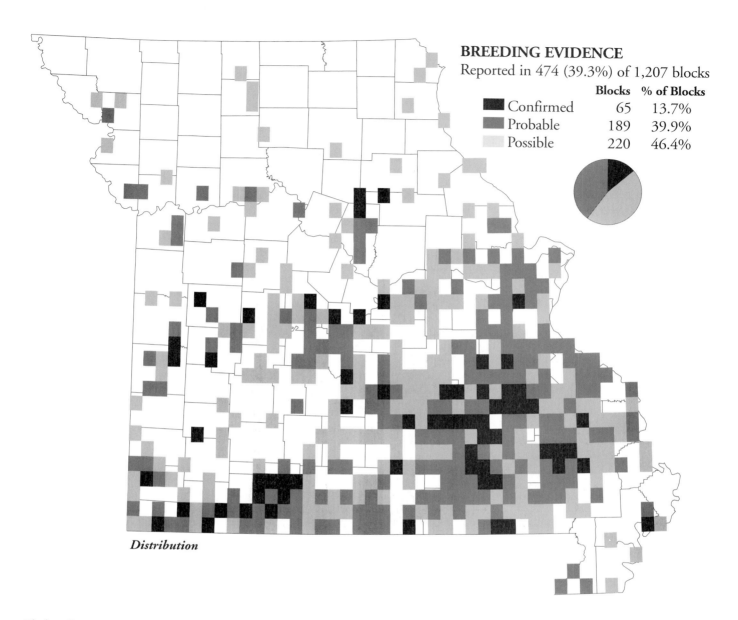

Distribution

Phelps County suggested that nesting activity likely occurred earlier for some pairs. Nest-related activity ended by July 15, but Atlasers observed fledglings until August 26.

Notes

Atlasers documented four records of Brown-headed Cowbird brood parasitism. According to Hopp et al. (1995) cowbirds parasitize almost half of all White-eyed Vireo nests. Ehrlich et al. (1988) also lists this species as a common host.

Bell's Vireo
Vireo bellii

Rangewide Distribution: East central & southwestern
   United States to central Mexico
Abundance: Common, declining in California, expanding in
   northeast
Breeding Habitat: Thickets, dense brush, hedgerows &
   mesquite
Nest: Deep rounded nest of leaves, bark & cocoons, lined
   with grass, down & hair, in shrub or bush
Eggs: 4 white with scattered brown spots on large end,
   occasionally unmarked
Incubation: 14 days
Fledging: 11–12 days

Bell's Vireos occupy the shrub stage of successional habitats, especially old fields, fence rows and upland brush patches in open fields (Brown 1993). In the eastern part of their range, they apparently avoid damp areas such as brushy draws (Peterjohn and Rice 1991). Bell's Vireos are exhibiting one of the greatest population declines in Missouri. Breeding Bird Survey data indicate a 5.1 percent average annual decline in Missouri from 1967 through 1989 (Wilson 1990).

Code Frequency

The unique, easily-identified song of this vireo makes it easy to locate when it sings most, early in the nesting season. If not detected audibly, however, Bell's Vireos can be difficult to find due to their subtle coloration, skulking behavior and the concealing, shrubby habitat where they nest. Addition-ally, unfamiliarity with this species may have resulted in fewer reports. Nests and other breeding evidence were difficult to observe because Bell's Vireos reside in dense foliage. Thus, nesting likely occurred in most of the blocks in which breeding was recorded as probable and in many where it was recorded as possible.

Distribution

Bell's Vireos were distributed primarily through the more open habitats in Missouri and they were essentially absent from the Ozarks. The greatest concentration of blocks where Bell's Vireos were recorded occurred on the Osage Plains, Springfield Plateau and through much of northern Missouri. They were nearly absent from the Mississippi Lowlands, presumably because of the intensive agriculture and lack of shrub cover in that natural division. Atlasers recorded them in three blocks near the St. Francis River in southern Butler County, just south of Dexter in Stoddard County and near Hornersville in southern Dunklin County. In these areas, they were generally associated with shrubby vegetation along drainage ditches.

Abundance

Abundance surveys mirrored block distribution for this species. Bell's Vireos were most abundant in the Glaciated Plains and Osage Plains natural divisions. Presumably, the appropriate combinations of foraging and nesting habitats are more prevalent in those regions.

Abundance by Natural Division
Average Number of Birds / 100 Stops

0.0   1.5
1.2   0.0
0.3   0.0

Breeding Phenology

EVIDENCE (# of Records)	MAR	APR	MAY	JUN	JUL	AUG	SEP
NB (1)			6/08	6/08			
NE (4)		5/25		6/20			
NY (2)			6/17	6/18			
FY (3)				6/28	7/07		

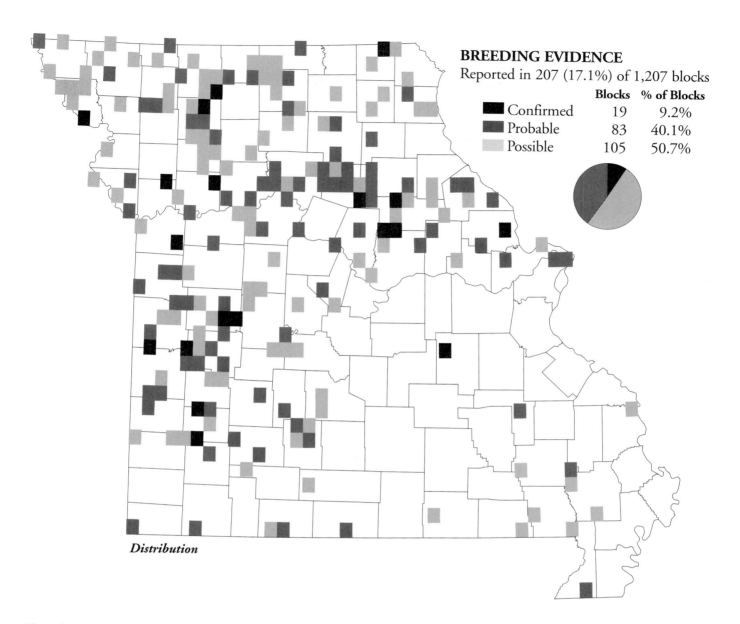

BREEDING EVIDENCE

Reported in 207 (17.1%) of 1,207 blocks

	Blocks	% of Blocks
Confirmed	19	9.2%
Probable	83	40.1%
Possible	105	50.7%

Distribution

Phenology

A nest with eggs was the earliest documentation of the onset of the breeding season. Evidence of the lateness of nesting included records of nest building, a nest with eggs and the feeding of young in their nest. Harrison (1975) reported that Bell's Vireos occasionally produce two broods in a season, which may explain the late nesting observed during the Atlas Project.

Notes

One-third of the blocks in which confirmed breeding was recorded involved evidence of Brown-headed Cowbird brood parasitism. Brood parasitism of Bell's Vireos is often severe and appears to vary regionally. In Indiana, parasitism occurred in seven of 13 nests, or 54 percent (Mumford 1952); in Kansas, in 24 of 35 nests, or 69 percent (Barlow 1962) and in Oklahoma in 18 of 61 nests, or 30 percent (Overmire 1962). Brown (1993) reported that a parasitism rate of 30 percent of Least Bell's Vireos in California has led to unstable populations that chance events could extirpate. High parasitism rates may contribute to the Bell's Vireo's rapidly declining population in Missouri.

Yellow-throated Vireo
Vireo flavifrons

"THREE-eight," best describes the part of the Yellow-throated Vireo's song that distinguishes it from that of the Red-eyed Vireo. This species, like the Philadelphia Vireo, is frequently mistaken as an unusual-sounding Red-eyed Vireo. The pronounced three-eight song, with the accent on the three, is harsh and loud, standing out at regular intervals during the song. Yellow-throated Vireos prefer treetops in open bottom land forests, mixed pine-oak forests and forested city parks.

Code Frequency

Because it was difficult to view this treetop warbler, most of the observations used to document this species were those obtained by sound. It appeared that if sufficient time was spent observing Yellow-throated Vireos, behaviors would eventually be detected that would confirm breeding. Therefore, Yellow-throated Vireos likely nested in most blocks where recorded.

Distribution

Yellow-throated Vireos were distributed statewide but fewer observations were recorded outside the more-forested parts of Missouri. Atlasers obtained no records from the southern two counties of the Mississippi Lowlands. Also, Yellow-throated Vireos were not detected in several counties in each of the other regions, most notably in the Springfield Plateau Natural Section.

Abundance

This species is reportedly fairly numerous in the Ozark Border Natural Division (Robbins and Easterla 1992), but Atlasers did not detect any on abundance routes.

Phenology

The Yellow-throated and White-eyed vireos were the earliest of the vireos to return to Missouri. The early observation of nest building occurred barely two weeks after the March 30 earliest-ever record (Robbins and Easterla 1992). Most nest-related activity occurred in late May and early June. Active birds were observed near a suspected nest site on May 25.

Notes

This species is a well-known host for Brown-headed Cowbirds (Ehrlich et al. 1988). Atlasers recorded two occasions of brood parasitism for Yellow-throated Vireos.

Abundance by Natural Division
Average Number of Birds / 100 Stops

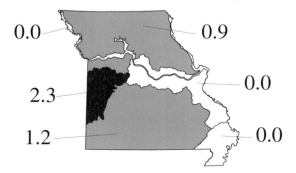

0.0   0.9

2.3

0.0

1.2   0.0

Breeding Phenology

EVIDENCE (# of Records)	MAR	APR	MAY	JUN	JUL	AUG	SEP
NB (6)		4/18		6/22			
NY (4)			6/12		7/23		
FY (8)			6/09			8/03	

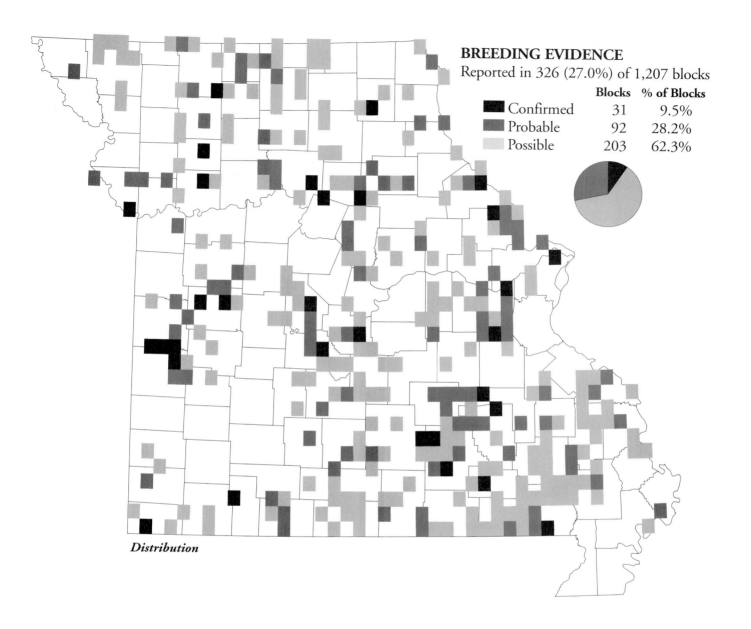

BREEDING EVIDENCE

Reported in 326 (27.0%) of 1,207 blocks

	Blocks	% of Blocks
Confirmed	31	9.5%
Probable	92	28.2%
Possible	203	62.3%

Distribution

Warbling Vireo
Vireo gilvus

Rangewide Distribution: Southwestern to south central Canada, most United States to central Mexico
Abundance: Common & widespread
Breeding Habitat: Open woods & thickets
Nest: Compact cup of bark, leaves, vegetation & grass, usually high in tree
Eggs: 4 white with brown or black spots
Incubation: 12 days
Fledging: 16 days

These plain, gray-colored vireos clearly make up in song what they lack in plumage. Close observation will often reveal the male singing while on the nest. Widmann (1907) described the Warbling Vireo's habitat as primarily restricted to trees along rivers and lakes, but noted it was adapting to orchards, gardens, parks and big shade trees in cities.

Code Frequency

No nests with eggs or young were documented, perhaps because Warbling Vireos usually nest on small branches, 6–9

Abundance by Natural Division
Average Number of Birds / 100 Stops

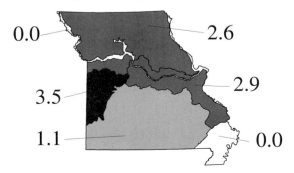

0.0   2.6
3.5
2.9
1.1   0.0

meters or higher above the ground. Due to the ease of locating birds by song, 82 percent of all records were based on sound.

Distribution

Warbling Vireos were distributed statewide with fewer records from the Ozark and Mississippi Lowlands natural divisions. The preference for large cottonwoods along large rivers established a pattern for this species' distribution in Missouri. In the Big Rivers Natural Division, where riverfront forests contain many cottonwoods, this species was recorded in 68 percent of blocks. In the Ozark and Mississippi Lowlands natural divisions, it was recorded in only 29 and 31 percent, respectively. In the Glaciated Plains, Ozark Border and Osage Plains natural divisions, Warbling Vireos were recorded in 59 percent, 54 percent, and 43 percent of blocks, respectively.

Abundance

Based on roadside point counts, the Osage Plains had the highest relative abundance. Only a few birds were found in the Ozark Natural Division where much more upland than bottomland exists. However, these low values may be due to roads crossing rather than paralleling appropriate habitat along waterways. Riparian counts would sample more appropriate habitat.

Phenology

Most eggs were probably laid beginning in mid-May. A May 15 record of an adult carrying a fecal sac would indicate Warbling Vireos lay some eggs in late April or early May.

Notes

Although no nests were observed, this species is considered a frequent host of Brown-headed Cowbirds. It is a host that makes no attempt to remove cowbird eggs (Ehrlich et al. 1988).

Breeding Phenology

EVIDENCE (# of Records)	MAR	APR	MAY	JUN	JUL	AUG	SEP
NB (9)			5/15		6/30		
FY (15)			6/01		7/19		

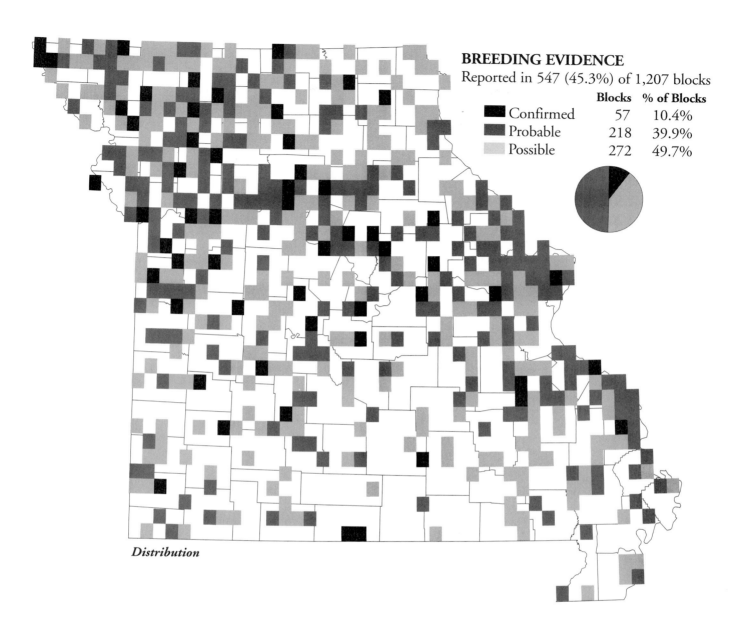

BREEDING EVIDENCE
Reported in 547 (45.3%) of 1,207 blocks

	Blocks	% of Blocks
Confirmed	57	10.4%
Probable	218	39.9%
Possible	272	49.7%

Distribution

Red-eyed Vireo
Vireo olivaceus

Rangewide Distribution: Southern Canada, north to eastern & southeastern (not southwestern) United States
Abundance: Abundant
Breeding Habitat: Woods, suburbs with large deciduous trees
Nest: Deep cup of vines, bark, grass, forbs & spider webs in deciduous tree
Eggs: 4 white with spots of browns or blacks at large end of egg
Incubation: 11–14 days
Fledging: 10–12 days

In 1907, Widmann suggested that this was the "most evenly distributed woodland summer resident from April 'til October." Today we still hear its repetitive song throughout the state. Its nest is an intricately woven cup that hangs by its rim from a horizontal forked branch. The nests of this and most vireos are most easily identified after the leaves have fallen. Only two nests were observed, one of which was seen outside the breeding season.

Code Frequency

Eighty-two percent of all observations were of song-related activities. These incessant vocalists made it easy to document their presence. Therefore, where they were not recorded they likely were scarce or absent. Because nests are usually well concealed in foliage, Atlasers found nests with eggs or young in only five blocks. However, the Red-eyed Vireo probably bred undetected in most of the blocks where it was found.

Distribution

Found statewide in forested areas, this species was recorded more frequently in the Ozark and Ozark Border than in other natural divisions. The sheer number of individuals in the Ozarks probably accounted for the numerous probable and confirmed records. Red-eyed Vireos were less frequently recorded in areas of open grasslands and agricultural regions of the Glaciated Plains and Mississippi Lowlands natural divisions and in the southern part of the Springfield Plateau Natural Section.

Abundance

The greatest abundance was found in the Ozark Natural Division where extensive forest habitat is present. In the Mississippi Lowlands where the second highest relative abundance was recorded, all observations were obtained on three stops on the Dexter Miniroute. This may have distorted this species' abundance.

Phenology

Nests and eggs were found by Atlasers between mid-May and mid-June, about a month later than Robbins and Easterla (1992) stated the first migrants had arrived. Most observations of fledglings and food delivery were made in the last week of June and the last week of July, which may indicate this is a two-brooded species.

Abundance by Natural Division
Average Number of Birds / 100 Stops

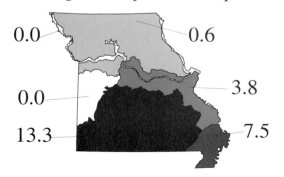

0.0   0.6

0.0   3.8

13.3   7.5

Breeding Phenology

EVIDENCE (# of Records)	MAR	APR	MAY	JUN	JUL	AUG	SEP
NB (9)			5/11		7/15		
NE (2)			5/18	6/17			
NY (3)			6/07	6/26			
FY (22)			5/18		7/30		

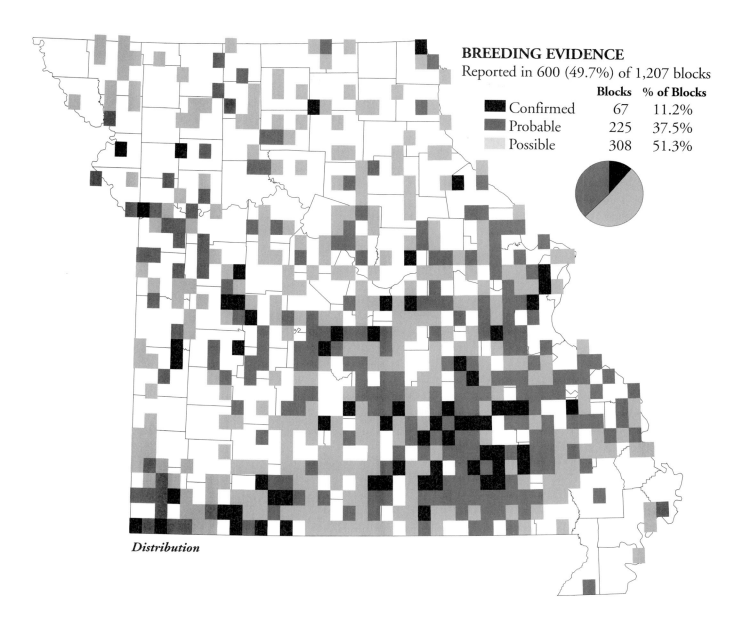

BREEDING EVIDENCE

Reported in 600 (49.7%) of 1,207 blocks

	Blocks	% of Blocks
Confirmed	67	11.2%
Probable	225	37.5%
Possible	308	51.3%

Distribution

Notes

Three records of Brown-headed Cowbird brood parasitism were noted for this species. Atlasers noted brood parasitism in 11.1 percent of observations of food for young or nest contents. Ehrlich et al. (1988) considered this species to be one of the most frequent hosts.

Blue-winged Warbler
Vermivora pinus

> **Rangewide Distribution:** Northeastern United States
> **Abundance:** Locally common, and expanding northeast & west
> **Breeding Habitat:** Small trees, scrubs & forest-stream edge
> **Nest:** Bulky cup of leaves, grass & grapevine bark, lined with grapevine fiber, in grass & vines
> **Eggs:** 5 white finely spotted with brown, mostly at large end
> **Incubation:** 10–11 days
> **Fledging:** 8–10 days

These splendid blue and yellow warblers are characteristic of brushy landscapes throughout the Ozarks. Their insect-like buzzing songs can be heard in overgrown pastures and forest openings from late April through June.

Code Frequency

Blue-winged Warblers are easily detected due to their unique, often-repeated song. Despite their brilliance, they are often hard to see and finding their nests is extremely difficult. Blue-winged Warblers reside and nest in dense cover and thickets that are inhospitable to birders. As a result, it is likely that nest sites did occur in many of the blocks in which evidence of possible and probable breeding was observed. Blue-winged Warblers were most easily recorded as probable or confirmed breeders in regions where they were most abundant. Individuals recorded in the less populated blocks with possible records may have been unmated or otherwise nonproductive.

Distribution

The Blue-winged Warbler's distribution is strongly associated with the eastern Ozarks, which falls well within its North American breeding range. This region affords the forest openings and brushy draws in which they nest. Blue-winged Warblers were not recorded in many of the blocks along the Arkansas line. The southern edge of their breeding range ends in Arkansas, and they may become scarcer as they approach that terminus. In the Mississippi Lowlands, Atlasers recorded Blue-winged Warblers only near Donaldson Point Conservation Area. Scattered locations occurred within the Osage Plains and in central and northeastern Missouri. They were essentially absent from northwestern Missouri except for two unexpected locations in Daviess and Holt counties. Robbins and Easterla (1992) stated that it was not known to breed north of the Missouri River and west of approximately 93°00'W, a north/south line running near the Swan Lake National Wildlife Refuge.

Abundance

Blue-winged Warblers were most abundant in the eastern Ozarks, which was also the region in which they were recorded in the greatest number of blocks. Presumably the appropriate combination of forest edge and brushy cover contributed to this pattern. Abundance dropped in the western Ozarks, which approaches the edge of their North American breeding range.

Abundance by Natural Division
Average Number of Birds / 100 Stops

0.0   0.2

0.0   1.6

2.4   0.0

Breeding Phenology

EVIDENCE (# of Records)	MAR	APR	MAY	JUN	JUL	AUG	SEP
NB (2)		5/01			6/19		
NY (4)			5/21	6/03			
FY (26)			5/21			7/28	

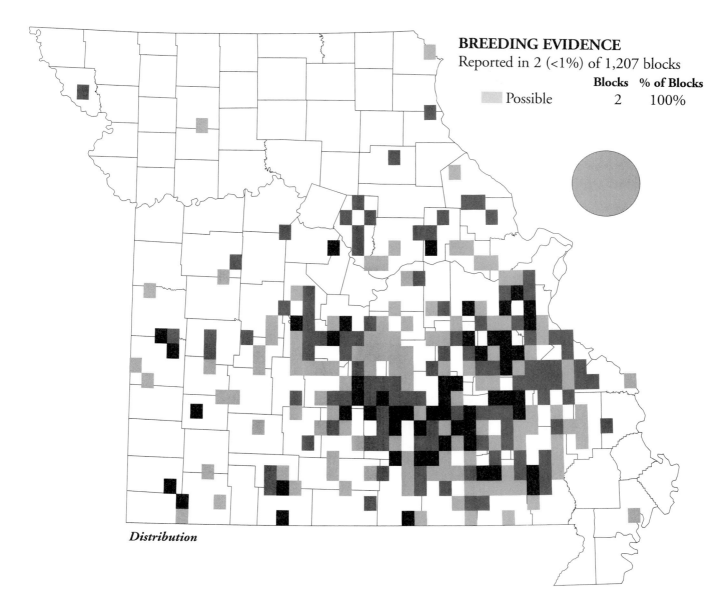

Distribution

Phenology

Blue-winged Warblers arrive in Missouri around the second week of April (Robbins and Easterla 1992) and apparently initiate nesting almost immediately. The observation of fledglings on May 4 in an area northeast of Salem in Dent County suggested a surprisingly early nesting. Most nesting occurred from May into June.

Notes

Brown-headed Cowbird parasitism of Blue-winged Warblers is well known (Ehrlich et al. 1988). In this project, two of 27 potential blocks recorded brood parasitism.

Northern Parula
Parula americana

Rangewide Distribution: Southeastern Canada &
  eastern United States
Abundance: Common
Breeding Habitat: Deciduous or coniferous woods,
  usually with tree lichens
Nest: Hanging pocket of lichens & conifer twigs, lined
  with fine material in tree
Eggs: 4–5 white to creamy with brown or reddish-
  brown markings
Incubation: 12–14 days
Fledging: 10–11 days

The diminutive Northern Parula is one of Missouri's
common warblers. Its nests are usually well hidden in the
foliage within or near upper tree canopies, making observation difficult. The Northern Parula's ascending buzzy song
can be heard along most bottomland and upland streams.
The name Parula comes from the chickadee- and titmouse-
like behavior this species employs as it forages along twigs
and hangs upside down on leaf bunches looking for insects.
Parus is the scientific or Latin name for the genus of titmice
and chickadees. Parula means little titmouse (Terres 1987).

Code Frequency

Although this species is easily detected by song, the concealed nests make it difficult to obtain codes associated with
nests and nest care. Atlasers observed only four nests during
the seven-year Atlas Project. It is likely that Northern Parulas
nested in many of the blocks where they were recorded as
possible breeders. A very low 10 percent of all observations
confirmed breeding of this species. Observations of fledglings and food being carried to young constituted 85 percent
of all confirmed records.

Distribution

Northern Parulas were common in all natural divisions
except the Glaciated Plains and the Mississippi Lowlands,
presumably because of their association with riparian woodlands. While this species was documented in 76 percent of
the blocks in the Ozark Natural Division, it was recorded in
only 23 percent of blocks in the Glaciated Plains and in 17
percent of blocks in the Mississippi Lowlands. Riparian surveys in the latter two regions would likely show this species
to be present along most wooded riparian corridors.

Abundance

Because this is a riparian species, and Atlas Project abundance maps are based on roadside counts, a complete picture of abundance is not available. This species is likely very
common along all wooded streams and rivers in the Ozark
and Ozark Border natural divisions.

Phenology

Northern Parulas arrive early in the spring beginning in
late March (Robbins and Easterla 1992). Nesting commenced in early May as determined by observations of food
being carried to young. Atlasers observed many fledglings in
mid-May and again in mid- to late-July, which may indicate
up to three broods for this species. Moldenhauer and

Abundance by Natural Division
Average Number of Birds / 100 Stops

0.0   1.2
1.2   4.1
9.8   0.8

Breeding Phenology

EVIDENCE (# of Records)	MAR	APR	MAY	JUN	JUL	AUG	SEP
NB (3)		5/01			6/22		
NE (1)					7/12 7/12		
NY (3)			6/03			7/23	
FY (28)			5/18				8/15

BREEDING EVIDENCE

Reported in 797 (66.0%) of 1,207 blocks

	Blocks	% of Blocks
Confirmed	60	9.9%
Probable	281	46.5%
Possible	264	43.6%

Distribution

Regelski (1996) reported at least two broods in southern states.

Notes

Brown-headed Cowbird brood parasitism was observed in 13 percent of observations of Northern Parula nest contents. Ehrlich et al. (1988) listed them as an uncommon cowbird host.

Yellow Warbler
Dendroica petechia

> **Rangewide Distribution:** Most of Canada, Alaska,
>   United States & Mexico
> **Abundance:** Common
> **Breeding Habitat:** Wet second growth woods, gardens,
>   & thickets
> **Nest:** Compact & neat, of weeds, bark, grass, lined
>   with fine material in tree
> **Eggs:** 4–5 off-white, occasionally pale green, with
>   brown, olive or gray marks
> **Incubation:** 11–12 days
> **Fledging:** 9–12 days

Yellow Warblers, the most widespread wood warblers, are habitat generalists in eastern North America but are closely associated with riparian thickets in the west. Missouri's Yellow Warblers are usually found in willow trees of wetland areas or along streams, especially in the Ozark Natural Division. According to Robbins and Easterla (1992), Yellow Warblers are associated with second-growth woodlands and thickets with willows, especially near water.

Code Frequency

This species' close association with wetland areas and riparian forests made it difficult to confirm. Only one nest was found and the 23 other reported confirmations were distributed among four other codes. Most codes related to individuals entering a possible nest site or carrying food to a possible nest site or fledglings.

Distribution

This species was distributed statewide in appropriate habitat. Locations seemed closely aligned with willow wetlands in the Glaciated Plains and Osage Plains natural divisions, cottonwood-willow riparian forest in the Big Rivers Natural Division, and moist, riparian zones along streams in the Ozark and Ozark Border natural divisions. In the Mississippi Lowlands, this species was found in forests outside river levees or in overgrown thickets along big ditches.

Abundance

Roadside count data poorly estimated abundance for this species because most wetland habitat and riparian zones were not easily surveyed from roads. Within appropriate habitat, Yellow Warblers were one of the more-commonly observed species.

Phenology

The first records of nest-building were noted in mid-May with incubation and fledging in late May and early June. Widmann (1907) suggested that most individuals move south and leave the breeding ground after mid-July. Robbins and Easterla (1992) asserted that most individuals have left the state by late August and early September.

Abundance by Natural Division
Average Number of Birds / 100 Stops

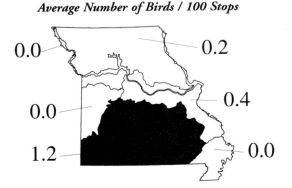

0.0   0.2

0.0   0.4

1.2   0.0

Breeding Phenology

EVIDENCE (# of Records)	MAR	APR	MAY	JUN	JUL	AUG	SEP
NB (4)			5/20		6/19		
NY (1)				6/11 6/11			
FY (11)				6/03		8/03	

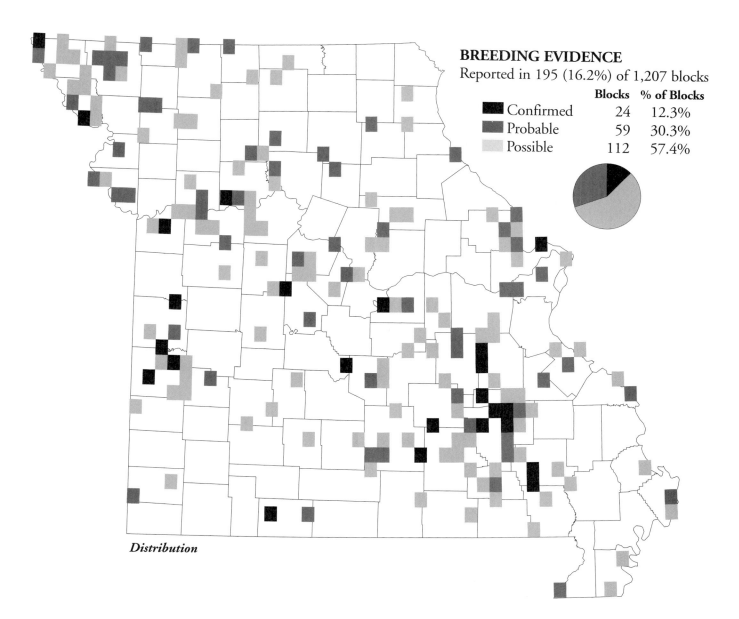

BREEDING EVIDENCE
Reported in 195 (16.2%) of 1,207 blocks

		Blocks	% of Blocks
■	Confirmed	24	12.3%
■	Probable	59	30.3%
■	Possible	112	57.4%

Distribution

Notes

Atlasers observed one case of brood parasitism out of 12 observations of nest contents. Yellow Warblers are one of the three most-frequently parasitized species (Ehrlich et al. 1988).

Chestnut-sided Warbler
Dendroica pensylvanica

Rangewide Distribution: Central to southeastern Canada, northeastern United States & southern Appalachian

Abundance: Fairly common

Breeding Habitat: Brush thickets, open-border wood & second growth

Nest: Cup of fine plant material, lined with finer material, in bush or shrub

Eggs: 4 white to off-white, with brown marks

Incubation: 12–13 days

Fledging: 10–12 days

This attractive, uniquely marked, warbler is one of the most common resident warblers in second-growth forests within its breeding range in the northeastern United States (Morse 1989). Although primarily a migrant in Missouri, a few historical breeding records have been obtained in the eastern Ozarks, ranging from the 19th century (Widmann 1907) to recent years (Evans 1980). Robbins and Easterla (1992) stated: "this species undoubtedly is a more regular breeder in the eastern section of the Ozarks and Ozark border than the few records indicate."

Code Frequency

Chestnut-sided Warblers are easy to detect and identify due to their well-known song. They tend to be noticed as they flit about the understory. Despite their relative conspicuousness, Chestnut-sided Warblers were recorded in only two blocks, suggesting that they are rare, although they do summer in the state and may breed. In one Atlas block, the observation of an agitated individual suggested probable breeding. In the other, a summering individual occurred in habitat appropriate for breeding, indicating possible breeding.

Distribution

The two locations discovered by the Atlas Project provide insufficient data to evaluate the distribution of the species. They do not support or refute the statement (Robbins and Easterla 1992) that Missouri's few summering Chestnut-sided Warblers are more associated with the eastern Ozarks and Ozark Border.

Phenology

The observations of Chestnut-sided Warblers, on July 1 and 29, were within the season when nesting is likely.

Notes

Although this species is a known Brown-headed Cowbird host (Ehrlich et al. 1988), Atlas Project participants obtained no evidence of brood parasitism.

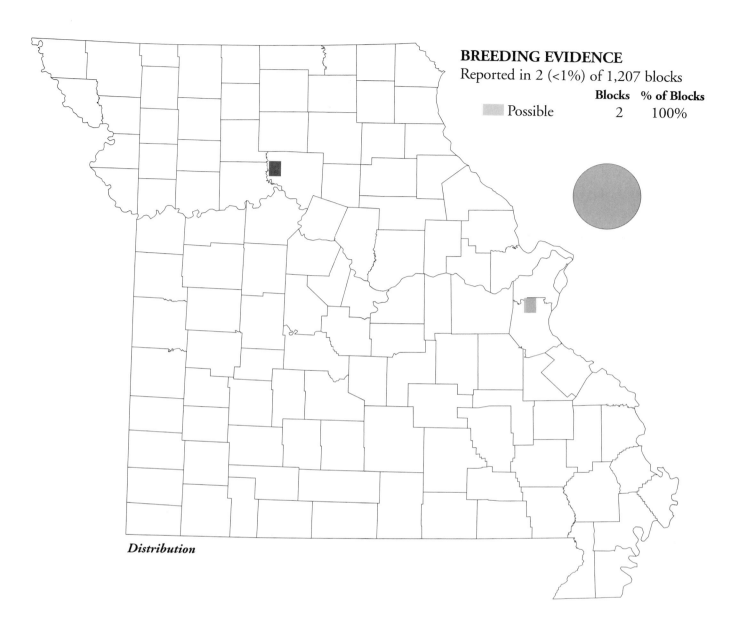

BREEDING EVIDENCE
Reported in 2 (<1%) of 1,207 blocks

	Blocks	% of Blocks
Possible	2	100%

Distribution

Yellow-throated Warbler
Dendroica dominica

Rangewide Distribution: Southeastern United States
Abundance: Fairly common to common
Breeding Habitat: Oak, pine woods, baldcypress & sycamore swamps
Nest: Grass, bark, weeds & caterpillar silk, lined with down & feathers; in southeast, of Spanish moss, inland in pine tree
Eggs: 4 greenish gray-white with purple, red or brown marks & wreathed
Incubation: 12–13 (?) days
Fledging: days to flegling not known

Known historically as the "Sycamore" Warbler in the Mississippi River Valley (Widmann 1907), this species is most often located while it sings from the upper branches of a Sycamore *(Platanus occidentalis)* tree. An alternative habitat is the mixed pine and hardwood forests of the Ozark Natural Division.

Code Frequency

Because this species nests in treetops, Atlasers usually based breeding confirmations on observations other than nest contents. The scarcity of observations of nest building, fledglings, suspected nest sites, fecal sacs carried by adults from nests and food being carried to young, are due to the difficulty of observing nests from the ground.

Distribution

The Yellow-throated Warbler's main distribution occurs within the forested regions of the Ozark, Ozark Border and Osage Plains natural divisions. The many forested riparian areas along Ozarks streams were the most frequent habitat in which this species was found. A few observations were recorded within the Eastern Glaciated Plains Natural Section north to the Iowa line. Although records of nesting activity exist north to Buchanan County in the Western Glaciated Plains Natural Section (Robbins and Easterla 1992), no Yellow-throated Warblers were observed within Atlas blocks in that section.

Abundance

This species was recorded on abundance counts only in the Ozark Natural Division. Special searches of riparian and mixed pine-hardwood forest areas in the Ozarks would likely provide a better picture of this species' abundance within appropriate habitat.

Phenology

Although first arrivals to the state begin appearing in March (Robbins and Easterla 1992), nest building was first observed on April 30. Food carried to young may be the best key to timing of this possibly two-brooded species (Ehrlich et al. 1988). Fledglings were observed from July 3 to August 21, certainly late enough to be from a second brood.

Abundance by Natural Division
Average Number of Birds / 100 Stops

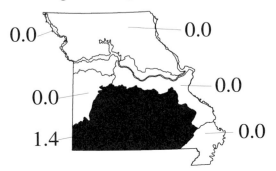

Breeding Phenology

EVIDENCE (# of Records)	MAR	APR	MAY	JUN	JUL	AUG	SEP
NB (7)		4/30		6/08			
FY (15)			5/20		7/13		

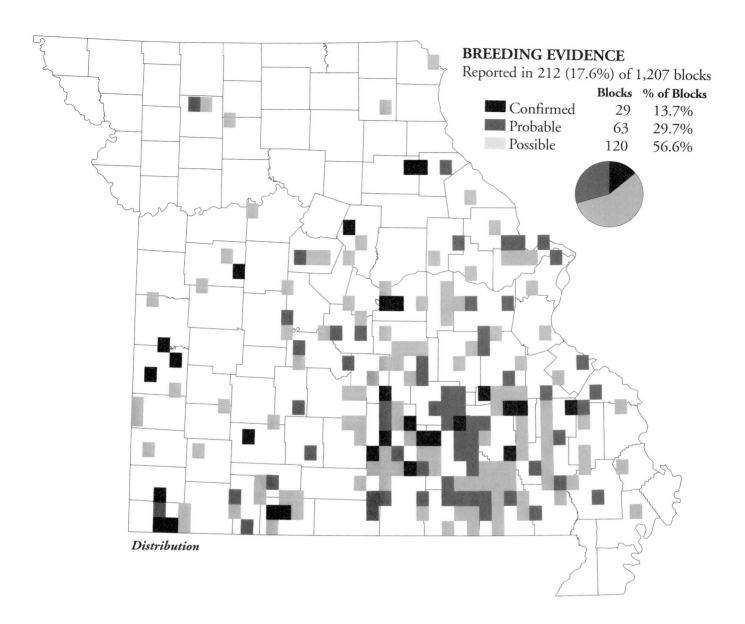

BREEDING EVIDENCE
Reported in 212 (17.6%) of 1,207 blocks

	Blocks	% of Blocks
Confirmed	29	13.7%
Probable	63	29.7%
Possible	120	56.6%

Distribution

Notes

Although Friedmann (1963) documented only one case of Brown-headed Cowbird brood parasitism with this host, three incidents were recorded during the seven-year Atlas Project.

Pine Warbler

Dendroica pinus

Rangewide Distribution: Southeastern Canada & eastern United States

Abundance: Common

Breeding Habitat: Pine forests with large stands & open canopies

Nest: Concealed, compact cup lined with feathers in pine tree

Eggs: 4 white to off-white with brown marks on large end

Incubation: 10(?) days

Fledging: 10(?) days

Perhaps more than any other Missouri breeding bird species, Pine Warblers are closely associated with specific types of vegetative cover. They are common in the Ozark Natural Division and closely associated with mixed hardwood and Short-leaf Pine *(Pinus echinata)* forests. They also can be found in mature plantations of other pine species outside the native Short-leaf Pine range. The song of the Pine Warbler is slower, more musical and less repetitive than that of the Chipping Sparrow with which it is sometimes confused.

Code Frequency

Pine Warblers characteristically move about slowly when foraging and singing in the upper half of large pine trees. With time, one can usually locate the bird and, if patient, may see it carry food to a nest. Most records (76 percent) were established from song evidence or by observing a bird making repeated visits to the same site. Nests were seldom found. Atlasers did confirm the species in 26 blocks by observing activities around nest sites.

Distribution

Pine Warblers were found almost exclusively within the Short-leaf Pine range of the Ozark and Ozark Border natural divisions. The main population may be in the heart of the Short-leaf Pine range, with a second area in the Elk River and the White River natural sections. Outside that range, an immature Pine Warbler was observed in late July in a 50–60 year-old Short-leaf Pine plantation in Knob Noster State Park. Apparently, Pine Warblers can nest in open, mature pine plantations anywhere in the state.

Abundance

The only records from Miniroutes or Breeding Bird Surveys were obtained in the Ozark Natural Division. According to Robbins and Easterla (1992), the highest yearly average of 3.4 birds was recorded on a 50-stop Breeding Bird Survey in Crawford and Dent counties. The Atlas Project recorded an average of 1.7 birds per 50 stops on surveys in the Ozark Natural Division. In routes where Pine Warblers were recorded, Atlasers logged 6.6 birds per 50 stops.

Phenology

Although some birds are present in late February, the earliest nest activity documented by Atlasers was in early May; the latest was in early August. The latter observation

Abundance by Natural Division
Average Number of Birds / 100 Stops

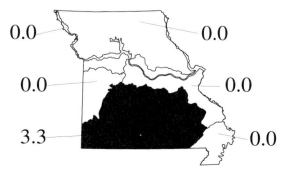

EVIDENCE (# of Records)	MAR	APR	MAY	JUN	JUL	AUG	SEP
NB (1)				6/05	6/05		
NY (1)		5/03	5/03				
FY (11)				6/06		8/05	

Breeding Phenology

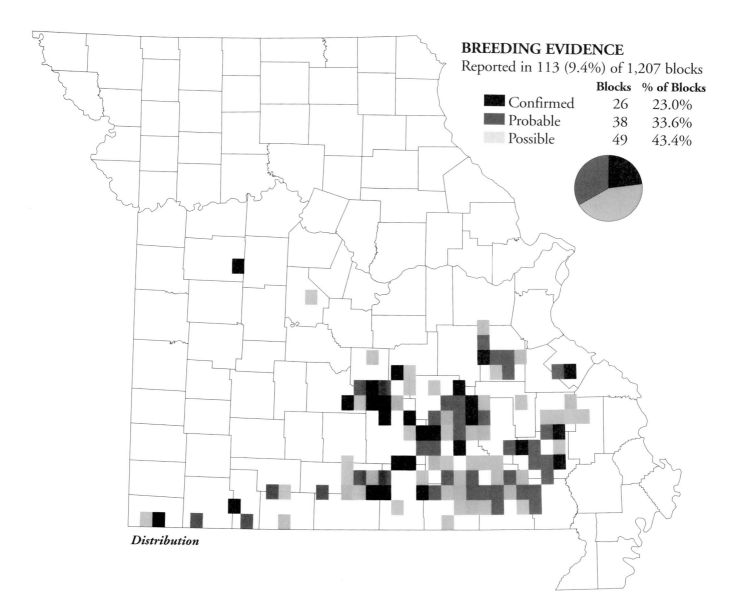

BREEDING EVIDENCE
Reported in 113 (9.4%) of 1,207 blocks

	Blocks	% of Blocks
Confirmed	26	23.0%
Probable	38	33.6%
Possible	49	43.4%

Distribution

may indicate a second brood because northern Arkansas birds begin nesting in late March (James and Neal 1986).

Notes

Atlasers recorded no Brown-headed Cowbird eggs or young associated with Pine Warblers. Pine Warblers foil Brown-headed Cowbird brood parasitism attempts by burying cowbird eggs in the bottom of nests (Ehrlich et al. 1988).

Prairie Warbler
Dendroica discolor

Rangewide Distribution: Eastern & southeastern
  United States including Gulf Coast
Abundance: Common
Breeding Habitat: Brushy hills, abandoned fields &
  open woods
Nest: Compact, of closely-felted plant material in tree
Eggs: 4 white to off-white usually wreathed with
  brown marks
Incubation: 12 days
Fledging: 9–10 days

Widmann (1907) called the Prairie Warbler the "hillside" or "glade" warbler in Missouri because of its abundance in glades in the White River Natural Section. Old fields with scattered red cedar trees or cedar glades are the best place to find this species in Missouri. As their preferred habitat succeeds to forest, this species moves to other suitable sites. Large, early-successional openings in forests also supply an excellent habitat for this species and others such as the Blue-winged Warbler and Yellow-breasted Chat. As large areas of the state succeed to forest, the Prairie Warbler population will shrink accordingly.

Code Frequency

Atlasers confirmed breeding of this species in 7 percent of blocks. Territorial behavior was recorded in 28 percent of the blocks in which it was found. The ventriloquistic voice of this species made it hard to locate. The roughness of the terrain and ticks and chiggers abundant in glades and brushy fields likely added to the difficulties Atlasers faced.

Distribution

Distribution was restricted to the Ozark Border and Ozark natural divisions except for an extension into the eastern Osage Plains Natural Division. In the St. Francois Mountains and White River natural sections, this species was recorded in the highest percentage of blocks, 68 percent and 49 percent, respectively.

Abundance

Abundance surveys recorded this species only in the Ozark Natural Division. The patchiness of the Prairie Warbler's habitat and the infrequency with which roads pass through it may contribute to low counts.

Phenology

Prairie Warblers are usually single-brooded. The fledglings observed August 21 may have been from a second brood or a renesting, or may have been immature birds beyond the stage of requiring parental care.

Notes

Atlasers reported no nests parasitized although this species is considered a highly parasitized species. Adults frequently abandon parasitized nests (Ehrlich et al.1988). However, only eight opportunities to record Brown-headed Cowbird parasitism occurred.

Abundance by Natural Division
Average Number of Birds / 100 Stops

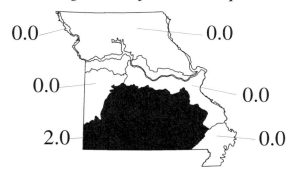

0.0   0.0

0.0   0.0

2.0   0.0

Breeding Phenology

EVIDENCE (# of Records)	MAR	APR	MAY	JUN	JUL	AUG	SEP
UN (1)	4/16	4/16					
NE (2)		5/12	5/27				
NY (1)		5/21	5/21				
FY (5)			6/03		7/15		

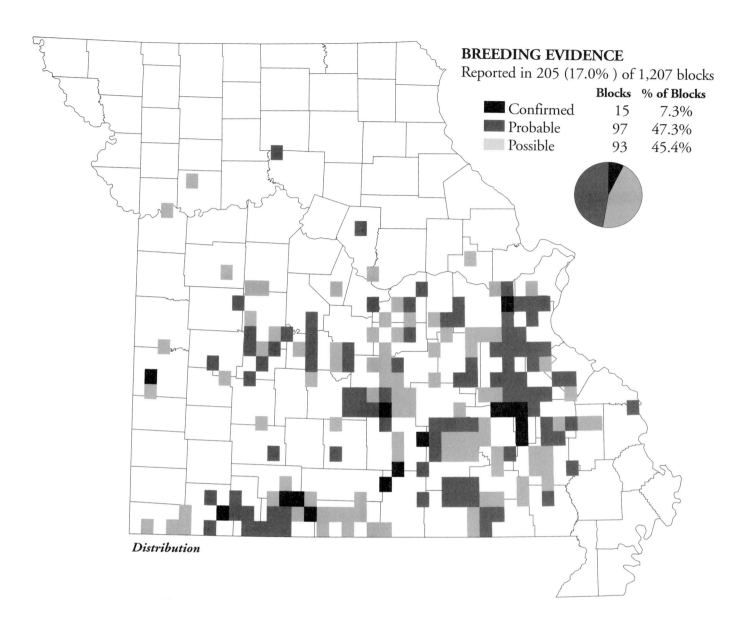

BREEDING EVIDENCE
Reported in 205 (17.0%) of 1,207 blocks

	Blocks	% of Blocks
Confirmed	15	7.3%
Probable	97	47.3%
Possible	93	45.4%

Distribution

Cerulean Warbler
Dendroica cerulea

Rangewide Distribution: Eastern & coastal United States, expanding northeast & south
Abundance: Fairly common but local
Breeding Habitat: Bottomland forests with tall mature deciduous trees
Nest: Small, shallow & compact, of bark, weeds, lichen & moss, lined with moss & hair, high in tree
Eggs: 4 gray, creamy or green-white, with variable brown marks
Incubation: 12–13(?) days
Fledging: Days to fledgling is unknown

The buzzy song of this treetop warbler can be heard in south Missouri's rich, riparian forests from late spring through June. Cerulean Warblers are named for the deep blue upper parts of the male. The female, which is difficult to spot, is turquoise above and lacks the male's black necklace and breast streaks. Once common, Cerulean Warblers bred in both upland and bottomland forests. The current Cerulean Warbler population is apparently a fraction of that of the late 1800s (Hamel et al. 1996). In Missouri, their population has declined at an average annual rate of 3.1 percent from 1967 to 1989 (Wilson 1990).

Code Frequency

The Cerulean Warbler is difficult to see and most easily detected by its call, although it is easily confused with calls of Northern Parulas and other warblers. Some probable breeding records were based on observations of territoriality, also likely perceived by sound. Few breeding confirmations were recorded for this rare species which has well-concealed nests. Atlasers observed no young or eggs in nests during the project.

Distribution

Most blocks with confirmed breeding evidences were associated with the Black, Current and Jack's Fork rivers. Also, a group of blocks with confirmed breeding was located in Bates and Vernon counties near the Schell-Osage Conservation Area. Forested riparian zones in these counties may support a larger concentration of this species and consequently more opportunity to confirm breeding. Scattered locations with lower breeding evidence occurred throughout the eastern Ozark Natural Division to the St. Louis area.

Phenology

Only nine confirmed breeding records were available for evaluating breeding phenology. While the only nest building observed was on June 19, some nesting apparently began a week or two earlier as evidenced by food being delivered to young on June 6. Later observations of birds on the nest and food for young indicate a protracted nesting season. Ehrlich et al. (1988) reported that they are single brooded.

Notes

Although Cerulean Warblers have been reported to host Brown-headed Cowbirds (Ehrlich et al. 1988), no such evidence occurred during the seven-year Atlas Project.

Breeding Phenology

EVIDENCE (# of Records)	MAR	APR	MAY	JUN	JUL	AUG	SEP
NB (1)				6/19	6/19		
FY (3)			6/06		7/10		

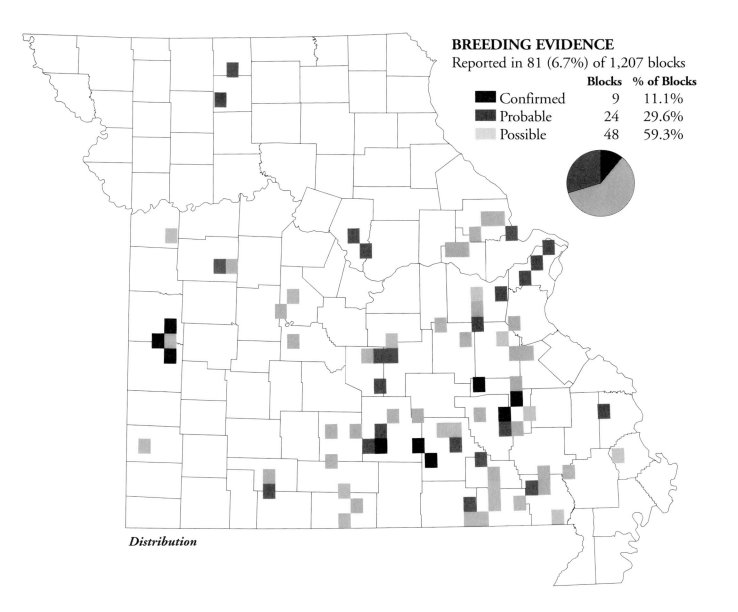

		Blocks	% of Blocks
■	Confirmed	9	11.1%
■	Probable	24	29.6%
■	Possible	48	59.3%

Distribution

Black-and-white Warbler

Mniotilta varia

Rangewide Distribution: Central & southeastern Canada & eastern United States

Abundance: Common

Breeding Habitat: Deciduous & mixed forests, hillsides & ravines

Nest: Leaves & coarse grass, lined with finer material, on ground

Eggs: 5 white to creamy, entirely flecked with brown or at large end, occasionally wreathed

Incubation: 10 days

Fledging: 8–12 days

These striking, bark-gleaning warblers nest on the ground in extensive tracts of mature upland forest. Dawson et al. (1993) hypothesized that they need a contiguous forest area of 116 hectares to breed. Additionally, because they nest on the ground, they likely have been impacted by woodland grazing, free-ranging pets and other predators. These factors should be considered when evaluating the data collected by the Atlas Project.

Code Frequency

This warbler is easy to detect and identify due to its conspicuous markings, bark-gleaning habits and recognizable song. Regions of the state with a scarcity of reports likely had a scarcity of Black-and-white Warblers. Their small nests, hidden beneath vegetation and litter on the ground, are difficult to find. Evidence of probable breeding, except for pairs and territoriality, was difficult to observe. Hence, many blocks in which Black-and-white Warblers were recorded as possible breeders likely contained nest sites.

Distribution

As an upland forest species that breeds in larger forest tracts, Black-and-white Warblers were primarily associated with the Missouri Ozarks. Scattered locations to the north and west may have constituted breeding localities, but there were only two confirmations of breeding north of the Missouri River. The only natural division with essentially no blocks reporting Black-and-white Warblers was the largely unforested Mississippi Lowlands.

Abundance

Abundance surveys verified the association of Black-and-white Warblers with the state's most forested lands. As expected, they were most numerous in the Ozarks.

Phenology

Black-and-white Warblers return from the south in mid-April (Robbins and Easterla 1992). Atlasers observed the earliest fledged young on May 25, indicating that nesting is initiated soon after arrival. The breeding season seemed protracted as birds were observed building nests as late as June 28 and evidence of young in nests was observed as late as July 10. Considering the Black-and-white Warbler's single-broodedness (Harrison 1975), these dates indicate a lengthy breeding season.

Abundance by Natural Division
Average Number of Birds / 100 Stops

0.0   0.0

0.0   0.0

1.8

0.0

Breeding Phenology

EVIDENCE (# of Records)	MAR	APR	MAY	JUN	JUL	AUG	SEP
NB (2)			6/04		6/28		
NE (1)			6/01	6/01			
FY (12)			5/18		7/17		

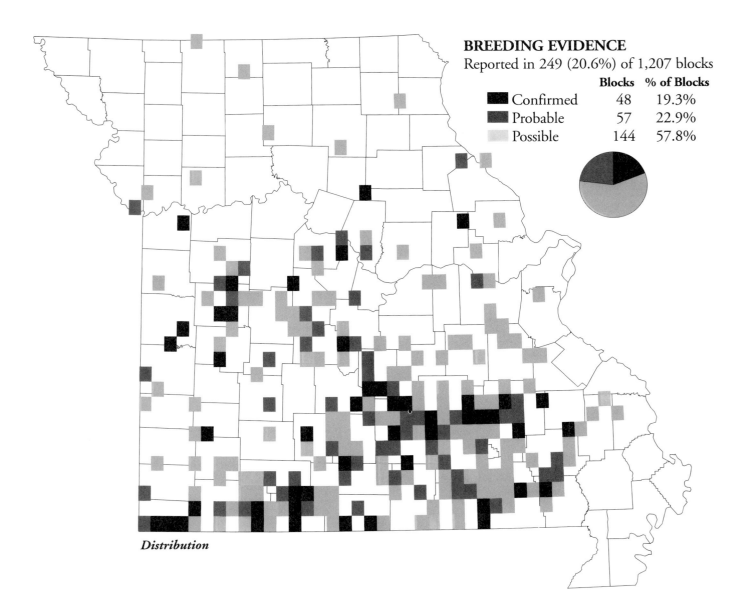

BREEDING EVIDENCE

Reported in 249 (20.6%) of 1,207 blocks

		Blocks	% of Blocks
■	Confirmed	48	19.3%
■	Probable	57	22.9%
■	Possible	144	57.8%

Distribution

Notes

Although Black-and-white Warblers are a known host to Brown-headed Cowbirds (Ehrlich et al. 1988), Atlasers rarely viewed nest contents and obtained no evidence of brood parasitism.

American Redstart
Setophaga ruticilla

Rangewide Distribution: Southern & northern central
  Canada, northwestern to southwestern United States
Abundance: Common
Breeding Habitat: Open mixed woods & forest edge with
  second growth
Nest: Cup of plant fibers, grass & roots, decorated with lichen,
  bark & feathers, lined with fine material in tree or shrub
Eggs: 4 white to off-white with brown marks, usually
  wreathed
Incubation: 12 days
Fledging: 9 days

These secretive, fly-catching warblers are associated with
young, second-growth deciduous woods (Bent 1953).
American Redstarts typically occupy forests where canopies
are sparse and the understory is dense (Peterjohn and Rice
1991). They seem most prevalent on moist hillsides and in
bottomlands in Missouri. Nests are placed up to ten meters
above ground in the fork of a tree or shrub (Harrison 1975).
They are often well-concealed in vegetation (Curson et al.
1994).

Code Frequency

Observing higher evidence than possible breeding was
apparently difficult for Atlasers. Redstarts were classified as
probable breeders in only one-fourth of the blocks in which
they were found, primarily through observation of pairs and
territorial behavior. In only four blocks were nests or nest
building observed. Because their nests are difficult to find,
American Redstarts likely bred in most of the blocks in
which they were recorded as possible and probable breeders.

Distribution

American Redstarts were absent in the northeastern and
southeastern corners of Missouri and were found most often
in the eastern Ozarks. The breeding range extended sparsely
into northwestern and southwestern Missouri. Redstarts
were recorded in a scattered pattern that does not allow con-
clusions regarding the species' true distribution.

Phenology

The observation of a bird on a nest on June 8 confirmed
breeding for American Redstart. Fledged young were discovered on June 24. American Redstarts produce only one
brood a season (Ehrlich et al. 1988).

Notes

Although Harrison (1975) has reported Brown-headed
Cowbird parasitism of American Redstarts, no such evidence was obtained during the Atlas Project

Breeding Phenology

EVIDENCE (# of Records)	MAR	APR	MAY	JUN	JUL	AUG	SEP
NB (1)				6/21	6/21		
FY (3)				6/17	7/04		

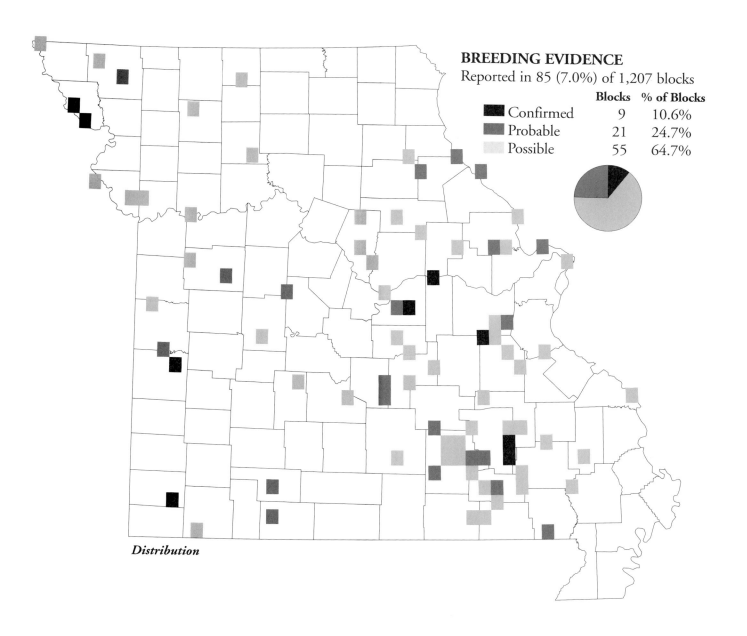

BREEDING EVIDENCE
Reported in 85 (7.0%) of 1,207 blocks

	Blocks	% of Blocks
Confirmed	9	10.6%
Probable	21	24.7%
Possible	55	64.7%

Distribution

Prothonotary Warbler
Protonotaria citrea

Rangewide Distribution: East Coast, Gulf Coast & eastern United States
Abundance: Fairly common & common in parts of south
Breeding Habitat: Snag or hollow tree in forest near water
Nest: Opening filled with moss, leaves, twigs & bark, lined with finer material in natural cavity
Eggs: 4–6 creamy with brown blotches
Incubation: 12–14 days
Fledging: 11 days

A golden-plumaged bird flashing through cypress swamps or along wooded river valleys is likely the "swamp candle" or Prothonotary Warbler. Its unusual name comes from the yellow color of the robes of the College of Prothonotaries Apostolic, a branch of the Roman Catholic Church. This is one of two wood warblers in North America that nests in cavities. The other is Lucy's Warbler from the wooded ravines in the American desert of the southwestern United States.

Code Frequency

About 21 percent of all Prothonotary Warbler records confirmed breeding. It was the fourth most-frequently confirmed of the 19 warblers recorded during the Atlas Project. Forty percent of confirmed records were of birds entering suspected nest cavities. Although it is generally easier to confirm the breeding of cavity nesters, Prothonotary Warbler nest sites are often situated in swamps and lakes and are difficult to approach. This species likely bred in most blocks where it was found.

Distribution

The most obvious distribution map feature is the abundance of blocks clustered in the Osage River basin, especially the lake areas of the Osage Plains Natural Division and the Springfield Plateau Natural Section. Numerous snags with holes excavated by woodpeckers stand in these lakes. The dead snags and cavities probably support a locally-elevated population of this species, making documentation easy. It is considered to be distributed statewide in appropriate forested wetland habitats where cavities for nesting occur.

Abundance

Abundance information was lacking on Miniroutes, probably because roads rarely pass through swamps and lake-flooded forests. The addition of artificial nesting cavities, such as bluebird boxes, may increase the number of local breeding pairs.

Phenology

Most individuals arrive in April, earlier in the south and later in the north. Atlasers' early observation of nests with young indicate that egg laying started by mid-April in some locations. The latest confirmed observations of this species exceeded the safe date by only one day.

Breeding Phenology

EVIDENCE (# of Records)	MAR	APR	MAY	JUN	JUL	AUG	SEP
NE (3)			5/15	6/18			
NY (3)		5/04		6/02			
FY (17)			5/30		7/19		

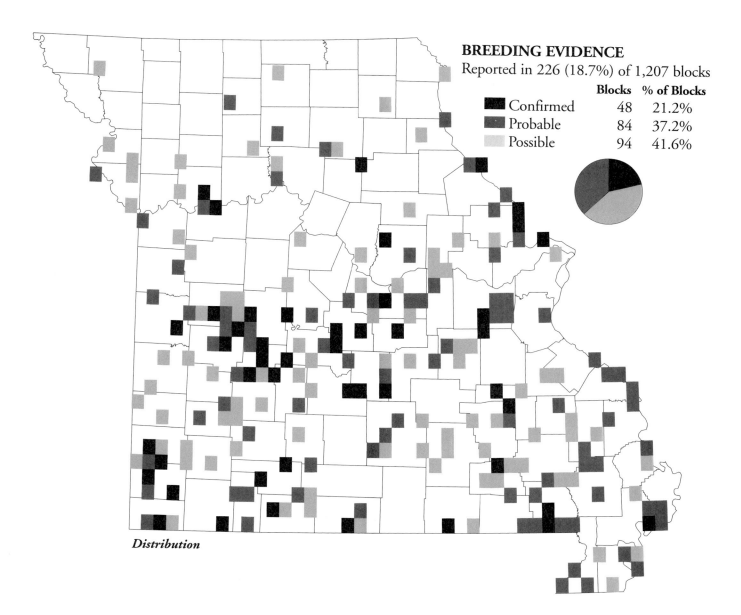

BREEDING EVIDENCE
Reported in 226 (18.7%) of 1,207 blocks

	Blocks	% of Blocks
Confirmed	48	21.2%
Probable	84	37.2%
Possible	94	41.6%

Distribution

Notes

Atlasers recorded only one parasitized nest. Prothonotary Warblers are considered a frequent host of Brown-headed Cowbird eggs and young according to Ehrlich et al. (1988). They report one example of a nest having seven cowbird eggs and no warbler eggs present.

Worm-eating Warbler

Helmitheros vermivorus

Rangewide Distribution: Eastern United States to northeastern Texas through northwestern Florida
Abundance: Common at the center of its range
Breeding Habitat: Brushy ravines, hillsides, dense understory
Nest: Skeletonized leaves, lined with fine material (always including mycelia of hair fungi) in low shrub on the ground
Eggs: 4–5 white with brown spots or blotches, usually wreathed
Incubation: 13 days
Fledging: 10 days

Dense patches of saplings on wooded slopes, especially near flowing water, are locations in which one frequently hears the dry, seedy trills of Worm-eating Warblers. Songs of this warbler are ventriloquistic making them difficult to locate in the dense understory of a forest. Worm-eating Warblers usually frequent large, forested landscapes with extensive forest interiors. According to Ehrlich et al. (1988), this species is very sensitive to forest fragmentation.

Code Frequency

Only 15.2 percent of records confirmed that Worm-eating Warblers breed in Missouri, about average for the 19 species of warblers located. Only one nest was found. Most confirmations were observations of fledging activities and fledglings. Due to the difficulty in confirming breeding, this species likely nested in most blocks where found. However, as a forest interior species, it may have been less successful in more fragmented forests.

Distribution

This species was most often found in the Ozark and Ozark Border natural divisions. The Springfield Plateau contained only 2.3 percent of the records for the Ozark Natural Division reflecting the lack of extensive forest cover in this natural section. The Ozark Border contained 20 percent of all the records probably due to extensive forested bluffs in this natural division. No Worm-eating Warblers were observed in the Mississippi Lowlands where appropriate habitat is nonexistent. Worm-eating Warblers were not reported from most of the northwestern one-third of Missouri.

Abundance

As a forest interior species, they were not well detected by roadside counts. However, relative abundance appears higher in the Ozark Natural Division, especially the White River and Lower Ozark natural sections. These sections have many wooded bluffs and hillsides in a forested landscape.

Phenology

The earliest breeding record was of a pair on May 11. Fledglings were observed between June 6 and July 8. This single-brooded species' breeding season in Missouri is short, with peak arrival in mid-May and most individuals departed by late August (Robbins and Easterla 1992).

Abundance by Natural Division
Average Number of Birds / 100 Stops

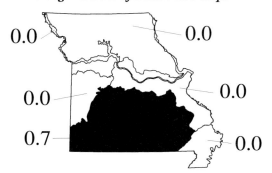

Breeding Phenology

EVIDENCE (# of Records)	MAR	APR	MAY	JUN	JUL	AUG	SEP
NB (1)				6/19	6/19		
NY (1)					7/10 7/10		
FY (15)			6/02			8/03	

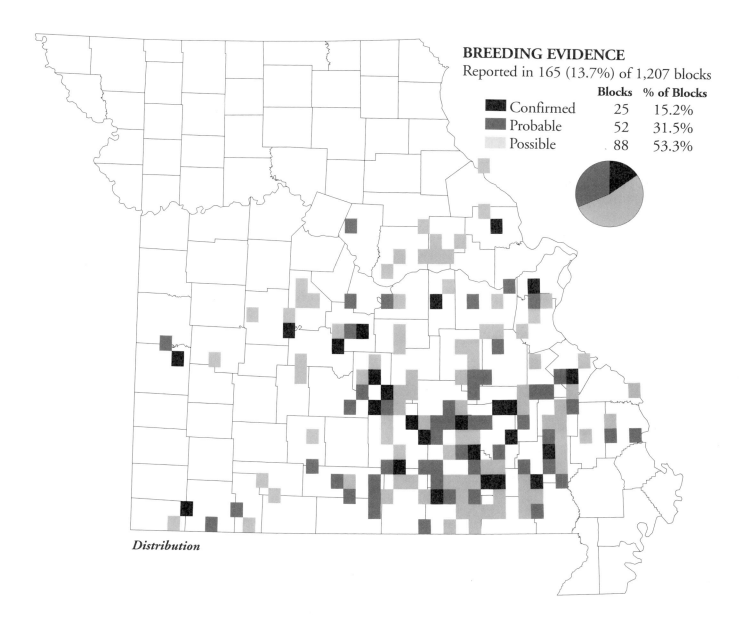

BREEDING EVIDENCE
Reported in 165 (13.7%) of 1,207 blocks

		Blocks	% of Blocks
	Confirmed	25	15.2%
	Probable	52	31.5%
	Possible	88	53.3%

Distribution

Notes

Although this species is considered a very rare Brown-headed Cowbird host (Ehrlich et al. 1988), Atlaser Dan Hatch found an adult feeding a recently-fledged Brown-headed Cowbird on July 6, 1987 in Texas County.

Swainson's Warbler

Limnothlypis swainsonii

> **Rangewide Distribution:** Southeastern United States
> **Abundance:** Uncommon & local
> **Breeding Habitat:** Canebrake, swamps & thickets in
>   moist woods
> **Nest:** Bulky, of leaves, lined with finer material in
>   vines & canes
> **Eggs:** 3 white unmarked, occasionally some faint
>   specks
> **Incubation:** 13–15 days
> **Fledging:** 10–12 days

Swainson's Warblers are listed as Endangered in Missouri. They have never been recorded in large numbers in Missouri and in occasional years have not been detected at all. They are closely associated with stands or "canebrakes" of giant cane *(Arundinaria gigantea)*, within extensively-forested landscapes along stream and river flood plains. The restricted local distribution of cane, and removal of this species' giant cane habitat, have been suggested as causes for low populations (Eddleman 1978). Tower sites kill many birds during migration which could result in a significant loss of individuals (Brown and Dickson 1994).

Code Frequency

The few Swainson's Warblers detected were usually revealed by songs emanating from canebrakes along rivers in the Lower Ozark Natural Section. Finding nests or observing pair activity was time consuming and difficult in dense vegetation. Because of their Endangered status, searches for Swainson's Warblers were conducted in areas in addition to Atlas blocks. Most Swainson's Warbler observations occurred outside Atlas blocks.

Distribution

Although rarely detected in Atlas Project blocks, other breeding sites are known in areas along the upper White River and Black River drainages and in forests outside the levee along the Mississippi River in the Mississippi Lowlands. These warblers may occupy favorable habitat across the southern edge of the state.

Abundance

A 1992–1993 study conducted by a University of Missouri at Columbia graduate student, Brian Thomas, located Swainson's Warblers at 29 sites in giant cane along several rivers, including the Current and Eleven Point (Thomas 1994). Thomas calculated an average of 3.65 eggs per nest with 2.11 fledglings for each of the 17 nests he observed.

Phenology

Swainson's Warblers arrive in late April and depart by mid-September. Most nesting attempts occur from May through July (Terres 1987). No data were available from the Atlas Project.

Notes

Thomas detected no Brown-headed Cowbird brood parasitism. The distribution of this species does not match appropriate cowbird habitat, which contains open land, pas-

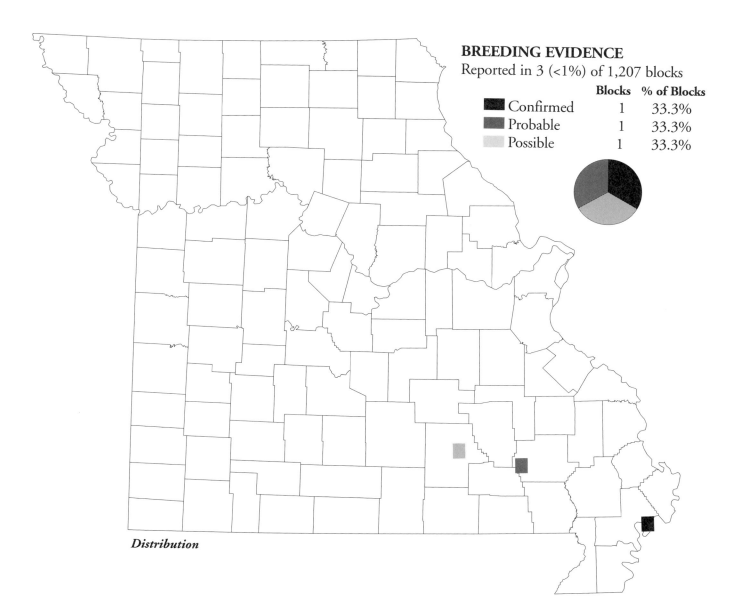

	Blocks	% of Blocks
Confirmed	1	33.3%
Probable	1	33.3%
Possible	1	33.3%

Distribution

tures and cropland for foraging areas. Ehrlich et al. (1988) considered Swainson's Warbler a locally common Brown-headed Cowbird host.

Ovenbird
Seiurus aurocapillus

Rangewide Distribution: Central & eastern Canada, northeastern & north central United States
Abundance: Common; abundant in eastern North America
Breeding Habitat: Large mature deciduous forests, rarely pine
Nest: Of dry grass, leaves & moss, often lined with hair, in the open on the ground
Eggs: 4–5 white with wreathed marks of brown or gray
Incubation: 11–13 days
Fledging: 8–10 days

Its loud ringing song announces the Ovenbird's presence long before it is seen. This forest-interior dwelling bird builds an oven-shaped nest on the ground, hence its name. Ovenbirds may be more successful raising young in larger forests than in isolated forest fragments. Donovan et al. (1995) suggested that long term viability of Ovenbird populations, as well as those of Wood Thrushes and Red-eyed Vireos, depend on the maintenance of heavily-forested landscapes throughout the breeding range. The Donovan et al. study suggested that the forested landscapes of the Ozark Natural Division support a "source" population wherein annual production of young exceed annual deaths. Researchers suggest that the many surrounding islands of

forest depend on Ozark forests to supply birds to support populations in which annual deaths exceed the production of young.

Code Frequency

Because locating the canopied nest of this ground-nesting species is difficult, nests were observed in only eight blocks. Ovenbirds likely nested in many locations where they were found but not confirmed. Eighty-two percent of all records were by sight or song during repeated visits.

Distribution

Seventy-seven percent of all records occurred in the Ozark and Ozark Border natural divisions. The remaining observations were scattered around the state except in extreme northwestern and southwestern areas. Absence of Ovenbirds from the Mississippi Lowlands is expected based on habitat needs. Range limits might explain absence from the most northwestern counties. Missouri is on the very southwestern periphery of the breeding range for this species.

Abundance

Ovenbirds are more abundant farther north in their range, the boreal transition forests of Wisconsin and Minnesota (Donovan et al. 1995).

Phenology

Ovenbirds usually arrive in mid- to late April (Robbins and Easterla 1992). Atlasers observed fledglings from June 1 until August for this sometimes two-brooded species. Occasionally Ovenbirds attempt three broods during outbreaks of spruce budworms in the boreal transition forests of the northern United States and Canada (Ehrlich et al. 1988).

Breeding Phenology

EVIDENCE (# of Records)	MAR	APR	MAY	JUN	JUL	AUG	SEP
NB (1)		5/04	5/04				
NE (2)			6/13		7/07		
NY (1)			5/29	5/29			
FY (3)				6/20	7/04		

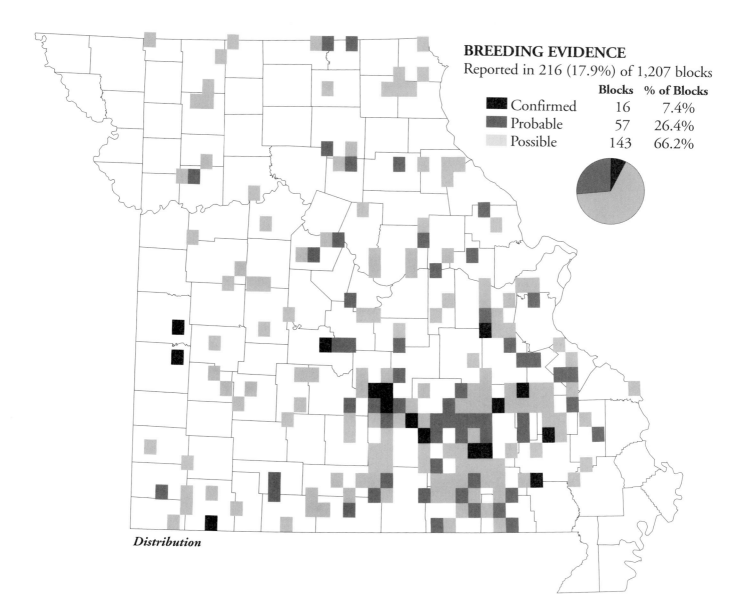

BREEDING EVIDENCE
Reported in 216 (17.9%) of 1,207 blocks

	Blocks	% of Blocks
Confirmed	16	7.4%
Probable	57	26.4%
Possible	143	66.2%

Distribution

Notes

No Ovenbirds were recorded as host to Brown-headed Cowbird young; however, Atlasers observed only three nests and three instances of food being carried to young.

Ovenbirds are considered a frequent host of Brown-headed Cowbirds (Ehrlich et al. 1988). Brood parasitism probably occurs more frequently in fragmented forests containing pasture areas where cowbirds can forage nearby (Thompson 1994). Landscapes like this are probably more common away from the forested Ozarks.

Louisiana Waterthrush

Seiurus motacilla

Rangewide Distribution: Eastern United States from Nebraska to eastern Texas to coast

Abundance: Uncommon but breeding in much of eastern United States

Breeding Habitat: Woodlands & forests bordering streams

Nest: Leaves, moss, twigs & bark, lined with fine material along bank near roots

Eggs: 5 white or creamy, with brown or purplish-gray blotches

Incubation: 13 days

Fledging: 10 days

Louisiana Waterthrushes head-bob and teeter along forested streams, their principal habitat. They flip leaves, creep along the shoreline under roots and jump around rocks in their search for invertebrates. Their loud ringing calls echo along Ozarks streams and their ventriloquistic song is difficult to trace to its source. Nests are usually well concealed next to a rock, exposed roots or leaves on a stream bank.

Code Frequency

The limited number of confirmed records was likely due both to the difficulty of traversing the area where nests are found and to the wariness of Louisiana Waterthrushes near a nest site. Watching a pair near a suspected nest site is usually more productive than actively searching along a stream bank. Eighty percent of confirmed breeding evidence was adults carrying food to young or fledglings.

Distribution

Eighty-one percent of blocks in which Louisiana Waterthrushes were reported occurred south of Latitude 38°30'00" (Jefferson City). Eighty-four percent of blocks occurred in the Ozark and Ozark Border natural divisions. The distribution is closely associated with forested riparian areas along streams. About 10 percent of the blocks are from the Glaciated Plains and 5 percent from the Osage Plains.

Abundance

Although poorly detected by road counts, Louisiana Waterthrushes did appear on a few Ozark, Ozark Border and Mississippi Lowland abundance counts. However, surveys that parallel rivers or streams are greatly needed to accurately assess relative abundance of most riparian species.

Phenology

One of the earliest warblers to arrive, Louisiana Waterthrushes appear by mid-March (Robbins and Easterla 1992). Atlas observations indicate nesting activities had commenced by early April. Perhaps better indicators of nesting were the numerous observations of food being carried to young and a fledgling observed Aug. 2. Most Louisiana Waterthrushes have left the state by early September (Robbins and Easterla 1992).

Abundance by Natural Division
Average Number of Birds / 100 Stops

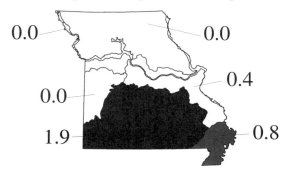

0.0   0.0

0.4

0.0

1.9   0.8

Breeding Phenology

EVIDENCE (# of Records)	MAR	APR	MAY	JUN	JUL	AUG	SEP
NB (1)				7/10	7/10		
NE (3)		4/27		6/13			
NY (4)			5/25	6/08			
FY (42)			5/08		7/26		

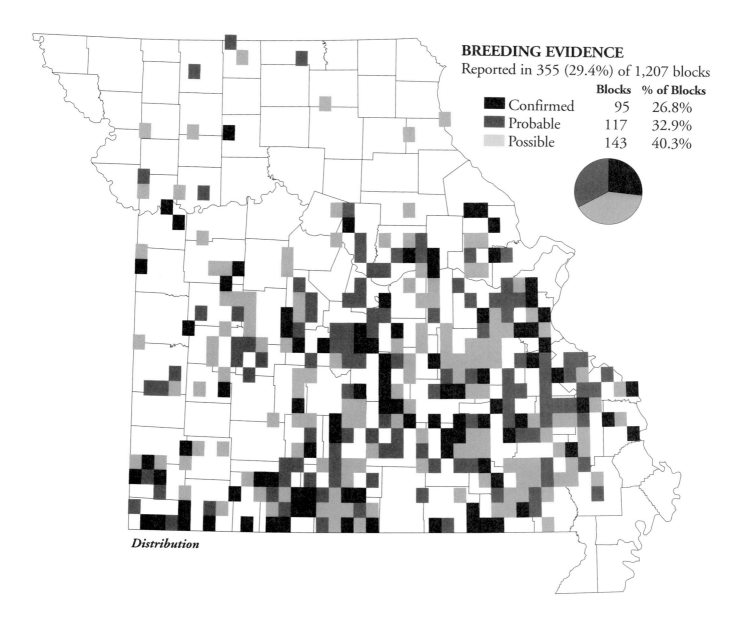

BREEDING EVIDENCE
Reported in 355 (29.4%) of 1,207 blocks

	Blocks	% of Blocks
Confirmed	95	26.8%
Probable	117	32.9%
Possible	143	40.3%

Distribution

Notes

Atlasers recorded seven instances of Brown-headed Cowbird parasitism in Louisiana Waterthrushes. Ehrlich et al. (1988) listed them as common cowbird hosts.

Kentucky Warbler
Oporornis formosus

Rangewide Distribution: Southeastern United States except Florida
Abundance: Common in dense woodlands
Breeding Habitat: Wooded thickets, dense tangled vegetation
Nest: Leaves, weed stems & grass, lined with rootlets at base of bush, on or near ground
Eggs: 4–5 white to creamy, marked with brown spots or blotches
Incubation: 12–13 days
Fledging: 8–10 days

Most people identify this bird of damp forest understory by its loud "churry churry churry" song rather than by its striking black face pattern and bright yellow plumage. Landscapes with continuous heavy forest cover are the best places to find these difficult-to-see warblers. However, many are found in smaller tracts of woods where damp understory shrubs are present.

Code Frequency

This vocal but secretive species was recorded in 42 percent of Atlas blocks. Because of dense undergrowth, it was confirmed in only 12 percent of blocks. Only 1 percent of records were observations of nests. Observations of food to young and fledglings, at 4 percent and 5 percent, respectively, were the most-frequently used breeding confirmations. Auditory observations likely accounted for nearly one-fourth of all records.

Distribution

About 76 percent of the records for Kentucky Warblers were from the Ozark and Ozark Border natural divisions. Seventeen percent of these confirmed breeding in these natural divisions, which include extensive forest cover with damp understory near streams and rivers. Kentucky Warblers were seldom found in small woodlots in the Glaciated and Osage Plains natural divisions. The Glaciated Plains Natural Division contained about 14 percent of the records (71), with only 7 percent of these (5) confirming breeding.

Abundance

Over eight times as many birds per 100 stops were recorded in the Ozark Natural Division as in the Glaciated Plains. The Ozark Natural Division also recorded about two to six times more than the Osage Plains, Ozark Border and Mississippi Lowlands natural divisions. The expansive, continuous-canopy forest of the Ozark Natural Division, with its flowing streams and moist conditions, may be important to sustaining large populations of this species.

Phenology

This species arrives in mid to late April (Robbins and Easterla 1992). Atlasers logged the first possible codes the first week of May. Most confirmation codes were recorded

Abundance by Natural Division
Average Number of Birds / 100 Stops

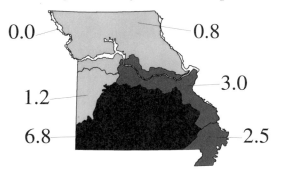

0.0   0.8

1.2   3.0

6.8   2.5

Breeding Phenology

EVIDENCE (# of Records)	MAR	APR	MAY	JUN	JUL	AUG	SEP
NB (3)			6/07		6/21		
NE (4)		5/24			7/06		
NY (3)			6/02		6/24		
FY (22)			6/01		7/24		

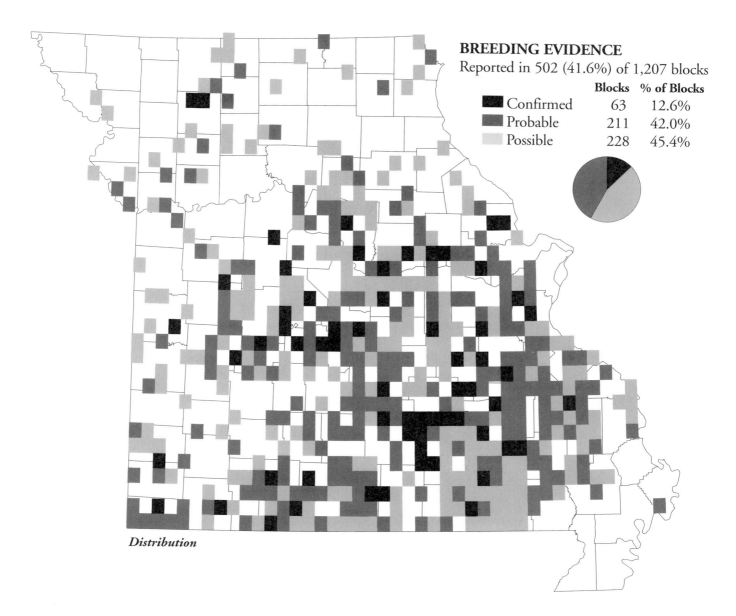

Distribution

between May 21 and June 15 for this single-brooded species, although nests and eggs were reported later. Pairs and territoriality were last observed on July 23 and 26 respectively, slightly before Atlasers reported the last fledglings on August 2.

Notes

Brown-headed Cowbird parasitism was recorded in two blocks, both in the Lake of the Ozarks area. The dense understory and scarcity of available cowbird habitat in that forested region likely accounted for Kentucky Warblers being infrequent Brown-headed Cowbird hosts. Ehrlich et al. (1988) considered this species a frequent host.

Common Yellowthroat
Geothlypis trichas

Rangewide Distribution: Southern Canada, southeastern Alaska & entire United States to southern Mexico
Abundance: Abundant & wide ranging
Breeding Habitat: Meadows, thickets & forest edges near water & marsh
Nest: Cup of bulky loose weeds, grass, bark & ferns, lined with fine material in bush
Eggs: 3–5 white to creamy, with brown or black marks, occasionally wreathed
Incubation: 12 days
Fledging: 10 days

Common Yellowthroats nest in early successional habitats, thickets and brushy fields. Yellowthroats especially associate with low-lying damp areas, including marshes and brushy draws. Best perceived by their distinctive "wichity-wichity-wich" song, the dramatically-marked males are usually easy to see when they sing from open perches. The drab olive females are much less noticeable. While pairs and nests are common, they are often in dense, brushy cover and are difficult to see.

Code Frequency

The detection of singing males resulted in most of the possible breeding records. Likewise, many of the probable breeding records resulted from observations of multiple singing males. Of the 1,105 blocks in which Common Yellowthroats were recorded, nests with eggs or young were observed in only four. Because of the difficulty in surveying the Common Yellowthroat's brushy habitat, and because the nests are well concealed (Harrison 1975) Common Yellowthroats likely nested in nearly all blocks in which they were recorded.

Distribution

The Atlas Project documented a solid statewide distribution for this species. Because no obvious geographical pattern is detectable from the map, Common Yellowthroats likely bred, and were missed, in those blocks where they were not recorded.

Abundance

Atlasers discovered Common Yellowthroats to be abundant in Missouri. It was one of few species that was relatively abundant in the Mississippi Lowlands Natural Division. It was the least numerous on routes through the more-forested regions of southern Missouri.

Phenology

Common Yellowthroats arrive between mid-April in the south and toward the end of April in the North (Robbins and Easterla 1992). Common Yellowthroats attempt two broods in a season (Harrison 1975). The late dates for nest building, nest with eggs, and fledglings (August 16) may be attributable to second broods.

Abundance by Natural Division
Average Number of Birds / 100 Stops

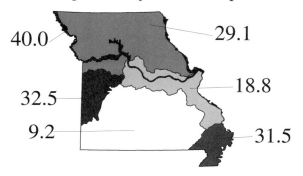

40.0   29.1

32.5

9.2   18.8

31.5

Breeding Phenology

EVIDENCE (# of Records)	MAR	APR	MAY	JUN	JUL	AUG	SEP
NB (15)		5/10			7/22		
NE (3)		5/16			7/15		
NY (1)				7/02	7/02		
FY (54)			6/01			8/29	

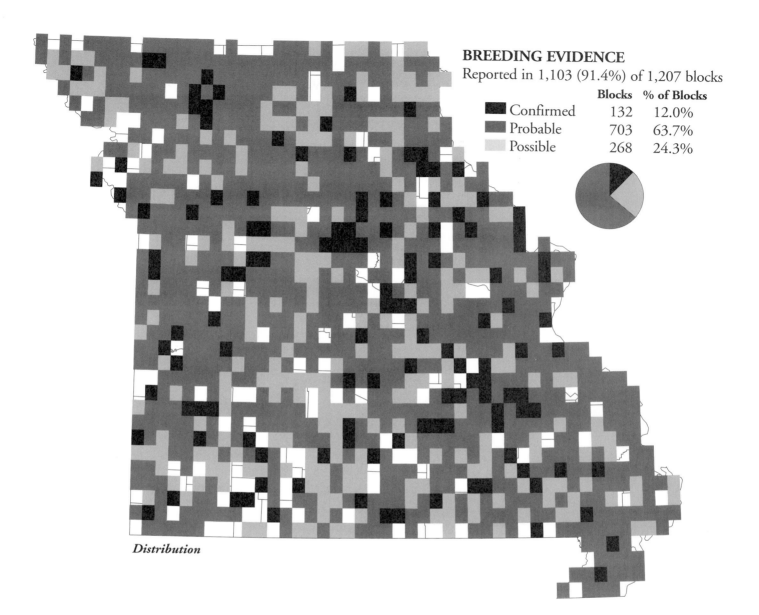

BREEDING EVIDENCE
Reported in 1,103 (91.4%) of 1,207 blocks

	Blocks	% of Blocks
Confirmed	132	12.0%
Probable	703	63.7%
Possible	268	24.3%

Distribution

Notes

Brown-headed Cowbirds frequently parasitize Common Yellowthroats (Ehrlich et al. 1988). The Atlas Project recorded seven instances of brood parasitism in four observations of nest contents and 54 observations of food being delivered to young.

Hooded Warbler

Wilsonia citrina

Rangewide Distribution: Southeastern Canada & eastern United States
Abundance: Fairly common
Breeding Habitat: Lowland woods with swampy brushy areas
Nest: Base of leaves, then bark & plant fiber, lined with finer material in shrub
Eggs: 3–4 creamy white with brown marks, occasionally wreathed
Incubation: 12 days
Fledging: 8–9 days

Hooded Warblers occupy and nest in understories of mature forests. They prefer moist deciduous communities (Ogden and Stutchbury 1994). Although often found in forested riparian zones, they avoid narrow corridors and are most associated with large forest tracts (Robbins and Easterla 1992).

Code Frequency

Most Hooded Warblers associate with dense forests and, therefore, are best detected by sound. Typically, they are glimpsed only briefly as they move among the understory vegetation. Atlasers had little success confirming breeding.

No nests were located, and in 79 percent of the blocks where they were found, birds were recorded as possible breeders. Breeding likely occurred undetected in most blocks in which the species was recorded, especially within the species' Ozark range. It is possible that some individuals sighted near the northwestern edge of their range were unmated, unsuccessful breeders.

Distribution

The Lower Ozarks represented the heart of the Hooded Warbler's range in Missouri, especially Reynolds and Shannon counties south to the Arkansas line. The range extended north to St. Francois and Washington counties. Unmated individuals may have been recorded in the outlying locations in Cole, Dallas and Ray counties.

Notes

Brown-headed Cowbirds routinely parasitize Hooded Warblers (Ogden and Stutchbury 1994). However, this phenomenon was not recorded during the Atlas Project due to lack of nest observations.

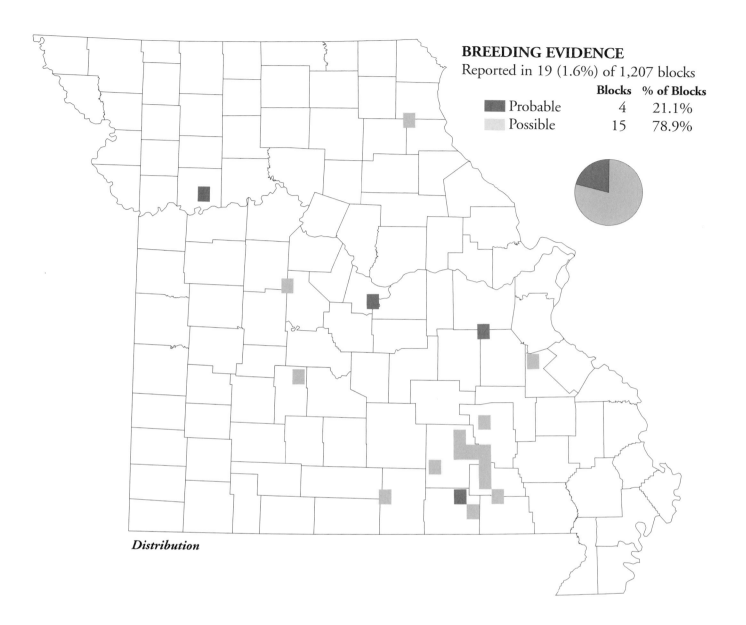

Yellow-breasted Chat
Icteria virens

Rangewide Distribution: Southern British Columbia, across southern Canada & United States
Abundance: Fairly common to common
Breeding Habitat: Brushy areas, thickets, vines & dense vegetation
Nest: Large cup of leaves, straw, weeds & woven bark, lined with fine weed stems & grass in shrub
Eggs: 3–4 white to creamy with brown marks usually at large end
Incubation: 11 days
Fledging: 8 days

Their large size and raucous calls make Yellow-breasted Chats the most unusual wood warblers in Missouri. They act and sound more like Northern Mockingbirds than warblers (Parkes 1964). Sometimes they hide among tangles of vines and shrubs only to burst forth in song, wings popping, flopping awkwardly and tail hanging limp.

Code Frequency

Eighty-three percent of the records were based on sight and sound rather than nest or fledgling observations. This is usual for a thicket-nesting species. In only 8.9 percent of blocks was breeding confirmed. Observation of food being carried to young and fledglings made up 66 percent of the confirmed records.

Distribution

Chats were found statewide with the Ozark and Ozark Border natural divisions containing the highest percentage of blocks in which this species was recorded. Recently cut-over areas in the Ozarks, usually more than four or five acres, make good chat habitat. Blocks with records gradually diminished northward and few observations were made in counties along the Iowa border. In the Mississippi Lowlands Natural Division thickets along flood ways and ditches were the only habitats available.

Abundance

The Ozark Natural Division had the greatest abundance of chat habitat in Missouri. The extensive shrub and sapling cover in the Ozark and Ozark Border natural divisions provides more quality habitat for this species than the Osage Plains and Glaciated Plains.

Phenology

Six July and August observations of food to young or a nest with young may have been second broods. Most chats depart Missouri by late August and early September (Robbins and Easterla 1992).

Notes

Ehrlich et al. (1988) consider Yellow-breasted Chats frequent Brown-headed Cowbird hosts. The Atlas Project recorded two instances of brood parasitism.

Abundance by Natural Division
Average Number of Birds / 100 Stops

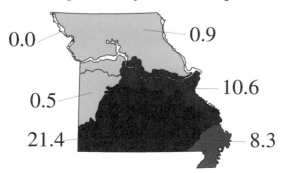

0.0   0.9
0.5   10.6
21.4   8.3

Breeding Phenology

EVIDENCE (# of Records)	MAR	APR	MAY	JUN	JUL	AUG	SEP
NB (7)			5/29	6/14			
NE (3)			6/05		7/06		
NY (4)		5/12			7/23		
FY (30)			5/27				8/25

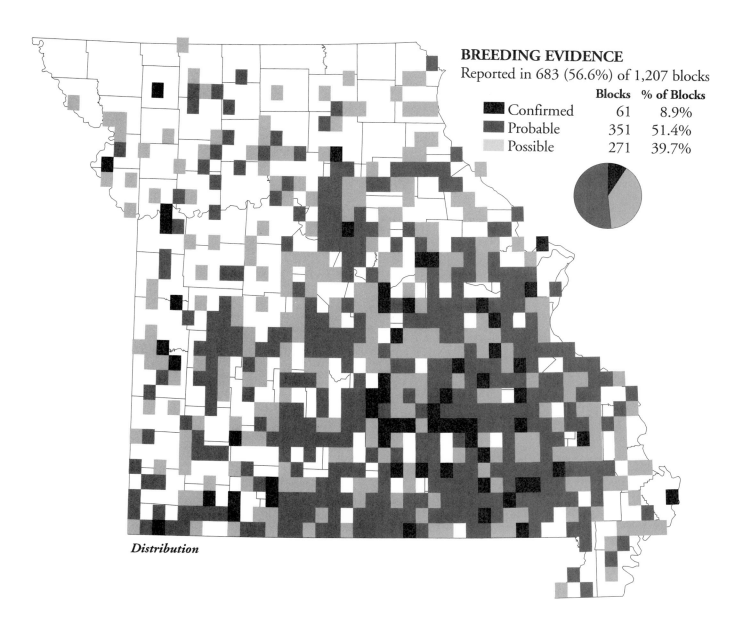

Distribution

Summer Tanager
Piranga rubra

> **Rangewide Distribution:** Southwestern United States through southeastern & east central United States
> **Abundance:** Common in pine-oak forests in the East & cottonwoods in the West
> **Breeding habitat:** Deciduous trees, pine-oak associations & riparian areas
> **Nest:** Loosely built of grass, feathers & moss, lined with fine grass, on horizontal branch of tree
> **Eggs:** 4 pale blue or pale green, marked with browns, occasionally wreathed
> **Incubation:** 12 days
> **Fledging:** 10 days

Like Northern Cardinals, Summer Tanagers are one of Missouri's red birds. Their secretive nature makes them hard to locate in the green leaves of forests and woodlots. However the "chicky-tucky-tuck" or "pity-tucky-tucky" calls and their robin-like song reveal their presence in the forest canopy. Male plumage emerges in spring as a blotchy green and red, which offers an identification challenge to the beginner.

Code Frequency

Summer Tanagers were confirmed to breed in only 13 percent of blocks. This likely resulted from the difficulty of seeing this species in trees after it has been heard. Only 20 percent of confirmations were based on observations of nests or carrying food to young. In southern Missouri, where this species is common, Atlasers observed much evidence of probable breeding. In the Glaciated Plains where this species is scarce and forested habitats are less abundant, possible codes were most frequent.

Distribution

The Summer Tanager was found throughout the state. In only six counties were no records obtained for this species. Observations were more frequent in the Ozark, Ozark Border and Osage Plains natural divisions. Observations were scattered throughout the Glaciated Plains Natural Division. No observations were recorded from the Mississippi Lowlands except in the forested areas along the Mississippi River and Crowley's Ridge Natural Section.

Abundance

The highest relative abundance was in the Ozark Natural Division which had nearly twice as many records as the Ozark Border Natural Division. As expected, Summer Tanager abundance was highest in regions of extensive forests. In the Mississippi Lowlands the number and location of routes precluded accurate sampling.

Phenology

The period during which eggs and nests were observed matches the period mentioned in Terres (1987) except that Atlasers documented nests and young over a longer period. Most food to young observations occurred between May 31 and July 31, with outlying dates of August 19 and 30.

Abundance by Natural Division
Average Number of Birds / 100 Stops

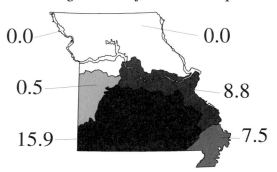

0.0   0.0

0.5   8.8

15.9   7.5

Breeding Phenology

EVIDENCE (# of Records)	MAR	APR	MAY	JUN	JUL	AUG	SEP
NB (12)			5/09		7/05		
NE (6)			5/20	6/11			
NY (12)				6/05		8/01	
FY (28)			5/31				8/30

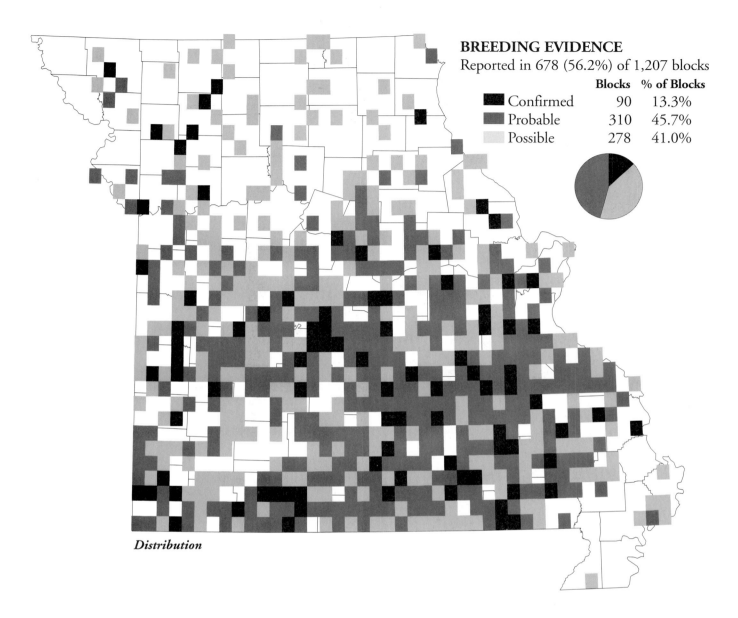

BREEDING EVIDENCE
Reported in 678 (56.2%) of 1,207 blocks

	Blocks	% of Blocks
Confirmed	90	13.3%
Probable	310	45.7%
Possible	278	41.0%

Distribution

Fledgling observation dates ranged from June 15 to August 19 with most found between June 15–26. Other FL records occurred between July 13 and 16, and August 19. The late observations indicate a second brood or renesting.

Notes

Terres (1987) documented only 18 occurences of Brown-headed Cowbird parasitism in Summer Tanagers. The Atlas Project documented two instances of brood parasitism despite few observation of nests.

Scarlet Tanager
Piranga olivacea

Rangewide Distribution: Southeastern Canadian border, northeastern to east central United States
Abundance: Common
Breeding Habitat: Large deciduous, coniferous trees in woods & parks
Nest: Loosely saucer shaped, of grass, roots, forbs & twigs, lined with finer material, in tree
Eggs: 4 bluish or greenish with brown marks, often wreathed
Incubation: 13–14 days
Fledging: 9–11 days

Few birds match the glamour and brilliance of Scarlet Tanagers and yet hide so easily in the forest canopy. Their burry robin-like song can be heard from the forested ridges in the Missouri and Arkansas Ozarks, which lie at the southwest corner of the Scarlet Tanager's North American breeding range. Scarlet and Summer tanagers' songs are sometimes difficult to distinguish. Listening to the songs one after the other points out obvious tonal differences.

Code Frequency

Observations of Scarlet Tanagers by songs, call notes and sight records accounted for 60 percent of all records ob-

tained for this species. In only four blocks did actual nest searches reveal eggs or young. Because nests were difficult to locate, Scarlet Tanagers likely nested in most blocks where they were found.

Distribution

Scarlet Tanagers were distributed statewide in forested areas, but were concentrated in the Ozark and Ozark Border natural divisions. Historically, their distribution was mainly in the prairie and Ozark border regions plus the valleys of the Ozarks (Widmann 1907). The presence of scattered wood lots in northern Missouri and larger stretches of riparian forests explain the thinly-scattered population in the Glaciated Plains Natural Division. This species is also scarce in the Mississippi Lowlands as noted by Widmann (1907), although perhaps for different reasons.

Abundance

The center of abundance for this species is in the Ozark and Ozark Border natural divisions where most of this state's large forested tracts occur. An extremely high number on a route in the Mississippi Lowlands may have misrepresented this species' abundance in that natural division.

Phenology

The late observation of nest building is likely a renesting for this single-brooded species. Nests with eggs and young were observed around the end of May until early July, with fledglings observed until August 19.

Notes

Atlasers reported no Scarlet Tanager nests parasitized by Brown-headed Cowbirds although Ehrlich et al.(1988) state that they are they commonly parasitized.

Abundance by Natural Division
Average Number of Birds / 100 Stops

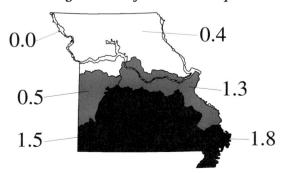

0.0   0.4

0.5   1.3

1.5   1.8

Breeding Phenology

EVIDENCE (# of Records)	MAR	APR	MAY	JUN	JUL	AUG	SEP
NB (5)			5/11		6/28		
NE (1)			5/31	5/31			
NY (83			6/04	6/18			
FY (6)				6/11	7/13		

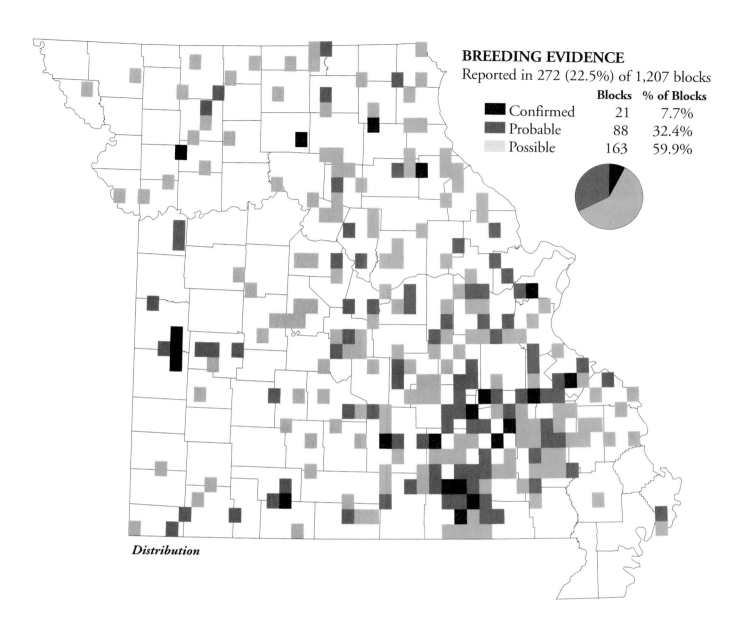

BREEDING EVIDENCE
Reported in 272 (22.5%) of 1,207 blocks

	Blocks	% of Blocks
Confirmed	21	7.7%
Probable	88	32.4%
Possible	163	59.9%

Distribution

Northern Cardinal
Cardinalis cardinalis

Atlasers used all but the physiological evidence code to record cardinals in blocks. Twenty-two percent of all observations recorded nests with eggs or young. Fifty-one percent of all records were confirmed, with 61 percent of these due to observations of fledglings or fledglings being fed.

Distribution

Northern Cardinals were distributed statewide. No regional difference in distribution was perceived, suggesting nesting habitat was available throughout the state.

Abundance

Cardinals were most abundant in the Ozark Border Natural Division as stated by Robbins and Easterla (1992). This is due to the abundance of appropriate habitat in that region. In the Glaciated Plains and Mississippi Lowlands natural divisions, this species was common, occurring at about 50 percent of stops.

Phenology

Courtship behavior and territoriality codes were first recorded when safe dates allowed, although these behaviors began in early to mid-February. Northern Cardinals obviously begin nesting early. Although Northern Cardinals usually produce two broods, Ehrlich et al. (1988) reported a potential for three and four broods.

Notes

With 40 reports of brood parasitism, Northern Cardinals were the most-frequently reported host of Brown-headed Cowbirds during the Atlas Project. Of 267 records for cardinals of food carried to young, nest and eggs, and nest with young, 15 percent involved Brown-headed Cowbirds. Ehrlich et al. (1988) stated that Northern Cardinals are common cowbird hosts, especially in the central portion of their range.

Rangewide Distribution: Eastern United States & expanding to southwestern United States
Abundance: Abundant throughout the East
Breeding Habitat: Thickets, dense shrubs & undergrowth
Nest: Stems, twigs, bark, grass & paper, lined with fine grass & hair, in sapling
Eggs: 3–4 grayish, bluish or greenish white; marked with brown, gray or purple
Incubation: 12–13 days
Fledging: 9–10 days

Of all the birds that nest in Missouri, Northern Cardinals may be the best known. They can be found in nearly every hedge, thicket or berry patch during the summer whether in rural areas, towns or suburbs. They sing from early February through August. Males whistle from the tops of saplings, and often, big trees.

Code Frequency

Northern Cardinals were observed in 1,200 blocks, more than any other species. Easy to find, observe and to confirm,

Abundance by Natural Division
Average Number of Birds / 100 Stops

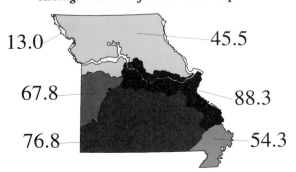

13.0   45.5

67.8   88.3

76.8   54.3

Breeding Phenology

EVIDENCE (# of Records)	MAR	APR	MAY	JUN	JUL	AUG	SEP
NB (33)		4/20				7/20	
NE (78)		4/09				8/02	
NY (56)		4/12					8/25
FY (137)		4/22					8/25

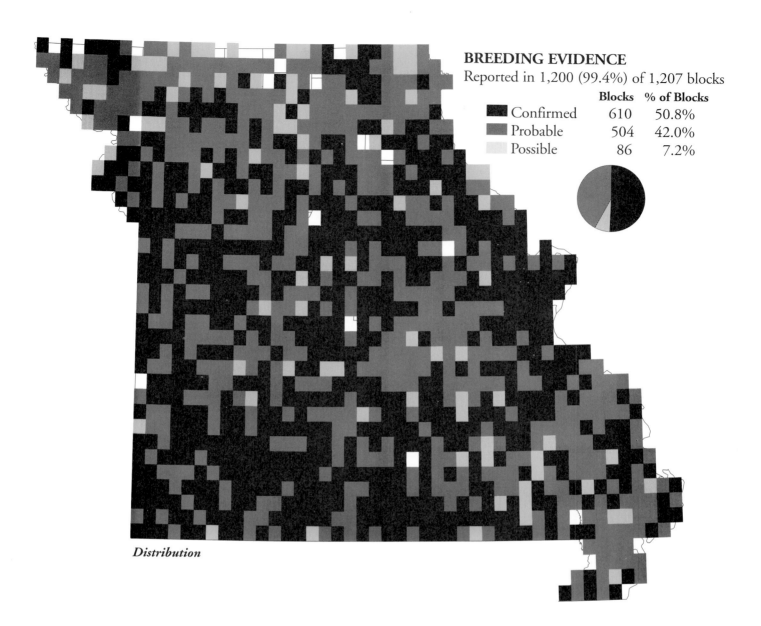

BREEDING EVIDENCE
Reported in 1,200 (99.4%) of 1,207 blocks

		Blocks	% of Blocks
■	Confirmed	610	50.8%
■	Probable	504	42.0%
▢	Possible	86	7.2%

Distribution

Rose-breasted Grosbeak

Pheucticus ludovicianus

Rangewide Distribution: Southern Canada & north-eastern United States
Abundance: Common in open woods
Breeding Habitat: Dense trees in second-growth woods & thickets
Nest: Twigs & coarse plant material, lined with roots, hair & fine twigs in tree or shrub
Eggs: 4 pale green, blue or bluish-green, brown or purple marks
Incubation: 13–14 days
Fledging: 9–12 days

The sweet, liquid, robin-like songs of Rose-breasted Grosbeaks are easily confused with several species. However, the black, white and rose colors of the males are unmistakable. Although the bright coloration would seem to stand out, these birds are often difficult to observe when singing from within foliage high in a tree. This species is associated with relatively open deciduous woods in both floodplains and uplands. It can be found in small woodland patches and forested corridors, both in rural areas and in well-forested urban neighborhoods.

Code Frequency

Rose-breasted Grosbeaks were most easily discovered by song. Fledglings were more frequently reported than any other breeding code and parents were easily observed feeding young.

Distribution

Eighty-eight percent of the records occurred north of the 38°30'00" parallel of latitude, roughly the northern two-thirds of the state. Most observations south of this latitude were recorded in the possible category and may represent migrants or unmated pairs rather than breeding individuals. There were sparse breeding records in both the eastern and western edges of the Osage Plains and Ozark Border natural divisions. The summer range of this species does not extend into Arkansas (James and Neal 1986).

Abundance

The abundance of this species peaks along the northern border of Missouri. Both the abundance and the distribution maps depict the southern periphery of the breeding range.

Phenology

Most nesting activity begins in mid- to late May. Second broods may have occurred, as suggested by nest and young observed from May 16 to July 10. Spring arrival begins in mid-April with departure by early October (Robbins and Easterla 1992).

Notes

Atlasers documented one instance of brood parasitism by Brown-headed Cowbirds. Cowbirds are common in the Glaciated Plains Natural Division so brood parasitism might be much higher than Atlas Project data indicated. Ehrlich et al. (1988) listed this species as a common host of cowbirds.

Abundance by Natural Division
Average Number of Birds / 100 Stops

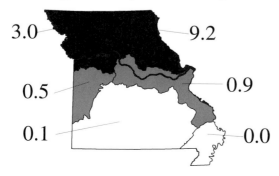

3.0   9.2
0.5   0.9
0.1   0.0

Breeding Phenology

EVIDENCE (# of Records)	MAR	APR	MAY	JUN	JUL	AUG	SEP
NB (11)			5/13		7/01		
NE (4)			5/26	6/22			
NY (3)				6/06	7/10		
FY (19)			5/16		7/20		

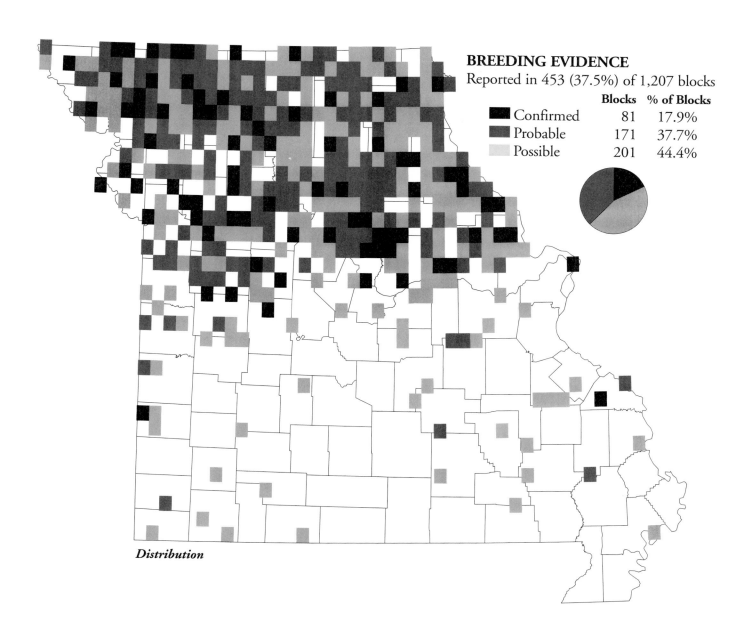

Reported in 453 (37.5%) of 1,207 blocks

	Blocks	% of Blocks
Confirmed	81	17.9%
Probable	171	37.7%
Possible	201	44.4%

Distribution

Blue Grosbeak
Guiraca caerulea

> **Rangewide Distribution:** Central & southern United States to central Middle America
> **Abundance:** Fairly common
> **Breeding Habitat:** Brushy areas, open woods, thickets & old fields
> **Nest:** Twigs, roots & bark, lined with finer material, inter-woven with snakeskin, leaves & paper, in tree or vine
> **Eggs:** 4 pale bluish-white & unmarked
> **Incubation:** 11–12 days
> **Fledging:** 9–10 days

Blue Grosbeaks typically associate with shrubby thickets interspersed with grassy fields, fence rows, roadsides and forest edges. Nests are usually low in small trees, shrubs, or tangles of vines, briars, and other vegetation and often near open areas or roads (Ingold 1993). Although Blue Grosbeaks are generally scarce (Ingold 1993), evidence suggests they are increasing in Missouri. The Missouri Breeding Bird Survey indicated a 1.2 percent average annual increase from 1967 through 1989 (Wilson 1990).

Code Frequency

Blue Grosbeaks are easy to locate wherever they occur because they tend to select open perches and their vocalizations are easily detected. Therefore, where Blue Grosbeaks were not recorded, they likely were not present. Because Blue Grosbeaks nest in brambles and other difficult-to-reach habitats, their nests and other confirming breeding evidence are difficult to observe. Therefore, it is likely that many probable breeding blocks contained nest sites. Most breeding was confirmed by the observation of fledglings or food for young. Of the 694 blocks where Blue Grosbeaks were found, nests with eggs or young were observed in only three.

Distribution

Blue Grosbeaks were widely distributed in southern Missouri, occurring in nearly 100 percent of the blocks near Missouri's southern border, except in the Mississippi Lowlands, where they are rare. In contrast, they were nearly 100 percent absent near the northern border of Missouri, which is near the northern edge of the Blue Grosbeak's breeding range.

Abundance

Blue Grosbeaks were most abundant in south central Missouri and portions of the southwestern corner of the state, which is the center of their southern Missouri range. These regions contain the mix of grasslands, brush and trees characteristic of Blue Grosbeak habitat.

Phenology

An accurate picture of the Blue Grosbeak's nesting phenology is afforded by 122 blocks in which breeding was confirmed. Second broods are common in the southern part of the breeding range (Ingold 1993), which may account for the late observations of nest building and a nest with young. In Oklahoma, Sutton (1967) found a nest with eggs on August 6 and recently fledged young on August 28.

Abundance by Natural Division
Average Number of Birds / 100 Stops

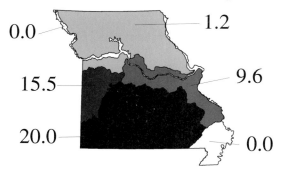

0.0   1.2

15.5   9.6

20.0   0.0

Breeding Phenology

EVIDENCE (# of Records)	MAR	APR	MAY	JUN	JUL	AUG	SEP
NB (20)			5/17		7/04		
NE (1)				7/04	7/04		
NY (2)				6/06		7/27	
FY (47)			5/22				8/26

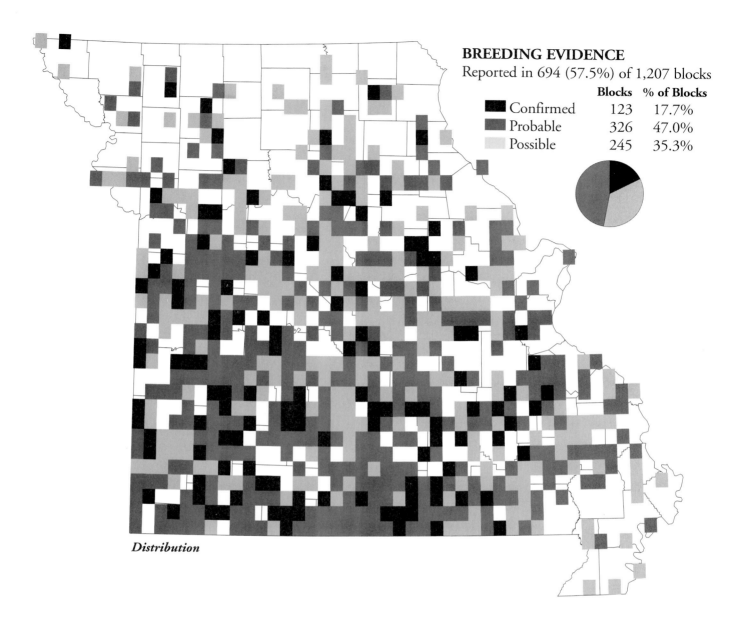

BREEDING EVIDENCE
Reported in 694 (57.5%) of 1,207 blocks

	Blocks	% of Blocks
Confirmed	123	17.7%
Probable	326	47.0%
Possible	245	35.3%

Distribution

Notes

Of the 45 blocks where Atlasers observed nest contents or fledglings, four instances of Brown-headed Cowbird parasitism were recorded. Ingold (1993) describes Blue Grosbeaks as a heavily-parasitized species.

Indigo Bunting
Passerina cyanea

> **Rangewide Distribution:** Southern Canada eastern
>   United States, some in southwestern United States
> **Abundance:** Common, but local & uncommon in the
>   southwest
> **Breeding Habitat:** Brushy vegetation, saplings &
>   weeds
> **Nest:** Grasses, leaves & weed stems, with finer lining,
>   in tree or tangles
> **Eggs:** 3–4 bluish to white eggs, usually unmarked
> **Incubation:** 12–13 days
> **Fledging:** 9–10 days

The "blue bird" of Missouri is a frequent misnomer
placed on this small, turquoise-colored, seed-eating bird.
Found throughout the state, its simple descending couplet
song is easily recognized. Indigo Buntings are most easily
observed along country roads as they fly to tree branches or
wires. Sometimes during their first spring, males have a
blotched blue and brown plumage, which is quite different
from the iridescent gem-like appearance of breeding males
when sunlight strikes their feathers.

Code Frequency

Atlasers recorded Indigo Buntings in all but 14 blocks.
All codes except UN were used to document breeding
behavior. Observations of fledglings (168) and food for
young (98) confirmed 22 percent of all breeding records.
Only 4.5 percent of all records were actual nest observations,
perhaps because nests were difficult to find in thickets, tangles and weed patches.

Distribution

Distributed nearly statewide, Indigo Buntings were
found in nearly all habitats. This species is common and
widespread and was the second most-widely distributed
species recorded during the Atlas Project. Indigo Buntings
were well distributed throughout the Ozark Natural
Division. This species has benefitted from shrubby habitat
created since 1900 by logging and old field succession
(Ehrlich et al. 1988).

Abundance

This species reached its highest relative abundance in the
Ozark, Ozark Border and Mississippi Lowlands natural divisions where it was two to three times more abundant than
in the Glaciated Plains. The profusion of edges between
forests and pastures in the southern and eastern parts of the
state probably offers more nesting and foraging habitat. The
Osage Plains averaged about 1.5 times as many birds per
route as the Glaciated Plains. Both agricultural regions likely
have fewer nest sites.

Phenology

Indigo Buntings arrive in mid- to late April (Robbins
and Easterla 1992). Atlasers observed the earliest instance of
nest building in Taney County, and of adults carrying food
to young in Osage County. A nest and eggs were observed
on May 7 in a block in the Festus area. Fledglings were

Abundance by Natural Division
Average Number of Birds / 100 Stops

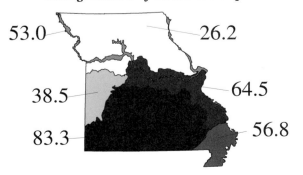

53.0

26.2

38.5

64.5

83.3

56.8

Breeding Phenology

EVIDENCE (# of Records)	MAR	APR	MAY	JUN	JUL	AUG	SEP
NB (32)			5/11			7/25	
NE (17)			5/15			7/31	
NY (18)				5/29			8/12
FY (98)			5/14				8/23

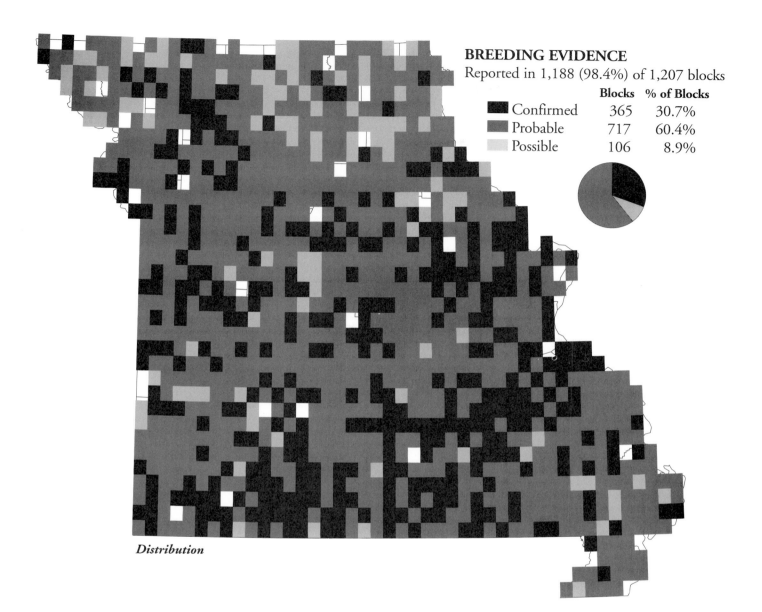

BREEDING EVIDENCE
Reported in 1,188 (98.4%) of 1,207 blocks

		Blocks	% of Blocks
Confirmed		365	30.7%
Probable		717	60.4%
Possible		106	8.9%

Distribution

observed through Aug. 29. Because most Atlasing activity had ceased by late August, September nesting activity went undocumented for this potentially four-brooded species (Payne 1992).

Notes

The Atlas Project reported 19 instances in which Indigo Buntings hosted Brown-headed Cowbird eggs and young. Ehrlich et al. (1988) report them as frequent hosts of Brown-headed Cowbirds.

Painted Bunting

Passerina ciris

> **Rangewide Distribution:** South central United States & Gulf Coast, southeastern coastal United States
> **Abundance:** Locally common
> **Breeding Habitat:** Scattered trees, brush, weeds & thickets
> **Nest:** Deep cup of grass, forbs & leaves, lined with fine grass & hair in small bush or vine
> **Eggs:** 3–4 pale blue-white or gray-white, spotted with red or brown
> **Incubation:** 11–12 days
> **Fledging:** 12–14 days

Although one of the most strikingly-colored species in Missouri, Painted Buntings often blend into tangles of vegetation. They associate with open woodlands and woodland edge, especially those adjacent to brushy fields. Nests are placed one to two meters above ground in bushes and low trees (Harrison 1975). Painted Buntings were not reported in Missouri at the turn of the century (Widmann 1907).

Code Frequency

Atlasers recorded Painted Buntings in only eight blocks. This species' habitat of rocky glades and thorny tangles is difficult to traverse and the single breeding confirmation reflects the scarcity of this species. Although not common anywhere, several birds have been reported outside Atlas blocks.

Distribution

Southwestern Missouri includes the northern edge of this species' range. The main range in Missouri is likely within the White River and Elk River natural sections. Most observations were locally distributed in glades and tangles with scattered cedars and honey locust. Several non-Atlas Project sightings were scattered north through the Osage Plains and east to Butler County. The authors know of no documented breeding records outside extreme southwestern Missouri.

Phenology

Atlas Project records suggest June and July as the best months to observe breeding evidence.

Notes

Painted Buntings are considered a frequent cowbird host (Ehrlich et al. 1988), but Atlasers did not report evidence of Brown-headed Cowbird parasitism.

Breeding Phenology

EVIDENCE (# of Records)	MAR	APR	MAY	JUN	JUL	AUG	SEP
FY (1)					7/25	7/25	

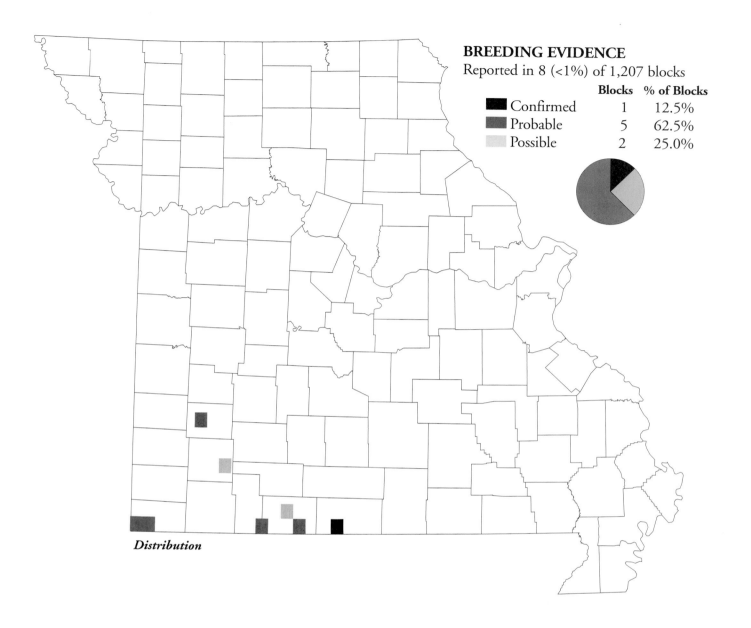

Reported in 8 (<1%) of 1,207 blocks

		Blocks	% of Blocks
■	Confirmed	1	12.5%
■	Probable	5	62.5%
▫	Possible	2	25.0%

Distribution

Dickcissel

Spiza americana

> **Rangewide Distribution:** South central Canada & eastern United States except for East Coast
> **Abundance:** Abundant, but population declining east of the Appalachians
> **Breeding Habitat:** Cultivated land, brushy fields & meadows
> **Nest:** Cup of forbs, grass & cornstalk inter-woven with leaves & grass, lined with finer material on or near the ground
> **Eggs:** 4 pale blue and unmarked
> **Incubation:** 12–13 days
> **Fledging:** 9 days

Dickcissels breed in weedy overgrown fields with dense grassy and herbaceous vegetation. Nests are typically placed one meter above ground in briars, shrubs, small trees and clumps of grass (James and Neal 1986). As with many grassland species, Dickcissel numbers are declining. They have experienced an average annual decline of 3.5 percent between 1967 and 1989 according to the Breeding Bird Survey (Wilson 1990). Dickcissel numbers also fluctuate dramatically from year to year and from place to place (Robbins et al. 1986). Reasons for these fluctuations are uncertain. The overall decline may be partly due to habitat changes and pesticides. Additionally, they may be declining because of control procedures on their Venezuelan wintering ground (Basili and Temple 1995).

Code Frequency

Typically, males perch in open situations. Their wiry "dick..dick-cissel" song permits easy detection. Therefore, where they were not found during the Atlas Project, they likely occurred in low numbers or not at all. Higher breeding evidence was hard for Atlasers to find, however. Most probable breeding records were based on the abundance of singing males and observations of territoriality. Food being delivered to young accounted for most confirmed breeding records. Because the nests were difficult to observe, Dickcissels likely bred in most blocks in which they were recorded as possible and probable breeders. Unmated, "floating" males may have accounted for some of these records.

Distribution

Dickcissels occurred throughout all but the forested regions of Missouri. They were observed in only 30 percent of blocks in the Ozarks Natural Division compared with 70 percent in the remainder of Missouri. The greatest frequency of breeding confirmations occurred in the Osage Plains.

Abundance

Correspondingly, Dickcissels were also most abundant in the Big Rivers and Mississippi Lowlands natural divisions.

Phenology

Dickcissels are late migrants to Missouri, arriving during the third week of April (Robbins and Easterla 1992). Atlasers observed nest building by May 7 and fledged young by May 25. Dickcissels typically raise two broods (Harrison 1975) and the late observation of a nest with young may

Abundance by Natural Division
Average Number of Birds / 100 Stops

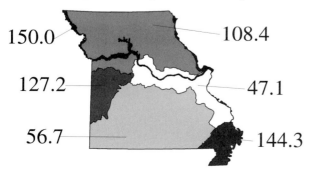

150.0   108.4
127.2   47.1
56.7   144.3

Breeding Phenology

EVIDENCE (# of Records)	MAR	APR	MAY	JUN	JUL	AUG	SEP
NB (30)			5/07			7/29	
NE (6)		5/14			7/22		
NY (12)			5/30		7/22		
FY (111)			5/24			8/16	

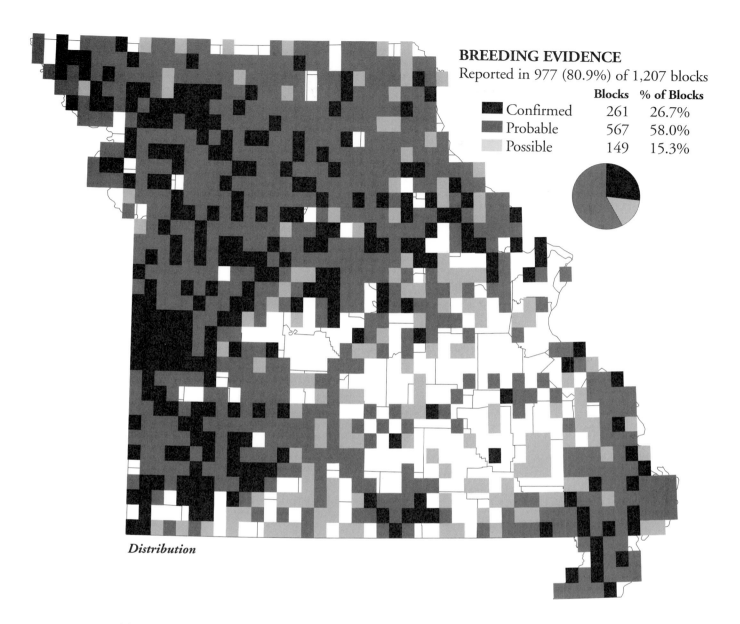

BREEDING EVIDENCE
Reported in 977 (80.9%) of 1,207 blocks

	Blocks	% of Blocks
Confirmed	261	26.7%
Probable	567	58.0%
Possible	149	15.3%

Distribution

represent a second brood. The observation of nest building July 29 was surprisingly late. Dickcissels typically begin to flock in late August in preparation for fall migration (Robbins and Easterla 1992).

Notes

Friedmann (1963) reported that parasitism of Dickcissels by Brown-headed Cowbirds was common. The Atlas Project recorded Brown-headed Cowbird parasitism in four of 18 blocks in which nest contents were observed.

Eastern Towhee
Pipilo erythrophthalmus

Rangewide Distribution: Southern Canadian border, most United States & Middle America
Abundance: Common
Breeding Habitat: Dense undergrowth, brushy areas & thickets
Nest: Cup of leaves, grass, bark, twig & roots, lined with hair & grass on ground
Eggs: 3–4 grayish or creamy white with brown spots, more on the large end
Incubation: 12–13 days
Fledging: 10–12 days

The Eastern Towhee was formerly known as the Rufous-sided Towhee or Red-eyed Towhee. The familiar "Drink you teeeee" song of this species is commonly heard during March and April throughout most of Missouri. Towhees sit at the top of saplings, or on branches of small trees, and sing their song repeatedly. This species nests on or near the ground, and the male has a black body while the female is a rich brown. Listening for the "cherwink" call, or seeing birds flash their white outer tail feathers as they fly off into the brush, is frequently the first indication of this species' presence.

Code Frequency

Breeding was confirmed in only 10 percent of the 889 blocks in which it was recorded. This might be because the Eastern Towhee nests in tangles and shrubs where it is difficult to observe or follow. Nests were reported in only 1 percent of Atlas blocks. About 87 percent of Atlas observations were birds singing in territories.

Distribution

This species was essentially distributed statewide with fewer records in the Mississippi Lowlands. The center of their distribution includes the Ozark and Ozark Border natural divisions, which apparently have an abundance of habitat. Widmann (1907) described their turn-of-the-century distribution as "locally common." The clearing of forest land for pastures and succession of pasture to brush land has created abundant habitat in the Ozarks for this species.

Abundance

As with distribution, the relative abundance of this species was greatest in the glades and brush lands of the Ozark and Ozark Border natural divisions. Agriculture in the Mississippi Lowlands and extreme northwestern Glaciated Plains natural divisions has likely restricted the available habitat for this species.

Phenology

Nesting activities commenced in early May. The late observation of a nest with young and the record of fledglings on August 19 may have been third broods for this normally two-brooded species. Three broods are suspected in the southeastern United States (Greenlaw 1996).

Abundance by Natural Division
Average Number of Birds / 100 Stops

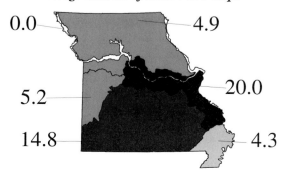

0.0   4.9

5.2   20.0

14.8   4.3

Breeding Phenology

EVIDENCE (# of Records)	MAR	APR	MAY	JUN	JUL	AUG	SEP
NB (5)		5/01			7/10		
NE (4)		5/11		6/11			
NY (4)			5/27			8/04	
FY (17)			5/21			8/05	

BREEDING EVIDENCE

Reported in 889 (73.7%) of 1,207 blocks

	Blocks	% of Blocks
Confirmed	90	10.1%
Probable	443	49.8%
Possible	356	40.1%

Distribution

Notes

Atlasers reported three records of Brown-headed Cowbird brood parasitism for this species. It is considered a frequent host by Ehrlich et al. (1988). The Spotted Towhee *(Pipilo maculatus)*, formerly considered a subspecies of Rufous-sided Towhee, could potentially nest in Missouri as it is a common nesting species in Kansas (Thompson and Ely 1989).

Bachman's Sparrow
Aimophila aestivalis

Rangewide Distribution: Southeastern United States, & southeastern Pennsylvania to Gulf Coast
Abundance: Uncommon, with northeastern range shrinking
Breeding Habitat: Open pine woods with dense brushy understory
Nest: Open & domed, of grass & forbs, lined with grass & hair on ground
Eggs: 3–5 white & unmarked
Incubation: 14 days
Fledging: 10–11 days

Historically, Bachman's Sparrows occupied glades and open pinewoods in Missouri. Their population has diminished because of the logging of pine forests in the early 1900s (Robbins and Easterla 1992) and the succession of glades and savannas to dense, woody vegetation (Hardin 1977). Their breeding range has retreated southward during this century. Until the mid-1960s they bred as far north as St. Louis (Robbins and Easterla 1992).

Code Frequency

Despite their small size and brown plumage, these sparrows are relatively conspicuous during the breeding season due to their occupation of open habitat and the male's loud, distinctive song. Thus, the Atlas Project likely provided an accurate portrayal of the status of the Bachman's Sparrow just as it did for similarly-detectable species such as Field Sparrows.

Distribution

Of the six blocks in which Atlasers observed Bachman's Sparrows, breeding was confirmed in only two. Confirmations were made by observations of fledged young and food deliveries to young. Bachman's Sparrow nests are placed on the ground and typically hidden beneath domes or existing vegetation (Chambers 1994), making nests difficult to find. The four blocks in which Bachman's Sparrows were recorded as possible or probable breeders may well have contained breeding individuals.

Abundance

Bachman's Sparrow records obtained from the Atlas Project were confined to the south central Ozarks. However, Chambers (1994) located additional birds in southwestern Missouri during 1992 and 1993.

Phenology

Atlasers obtained most evidence of breeding during June, including food deliveries that indicated there were either young in the nest or recently fledged young. A late season observation of fledged young on August 18, may be a second brood. Harrison (1975) observed that Bachman's Sparrows are double- or triple-brooded in the south.

Breeding Phenology

EVIDENCE (# of Records)	MAR	APR	MAY	JUN	JUL	AUG	SEP
FY (1)				6/25	6/25		

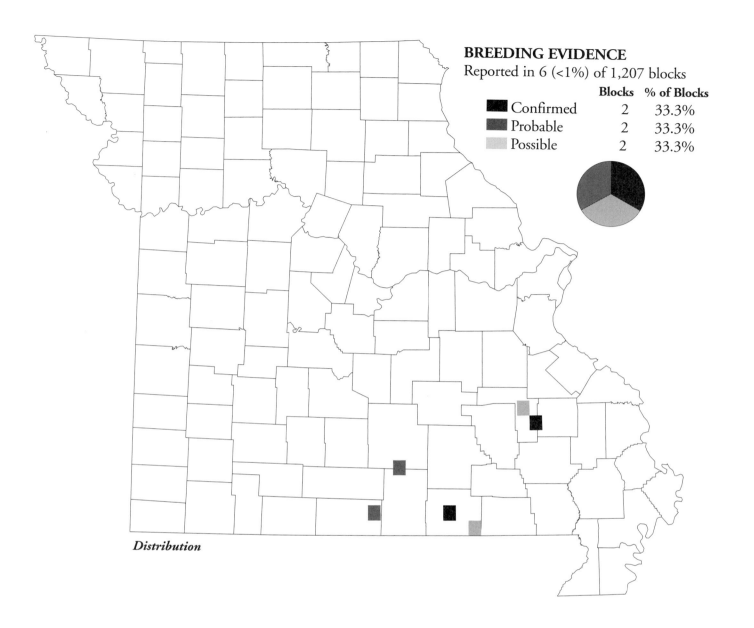

BREEDING EVIDENCE
Reported in 6 (<1%) of 1,207 blocks

	Blocks	% of Blocks
■ Confirmed	2	33.3%
■ Probable	2	33.3%
■ Possible	2	33.3%

Distribution

Chipping Sparrow
Spizella passerina

Rangewide Distribution: Northwestern through southern Canada, southeastern Alaska, all United States
Abundance: Widespread & common
Breeding Habitat: Open oak-pine woods, forest edge & thickets
Nest: Cup of grass, forbs & roots, lined with hair & fur in tree or vine
Eggs: 4 bluish-green with dark brown or black marks, occasionally wreathed
Incubation: 11–14 days
Fledging: 10 days

These small sparrows with a white eye-line and rusty cap are most often associated with suburban and rural yards where they search lawns for insects. Parks, cemeteries and golf courses also offer appropriate habitat. They occasionally occur distant from human habitation in open pine woodlands, orchards and brushy pastures. Chipping Sparrows are best recognized by their even trill of chip notes which they emit from elevated perches. Frequently, they construct their neat grass nests in ornamental shrubbery; however, they also nest in early successional brush, hedgerows and small trees.

Code Frequency

Most Atlasers knew where to locate and how to identify Chipping Sparrows. Therefore, where they did not record Chipping Sparrows, the birds likely were absent. Although confirmation of breeding should have been easy due to the association of nests with residential plantings, some Atlasers may have been reluctant to search for nests around homes. Chipping Sparrows, therefore, likely nested in most blocks in which they were found.

Distribution

Chipping Sparrows were found to have statewide distribution except for the Mississippi Lowlands. Their nearly complete absence from the Mississippi Lowlands seems inexplicable, as sufficient nesting habitat is apparently present. An interesting pattern of Chipping Sparrow scarcity extended from the southern Osage Plains through Carroll and Cooper counties. This is also difficult to understand because there is suitable habitat in these areas. It may be due to western Missouri being near the western limit of this species' range. Breeding Chipping Sparrows are absent from the southern Great Plains (Johnsgard, 1979) a short distance from western Missouri.

Abundance

Chipping Sparrows were most abundant in south central Missouri and the eastern Ozarks. Areas with the fewest Chipping Sparrows were the Mississippi Lowlands and western Missouri south of Kansas City.

Phenology

Chipping Sparrows arrive in southern Missouri in early March and a month later in northern Missouri (Robbins and Easterla 1992). The 236 Atlas Project blocks in which breeding was confirmed provided an excellent portrayal of nesting

Abundance by Natural Division
Average Number of Birds / 100 Stops

0.0
7.4
0.5
13.8
14.4
0.0

Breeding Phenology

EVIDENCE (# of Records)	MAR	APR	MAY	JUN	JUL	AUG	SEP
NB (32)	4/07				7/15		
NE (18)		4/28			7/15		
NY (21)			5/14			8/12	
FY (66)			5/15			8/10	

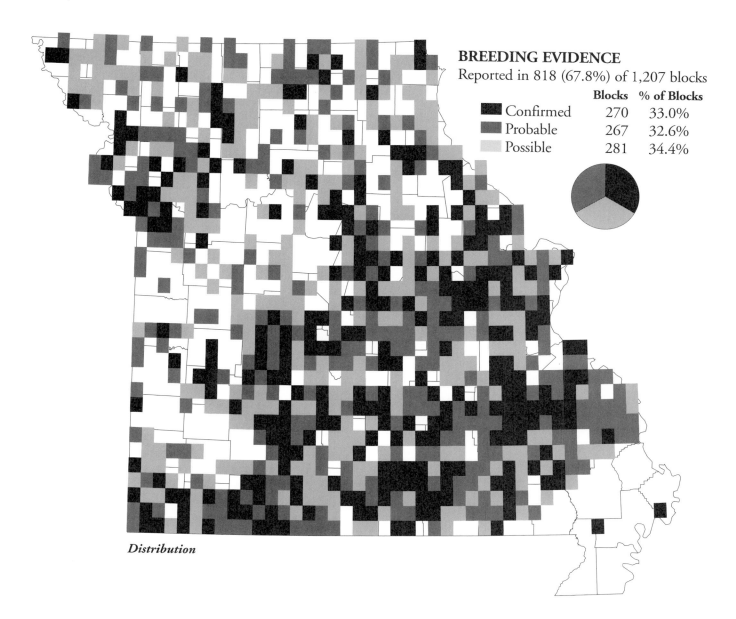

BREEDING EVIDENCE
Reported in 818 (67.8%) of 1,207 blocks

	Blocks	% of Blocks
Confirmed	270	33.0%
Probable	267	32.6%
Possible	281	34.4%

Distribution

phenology. Second broods are commonplace (Harrison 1975). Late dates for nest building and a nest with young may have been second broods or renesting attempts.

Notes

Brown-headed Cowbird parasitism of Chipping Sparrows is common (Ehrlich et al. 1988). Brown-headed Cowbird parasitism was recorded in seven out of 39 observations of nest contents.

Field Sparrow
Spizella pusilla

Rangewide Distribution: Southeastern Canada, northwest
to southeast Montana through eastern United States
Abundance: Fairly common
Breeding Habitat: Areas with scattered trees, brush &
thorn shrubs
Nest: Vine tangle of grass & forbs, lined with fine
material near or on the ground
Eggs: 3–5 creamy pale green-blue white, with brown
marks
Incubation: 12 (10–17) days
Fledging: 7–8 days

Field Sparrows are well known to birdwatchers by their
plaintive trill and easily-recognized field marks. They occupy
a variety of brushy, successional habitats, and are especially
associated with overgrown pastures containing weeds,
brushy tangles and small, scattered trees. Red cedar dominated habitats, including glades, are appropriate habitats for
this species. To a lesser extent, they are found along roadsides and in grassy fields with little woody vegetation, such
as native prairies (Carey et al. 1994).

Code Frequency

Field Sparrows are one of the easiest species to detect due
to their persistent singing. Singing commences upon their
arrival in spring and continues through most of the breeding
season, even during the heat of the day. Atlasers used sound
to record most of the probable breeding evidence, primarily
multiple singing males and territoriality. Breeding was confirmed in one-third of the blocks in which they were
observed. Although nests are low and sometimes easy to
locate, they were found in only 8 percent of Atlas blocks.
Field Sparrows are expected to have bred in most blocks in
which they were observed.

Distribution

Field Sparrows were distributed statewide. Regions with
fewer Field Sparrow observations were those that are more
extensively forested, such as the Lower Ozark Natural
Section, and those that lack brushy fields, such as the
Mississippi Lowlands Natural Division.

Abundance

Miniroute and Breeding Bird Survey data indicate that
Field Sparrows are apparently most abundant in southwestern Missouri where the Ozark Natural Division meets the
Osage Plains. They were generally less numerous in northern Missouri and the Mississippi Lowlands and appeared
least abundant in the Big Rivers Natural Division.

Phenology

Field Sparrows return to breeding territories in Missouri
during March (Robbins and Easterla 1992). Despite the
almost immediate onset of singing, Atlas observations of
eggs and nestlings indicate actual nesting events do not
commence until April. Carey et al. (1994) reported eggs as
early as April 29 in Missouri. Two or more broods are

Abundance by Natural Division
Average Number of Birds / 100 Stops

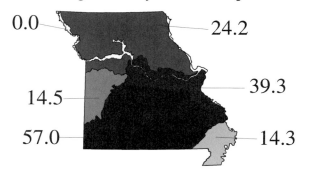

0.0

24.2

14.5

39.3

57.0

14.3

Breeding Phenology

EVIDENCE (# of Records)	MAR	APR	MAY	JUN	JUL	AUG	SEP
NB (18)		4/29				8/08	
NE (30)		5/07			7/29		
NY (19)		5/10			7/27		
FY (113)		5/07				8/14	

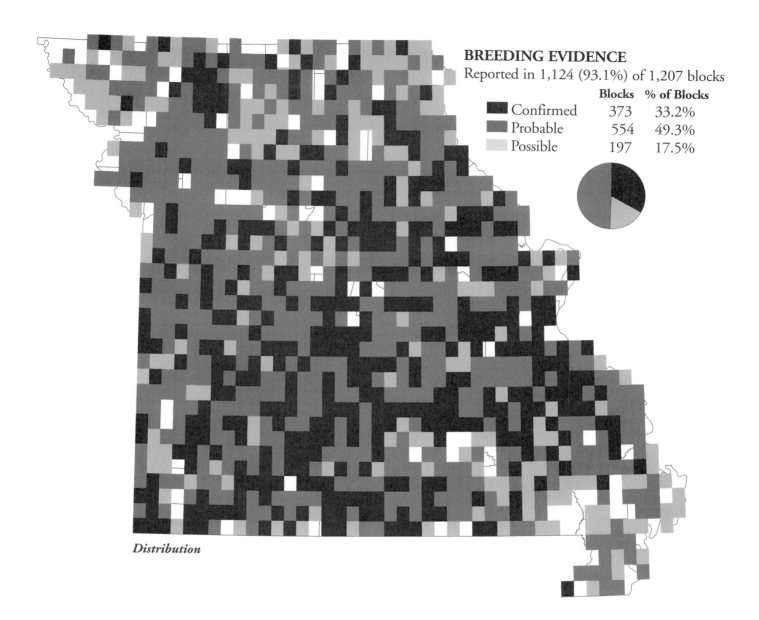

BREEDING EVIDENCE
Reported in 1,124 (93.1%) of 1,207 blocks

		Blocks	% of Blocks
■	Confirmed	373	33.2%
■	Probable	554	49.3%
■	Possible	197	17.5%

Distribution

reported for this species (Carey et al. 1994). Nests with eggs and young observed in late June and July may have been subsequent broods or renesting attempts. Carey et al. (1994) reported fledglings as late as August 15 in Pennsylvania and October 10 in Ohio.

Notes

Field Sparrows were one of the most common hosts of Brown-headed Cowbird parasitism discovered during the Atlas Project. The phenomenon was recorded in 16 of 160 blocks in which nest contents or tended young were observed. Cowbird parasitism varies geographically. Carey et al. (1994) reported a 13 percent parasitism frequency in Missouri, compared with 11 percent in Illinois, 32 percent in Ohio and 80 percent in Iowa.

Vesper Sparrow
Pooecetes gramineus

Rangewide Distribution: Southern Canada, northern
  & central United States
Abundance: Fairly common
Breeding Habitat: Dry scrub field, pastures with
  scattered trees
Nest: Bulky, loose, of grass & roots, lined with fine
  material in excavated depression on the ground
Eggs: 3–4 creamy white or pale greenish-white, with
  brown marks
Incubation: 11–13 days
Fledging: 9 (7–14) days

In Missouri, Vesper Sparrows are most likely to be found in open farming country. In areas where wood lots are scattered throughout the landscape, Vesper Sparrows frequently sing along the edge of woods or from atop a fence post in an open field. Their song and a flash of white outer-tail feathers will usually verify identification.

Code Frequency

Atlasers documented Vesper Sparrows in only 3 percent (88) blocks and only one nest with eggs was found. Most records were of an individual, a pair or a territorial individual.

Distribution

Vesper Sparrows were distributed almost entirely north of the Missouri River. Most sightings were restricted to the Western Glaciated Plains, Grand River and northern part of the Eastern Glaciated Plains natural sections. Scattered records south of the Missouri River suggest the species breeds locally at the southern edge of its range in the Midwest. However, these possible breeding records may have been observations of migrants.

Abundance

Vespers Sparrows were most abundant in the north central part of the Glaciated Plains Natural Division, where considerable pasture land is mixed with crop land and scattered wood lots. Surveyors averaged about five birds per 100 stops on certain routes in the Glaciated Plains Natural Division.

Phenology

The first birds were observed on May 15 with territorial birds recorded on May 31. Some fledglings were off the nest by June 11 and fledglings were documented through August 14. Although Terres (1987) reported that Vesper Sparrows produce two to three broods each year, Atlas Project observations provide little information on the timing of broods in northern Missouri.

Abundance by Natural Division
Average Number of Birds / 100 Stops

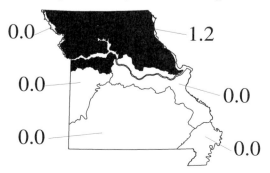

0.0   1.2

0.0   0.0

0.0   0.0

Breeding Phenology

EVIDENCE (# of Records)	MAR	APR	MAY	JUN	JUL	AUG	SEP
NE (1)				6/20	6/20		
FY (4)			6/12		7/10		

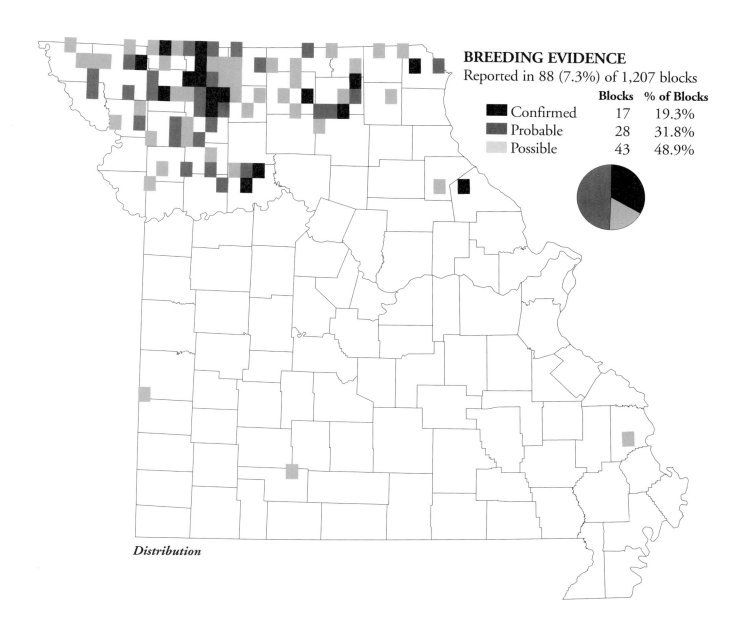

BREEDING EVIDENCE
Reported in 88 (7.3%) of 1,207 blocks

	Blocks	% of Blocks
■ Confirmed	17	19.3%
■ Probable	28	31.8%
■ Possible	43	48.9%

Distribution

Notes

Vesper Sparrows are considered a common host of Brown-headed Cowbirds (Ehrlich et al. 1988), but no instances of Brown-headed Cowbird parasitism were observed during this study.

Lark Sparrow
Chondestes grammacus

Rangewide Distribution: United States to Mississippi & Mexico, not East Coast
Abundance: Fairly common west of Mississippi
Breeding Habitat: Forest edge, grassy fields & brush areas
Nest: Fine grass, twigs & forbs on ground or in shrub or crevice
Eggs: 4–5 creamy to grayish-white, marked with browns or blacks
Incubation: 11–12 days
Fledging: 9–10 days

The Lark Sparrow appears as a typical brown-colored sparrow when viewed from a distance. On close inspection, the head plumage is ornately patterned with distinctive chestnut-colored cheek patches and crown stripes bordered in white and black. Lark Sparrows usually nest on or near the ground close to a sapling, shrub base, or in rocks that provide protection. Nests are frequently constructed in over-grazed pastures.

Code Frequency

Although Lark Sparrows were reported in 45 percent of all blocks, in only 28 percent of blocks did observations confirm breeding. Nests with eggs and young accounted for only 1 percent of all codes.

Distribution

Lark Sparrows were widely distributed in the northwestern half of Missouri and in agricultural fields and pastures scattered throughout the Ozark and Ozark Border natural divisions. They nest in overgrazed pastures, habitat that resembles western rangeland and native short-grass prairie. Only a few records occurred in the Mississippi Lowlands apparently due to lack of a suitable nesting habitat.

Abundance

The greatest abundance for this species occurred in the Big Rivers and Ozark natural divisions. Atlasers did not report this species on Miniroutes in the Mississippi Lowlands where distribution was sparse and patchy.

Phenology

Most Lark Sparrows arrive in late April and early May (Robbins and Easterla 1992). Fledglings were recorded through August 28. Most birds depart the state by mid-September (Robbins and Easterla 1992).

Notes

Atlasers reported only one Lark Sparrow nest parasitized by Brown-headed Cowbirds. Perhaps this species' nests are sufficiently concealed to prevent cowbird detection (Rothstein 1971). Ehrlich et al. (1988) list Lark Sparrows as an occasional cowbird host.

Abundance by Natural Division
Average Number of Birds / 100 Stops

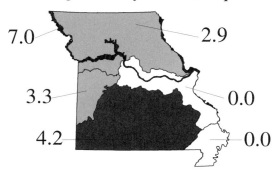

7.0   2.9

3.3   0.0

4.2   0.0

Breeding Phenology

EVIDENCE (# of Records)	MAR	APR	MAY	JUN	JUL	AUG	SEP
NB (11)			5/10		7/03		
NE (5)		4/28		6/09			
NY (3)			5/25		7/02		
FY (41)			5/16		7/27		

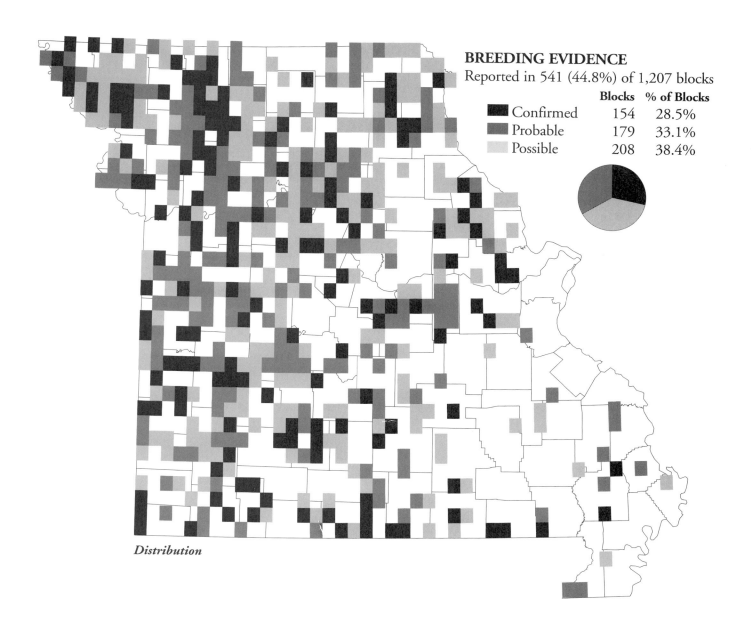

	Blocks	% of Blocks
Confirmed	154	28.5%
Probable	179	33.1%
Possible	208	38.4%

Distribution

Savannah Sparrow

Passerculus sandwichensis

Rangewide Distribution: All Canada & northern United States
Abundance: Common
Breeding Habitat: Open grassy areas, meadows & moderate height vegetation
Nest: Coarse grass, lined with fine material in depression on ground
Eggs: 3–5 pale greenish-blue, off-white, with brown marks & wreathed
Incubation: 12–13 days
Fledging: 7–10 days

A secretive, grassland species that prefers wet meadows, Savannah Sparrows are little-known breeders in the state. During migration, this species is extremely common in agricultural and rural areas, which provides an opportunity to learn their habits and plumage variations.

Code Frequency

Although they are common in states farther north, the Atlas Project did not obtain sufficient information to give an accurate picture of this species' distribution throughout the state. Their rarity, similarity to Song Sparrows and their sparse distribution despite abundant suitable habitat may have led to them being overlooked frequently. Further searches in Atlas blocks where this species was confirmed might provide insight into the species' habitat requirements.

Distribution

The patchy distribution and sparse breeding population of Savannah Sparrows in Missouri may be due to a lack of appropriate breeding habitat at the extreme southern limit of their breeding range. Little is known about the habitat characteristics of breeding sites or if the birds successfully reproduce every year. Atlas Project findings contradict Wheelwright and Rising (1993), who suggested that the southern edge of the breeding range for the Savannah Sparrow is found in the northern third of Iowa. Jackson et al. (1996) recorded breeding Savannahs in Iowa counties adjacent to the Missouri border, but found them to be much more numerous in the northern half of the state.

Abundance

Because this species was recorded in few Miniroutes and Breeding Bird Survey routes, little abundance information can be extrapolated from the data. It is likely a scarce nester in Missouri. Wheelwright and Rising (1993) suggested Savannah Sparrows have likely benefited from agriculture. Hayfields, pastures and open habitats may provide more nesting and foraging habitat. Declines in abundance may be attributed to the loss of open habitats resulting from successional changes.

Phenology

Savannah Sparrows are considered a two-brooded species (Ehrlich et al. 1988). Insufficient data exists to determine timing of breeding in Missouri. The peaks of migration in Missouri are mid-April and mid-October (Robbins and Easterla 1992).

Breeding Phenology

EVIDENCE (# of Records)	MAR	APR	MAY	JUN	JUL	AUG	SEP
NB (1)			6/14	6/14			
FY (2)				7/09	7/10		

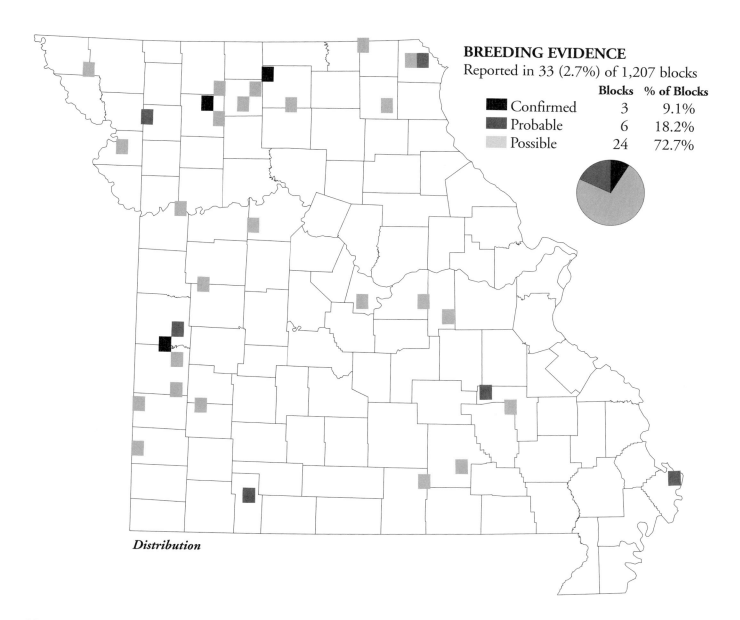

BREEDING EVIDENCE

Reported in 33 (2.7%) of 1,207 blocks

	Blocks	% of Blocks
■ Confirmed	3	9.1%
■ Probable	6	18.2%
■ Possible	24	72.7%

Distribution

Notes

Atlasers reported no records of brood parasitism for this species. Savannah Sparrows are considered uncommon Brown-headed Cowbird hosts by Ehrlich et al. (1988).

Grasshopper Sparrow

Ammodramus savannarum

Rangewide Distribution: Southern Canada, central &
eastern United States, also eastern Washington to
southern California

Abundance: Fairly common

Breeding Habitat: Prairie, fallow fields & short grasses

Nest: Grass, lined with fine material in depression or
rim at ground level

Eggs: 4–5 creamy white, red to brown marks, occasionally wreathed

Incubation: 11–12 days

Fledging: 9 days

Grasshopper Sparrows prefer medium height grasses such
as unmown hayfields and lightly-grazed pastures for breeding (Peterjohn and Rice 1991). Occasionally they are found
in extremely small, fragmented grasslands such as roadside
ditches and isolated fields. As is true with many grassland
species, Grasshopper Sparrow populations have declined
throughout the continent (Robbins et al. 1986). According
to Breeding Bird Survey data, an average annual decline of
3.3 percent occurred from 1967 to 1989 in Missouri
(Wilson 1990).

Code Frequency

Grasshopper Sparrows are most readily-detected by their
buzzy insect-like songs. Once located by sound, singing
individuals can usually be sighted perched on grass stems,
shrubs, fences or power lines. Some birders have difficulty
hearing this sparrow's high frequency song and it is therefore
conceivable that birds were missed in some blocks. Once
found, however, this species was easily elevated to the probable breeding level, based on observations of territoriality and
multiple singing males. Breeding confirmations were difficult as expected for a species whose nests are on the ground,
often well hidden in grass. Therefore, they likely bred undetected in most blocks in which they were found.

Distribution

Although ranging statewide, these sparrows were
extremely scarce throughout most forested parts of the state,
including the Ozark Natural Division and Lincoln Hills
Natural Section. They were also absent from blocks in
regions with few appropriate grasslands, such as the intensively row-cropped areas of the Mississippi Lowlands and
northern Carroll and Livingston counties. They were also
absent in large urban areas.

Phenology

Grasshopper Sparrows arrive in their Missouri breeding
areas during late April (Robbins and Easterla 1992). Atlasers
recorded a bird on the nest May 17. The first fledglings were
observed on June 10. Second and third broods have been
reported for this species (Ehrlich et al. 1988). Atlas Project
observations of nest building may have represented second
or third broods.

Abundance by Natural Division
Average Number of Birds / 100 Stops

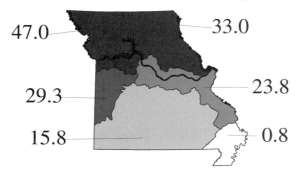

47.0   33.0

29.3   23.8

15.8   0.8

EVIDENCE (# of Records)	MAR	APR	MAY	JUN	JUL	AUG	SEP
NB (2)				6/25	7/27		
NY (6)			5/24		7/13		
FY (67)			5/25			8/18	

Breeding Phenology

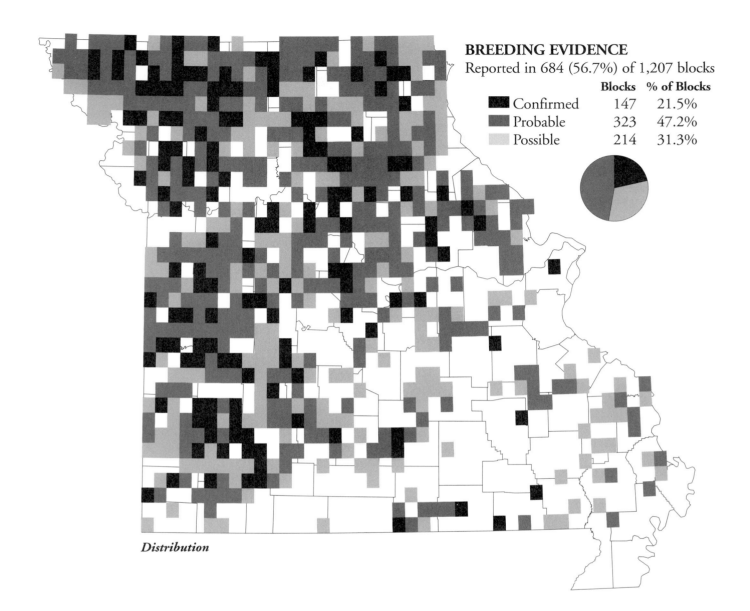

BREEDING EVIDENCE

Reported in 684 (56.7%) of 1,207 blocks

		Blocks	% of Blocks
■	Confirmed	147	21.5%
■	Probable	323	47.2%
■	Possible	214	31.3%

Distribution

Notes

Ehrlich et al. (1988) described Grasshopper Sparrows as uncommon Brown-headed Cowbird hosts. Only one instance of Brown-headed Cowbird parasitism was documented during the Atlas Project.

Henslow's Sparrow
Ammodramus henslowii

Rangewide Distribution: Southeastern Canada, & northeastern United States
Abundance: Uncommon, local & declining
Breeding Habitat: Areas with dense grasses, sedges & shrubs
Nest: Grass & forbs, lined with fine material on or near ground with overhead grass
Eggs: 3–5 creamy or green-white, with red or brown marks, wreathed
Incubation: 11 days
Fledging: 9–10 days

Historically, Henslow's Sparrows bred in native prairie habitats throughout the Midwest (Herkert 1994). Widmann (1907) considered them locally common summer residents in wet meadows throughout the Osage Plains and Ozark Border natural divisions. Currently, Henslow's Sparrows breed in remnant prairie tracts, hayfields, and pastures with grasses and dead residual vegetation. Standing dead vegetation (Zimmerman 1988) and accumulations of ground litter (Herkert 1994) are recognized as important components of Henslow's Sparrow habitat.

Code Frequency

Henslow's Sparrows are most easily seen when they sing from an elevated perch within their dense tall-grass habitat. Otherwise, they are secretive and tend to flit from one hiding place to another. Because some observers have difficulty hearing this species' weak, ventriloquistic song, Henslow's Sparrows may have occurred in more blocks than suggested on the map. Additionally, because of the difficulty in confirming breeding, Henslow's Sparrows likely bred in most blocks in which they were found. No nests were discovered and breeding was confirmed in only six instances.

Distribution

Henslow's Sparrows were sparsely distributed in western and northern Missouri. The locations of most records correspond to public prairies, including Bushwhacker, Diamond Grove, and Taberville conservation areas and Prairie State Park in southwestern Missouri. The records from northern Missouri were somewhat unexpected. Atlasers in northern Missouri made most of their observations in private hayfields and Conservation Reserve Program acreage. Several wet summers during the late 1980s prevented or delayed haying. This may have benefitted Henslow's Sparrows.

Phenology

Confirmed breeding evidence observed in six Atlas blocks provided insufficient information to evaluate this species' breeding phenology. Henslow's Sparrows migrate to Missouri in early April (Robbins and Easterla 1992) and probably commence breeding activities in late April. The earliest evidence of nesting was based on food being delivered to young. Atlasers observed the first fledglings on June 25. Henslow's Sparrows produce two broods (Ehrlich et al. 1988) and the late observation of fledglings on July 2 may have been associated with a second brood or a renesting attempt.

Breeding Phenology

EVIDENCE (# of Records)	MAR	APR	MAY	JUN	JUL	AUG	SEP
FY (3)				6/08	7/09		

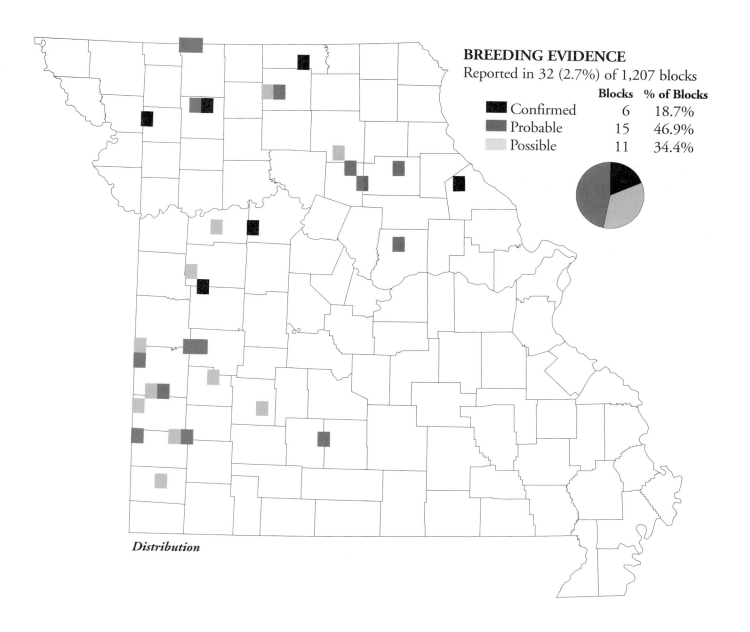

BREEDING EVIDENCE
Reported in 32 (2.7%) of 1,207 blocks

		Blocks	% of Blocks
■	Confirmed	6	18.7%
■	Probable	15	46.9%
■	Possible	11	34.4%

Distribution

Song Sparrow
Melospiza melodia

Rangewide Distribution: Southern Canada & southwestern, northern & northeastern United States
Abundance: Common & permanent in many areas
Breeding Habitat: Dense vegetation, watercourses, coasts & forest edge
Nest: Grass, forbs, leaves & bark strips, lined with fine material on the ground
Eggs: 3–4 pale blue to greenish-white, with reddish or brown marks
Incubation: 12–14 days
Fledging: 9–12 days

Song Sparrows vary considerably in plumage throughout their continental range. The long rounded tail, gray eyebrow line and streaked "stickpin-like" breast spot distinguish them from other Missouri sparrows. Frequented habitats include thickets along streams and ponds and roadside ditches dominated by willows and other woody vegetation.

Code Frequency

This is a noisy, easily-observed species that usually responds to human approach by uttering its distinctive call.

Seventy-six percent of all records were based on sight, song or territoriality. Observations of the Song Sparrow's well-known song likely were the basis for most possible records. Most confirmed records were sightings of fledglings and food being carried to young. Nests were apparently difficult to find with only six reported.

Distribution

Song Sparrows were found throughout the Glaciated Plains, Ozark Border, Big Rivers and Mississippi Lowlands natural divisions. Records scattered southward along Missouri's western border were usually associated with wetland complexes such as Schell-Osage Conservation Area. This species may be expanding its breeding range southward. The first breeding evidence was observed in Arkansas near Blytheville on June 22, 1993, when a singing male with a female were observed on a Breeding Bird Survey route in Mississippi County (Parker 1993).

Abundance

Song Sparrows were common in the Glaciated Plains and Big Rivers natural divisions. While there were scattered records in the Osage Plains, Atlasers did not record any on Miniroutes. Song Sparrows may be concentrated at wetland complexes in the Osage Plains and could have been missed by abundance surveys. In northern Missouri, Song Sparrows are distributed more evenly across the Glaciated Plains at the southern periphery of their main breeding range.

Phenology

Atlasers reported nesting activity from April 30 through August 4. Most observations of fledglings occurred from May 17 through August 30. This species is two-, three- and sometimes four-brooded (Ehrlich et al. 1988). In the southern part of Missouri, nesting attempts could commence as early as March.

Abundance by Natural Division
Average Number of Birds / 100 Stops

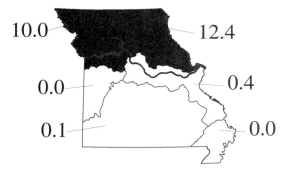

10.0   12.4
0.0   0.4
0.1   0.0

Breeding Phenology

EVIDENCE (# of Records)	MAR	APR	MAY	JUN	JUL	AUG	SEP
NB (7)		5/06			6/28		
NE (2)			5/30	6/03			
NY (4)			5/13		7/02		
FY (22)			5/20				8/24

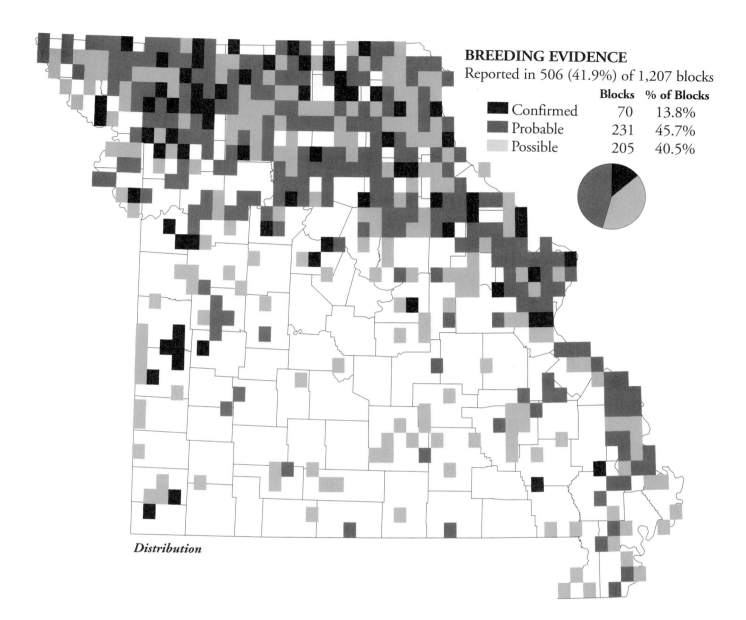

BREEDING EVIDENCE
Reported in 506 (41.9%) of 1,207 blocks

		Blocks	% of Blocks
■	Confirmed	70	13.8%
■	Probable	231	45.7%
■	Possible	205	40.5%

Distribution

Notes

Of the 28 blocks in which there were opportunities to detect Brown-headed Cowbird parasitism, parasitism was documented in three (10.7 percent). Although 11 other species were more frequently parasitized during in the Atlas Project, Song Sparrows (along with Yellow Warblers) are considered the most frequent host species for Brown-headed Cowbirds in North America (Ehrlich et al. 1988).

Bobolink

Dolichonyx oryzivorus

Rangewide Distribution: Southern Canada & northern
    United States
Abundance: Uncommon, with declining breeding range
    east & northeast
Breeding Habitat: Tall grass, wet meadows & cultivated
    cropland
Nest: Grass & forbs, lined with finer grass in depression
    on the ground
Eggs: 5–6 gray to pale reddish-brown, with brown or
    purple marks
Incubation: 10–13 days
Fledging: 10–14 days

Bobolinks reside in open country and often favor damp meadows (Martin and Gavin 1995). Typically a colonial nesting species, they place their nests on the ground in dense vegetation, usually in hayfields of grass, clover or alfalfa (Martin and Gavin 1995). Territorial males emit a jangling call, often while flying above a field they have selected for nesting. As with many grassland bird species, Bobolink numbers have declined significantly throughout their range (Peterjohn and Sauer 1995).

Code Frequency

Bobolinks are reasonably easy to detect and Atlasers often observed probable or confirmed breeding evidence for this species. These included territorial singing, courtship, fledged young and food for young. Bobolinks recorded as possible breeders might have been late migrants. Because the drab female approaches and departs the nest through thick vegetation (Martin and Gavin 1995), locating nests is extremely difficult. Only one nest was discovered during the Atlas Project.

Distribution

As expected, Bobolinks bred in northern counties. Locations in central Missouri south of the Missouri River were scattered and unexpected. Especially noteworthy were outlying breeding locations in Bates and Vernon counties. Certain years of the Atlas Project were especially favorable for Bobolink nesting because conditions were apparently too wet to mow hayfields. The resultant rank grasses are preferred by Bobolinks, and may have led them to nest in some of the outlying locations.

Abundance

Bobolinks were most abundant in the agricultural regions of northeastern and north central Missouri.

Phenology

Bobolink migration peaks in mid-May (Robbins and Easterla 1992) and nesting was well underway by mid-June as evidenced by food carried to young on June 16. Although Bobolinks are usually single-brooded, they can raise a second brood. Second broods in New York have been recorded between June 24 and July 1 (Martin and Gavin 1995). A nest with eggs on July 10 may represent a second brood, renesting attempt or an abandoned nest.

Abundance by Natural Division
Average Number of Birds / 100 Stops

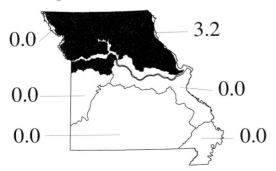

0.0   3.2

0.0   0.0

0.0   0.0

Breeding Phenology

EVIDENCE (# of Records)	MAR	APR	MAY	JUN	JUL	AUG	SEP
NE (1)					7/10	7/10	
FY (6)				6/16		7/26	

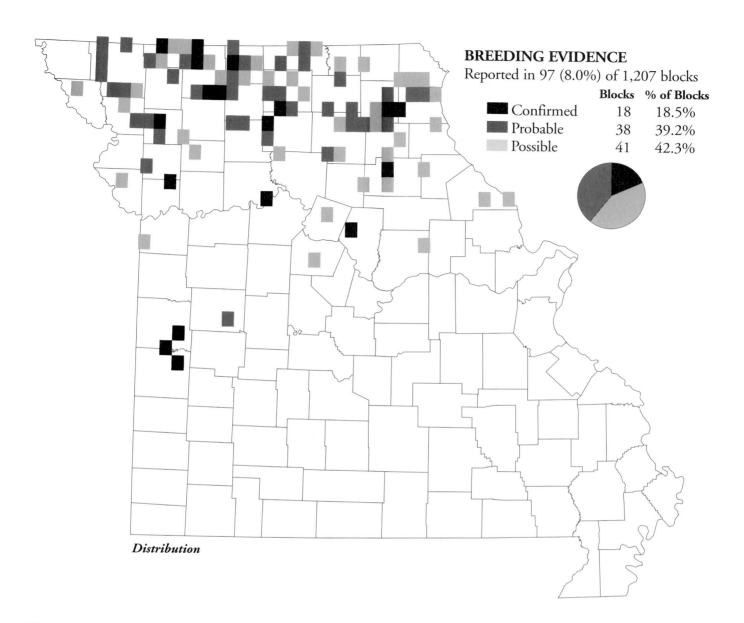

BREEDING EVIDENCE
Reported in 97 (8.0%) of 1,207 blocks

	Blocks	% of Blocks
Confirmed	18	18.5%
Probable	38	39.2%
Possible	41	42.3%

Distribution

Notes

Brown-headed Cowbird parasitism was recorded in one block. Cowbird parasitism of Bobolinks is infrequent except in Iowa (Friedmann 1963).

Red-winged Blackbird
Agelaius phoeniceus

> **Rangewide Distribution:** Southern Canada & entire United States & Middle America
> **Abundance:** Widespread & abundant
> **Breeding Habitat:** Brush, aquatic vegetated areaS by rivers, ponds & swamps
> **Nest:** Woven sedge & grass lined with fine grass & rushes in vegetation near or over water
> **Eggs:** 3–4 pale bluish-green with dark color markings
> **Incubation:** 10–12 days
> **Fledging:** 11–14 days

These crimson-shouldered residents of marshes, wet meadows and weedy roadside ditches are well known by most rural Missourians. The Red-winged Blackbird's "konk-a-ree" song likely emanates from every pond in Missouri. It is fascinating to watch males display their colorful shoulders and defend their marsh nesting territories. Assembled spring and late summer flocks may number in the hundreds or thousands of birds. Night roosts may occasionally reach into the millions. At 190 million birds in a winter estimate, this species likely was the most abundant bird species in North America (Nero, 1984, in Yasukawa and Searcy 1995).

Code Frequency

Red-winged Blackbirds were plentiful and easily detected. This species was recorded in 98 percent of blocks (1,185), ranking it fifth among breeding species in block frequency. An amazing 68 percent of records were confirmed. Where Atlasers did not record this species, they were likely absent. In blocks where they were not confirmed to breed there were likely few birds, thus offering Atlasers few opportunities to find a nest.

Distribution

This species was distributed statewide. With such a solidly-distributed species, certain features such as Truman and Table Rock lakes created distribution gaps, as did the larger forests in Wayne, Reynolds and Shannon counties. Although appropriate habitat does occur in the Ozark and Ozark Border natural divisions, most sites are small and patchily distributed, and could easily be overlooked.

Abundance

The Red-winged Blackbird was one-fifth as abundant in the Ozark Natural Division as in the Mississippi Lowlands. In the Glaciated Plains Natural Division, also an agricultural region, Red-winged Blackbirds were recorded about one-half as often as in the Mississippi Lowlands. Perhaps the Mississippi Lowland's many ditches, sloughs and the preponderance of waste grains enable it to support a larger population than the fields, pastures and croplands of the Glaciated Plains.

Phenology

Red-winged Blackbirds were on nests by April 17 and were feeding fledglings by May 13. This two- and three-brooded species continued nesting through early August as demonstrated by late dates for food being carried to young,

Abundance by Natural Division
Average Number of Birds / 100 Stops

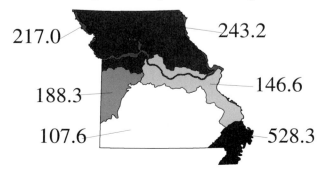

217.0   243.2

188.3   146.6

107.6   528.3

Breeding Phenology

EVIDENCE (# of Records)	MAR	APR	MAY	JUN	JUL	AUG	SEP
NB (53)	4/09				7/13		
NE (120)		4/30				7/23	
NY (60)		5/01				7/26	
FY (166)		5/09					8/08

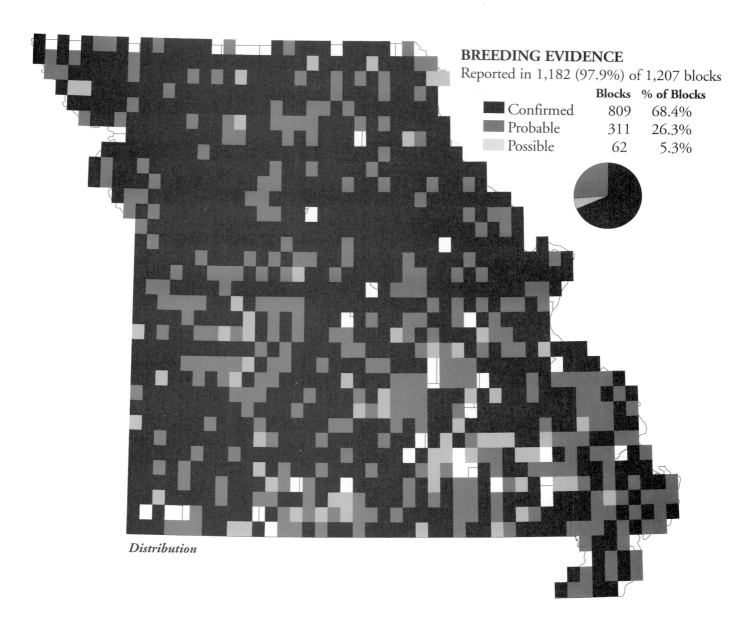

		Blocks	% of Blocks
	Confirmed	809	68.4%
	Probable	311	26.3%
	Possible	62	5.3%

Distribution

nest with eggs and nest with young. Some nesting likely continued well after these dates. However, with less Atlasing conducted in August, late season breeding activity was likely under represented.

Notes

Atlasers reported Brown-headed Cowbird eggs and young fledgling cowbirds being raised by Red-winged Blackbirds in 25 blocks. This is 7.3 percent of the 344 blocks in which Atlasers observed red-wing clutches or young being fed. These blackbirds are considered frequent cowbird hosts by Ehrlich et al. (1988).

Eastern Meadowlark
Sturnella magna

> **Rangewide Distribution:** Southeastern Canada, south central & eastern United States to South America
> **Abundance:** Common
> **Breeding Habitat:** Fallow fields, prairies & cultivated land
> **Nest:** Grass, lined with fine grass & hair, with a domed canopy of grass & side opening in depression on ground
> **Eggs:** 3–5 white with brown or purple marks
> **Incubation:** 13–15 days
> **Fledging:** 11–12 days

Eastern Meadowlarks are a familiar sight throughout the farmlands of the Midwest. Upland fields, especially native prairies, unmown hayfields, lightly-grazed pastures, golf courses and airports are appropriate habitat (Lanyon 1995). Although they do not require woody vegetation in nesting territories (Peterjohn and Rice 1991), they will occasionally nest in shrubby, overgrown fields. Eastern Meadowlarks will also nest in fallow fields and smaller tracts of grassland. As with many grassland birds, numbers of Eastern Meadowlarks are declining. Wilson (1990) reported an average annual decline in Missouri of 1.7 percent from 1967 through 1989 based on the Breeding Bird Survey.

Code Frequency

Vigorous singers from March through June, Eastern Meadowlarks are extremely easy to detect. Thus, where they do not appear on the map, they likely did not occur. Meadowlarks frequently construct roofed nests on the ground that are difficult to locate, hence few breeding confirmations were obtained. They likely bred in most of the possible and probable breeding locations shown on the map. Although easily identified by song or call, they can be confused with Western Meadowlarks if identification is based solely on sight. However, Atlasers provided documentation indicating they had accurately identified the two species.

Distribution

Eastern Meadowlarks were distributed statewide. A small gap in range occurred in the eastern Ozarks, extending through Crawford, Center, Reynolds, Shannon and Washington counties. This extensively-forested region apparently provides fewer nesting sites for Eastern Meadowlarks. Eastern Meadowlarks were found in the extensively row-cropped areas of the Mississippi Lowlands Natural Division although little breeding was confirmed there. Breeding confirmations were greatest in the Osage Plains. The distribution of Eastern Meadowlarks overlapped the range of Western Meadowlarks in northwestern Missouri. The two species may have selected slightly different habitats in the area of overlap although this is not apparent from the map. Lanyon (1956) found that where both occur in the northern plains states, Eastern Meadowlarks select more moist sites.

Abundance

Eastern Meadowlarks were most abundant in the Osage Plains Natural Division. They were less common in the Glaciated Plains, Ozark Border and Ozark natural divisions and least abundant in the Mississippi Lowlands. Robbins and Easterla (1992) suggest that numbers in the Ozark and Mississippi Lowlands natural divisions have increased during the 20th century.

Abundance by Natural Division
Average Number of Birds / 100 Stops

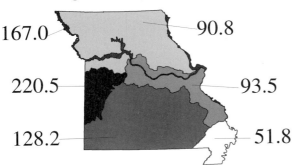

167.0   90.8

220.5   93.5

128.2   51.8

Breeding Phenology

EVIDENCE (# of Records)	MAR	APR	MAY	JUN	JUL	AUG	SEP
NB (10)		4/23			6/30		
NE (16)		5/01			6/28		
NY (7)			5/25	6/16			
FY (221)		4/21					8/20

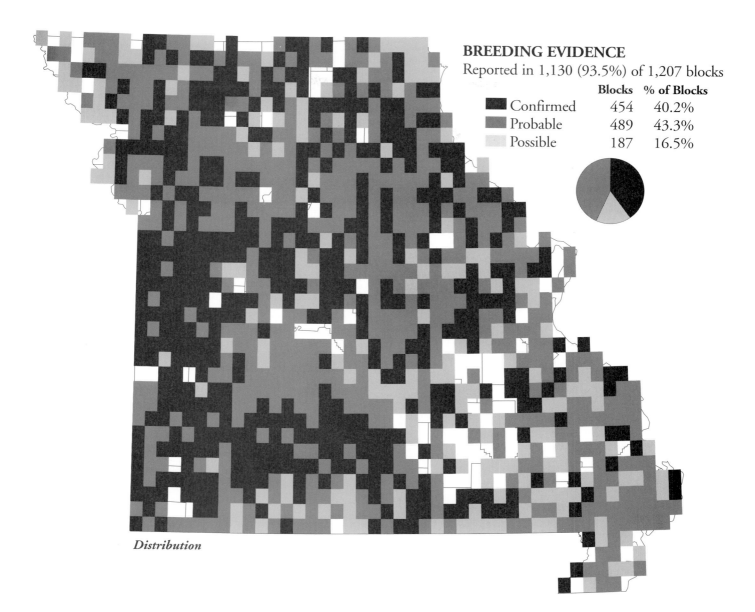

BREEDING EVIDENCE
Reported in 1,130 (93.5%) of 1,207 blocks

	Blocks	% of Blocks
Confirmed	454	40.2%
Probable	489	43.3%
Possible	187	16.5%

Distribution

Phenology

Eastern Meadowlarks migrate to Missouri primarily between late February through March (Robbins and Easterla 1992), although Atlasers did not document nest building until late April. Two broods per season are expected (Lanyon 1995). Late June dates for both eggs and nestlings may be attributable to second broods or renesting attempts.

Notes

Eastern Meadowlarks were found parasitized by Brown-headed Cowbirds in four of 244 blocks where Atlasers observed nest contents or deliveries of food to fledglings. Friedmann (1963) stated that Eastern Meadowlarks are uncommon cowbirds hosts.

Western Meadowlark
Sturnella neglecta

1994). Hybrids may infrequently occur where their ranges overlap (Robbins and Easterla 1992; Lanyon 1957 in Terres 1987).

Code Frequency

It was difficult for Atlasers to find nests of grassland birds and Western Meadowlarks presumably bred undetected in most blocks where they were discovered. Sight or song observations accounted for 73 percent of all records documenting their presence, multiple singing males or territoriality. Observers located two nests. Observations of food being delivered to young accounted for most confirmations. For only 24 percent of Western Meadowlark records was breeding confirmed compared with 40 percent for the Eastern Meadowlark.

Distribution

Missouri lies on the southeastern periphery of the breeding range of this widespread western species. Although native prairies and pastures were lacking in northwestern Missouri, Western Meadowlarks were found in a variety of habitats, from cropland and hedgerows to fallow fields. This species' distribution was concentrated in the Western Glaciated Plains Natural Section and decreased toward the east within the Glaciated Plains Natural Division. Scattered observations were noted throughout the Osage Plains Natural Division. The more easterly observations may represent breeders, migrants or misidentifications of Eastern Meadowlarks.

Abundance

Although they were fairly common throughout the Glaciated Plains, most Western Meadowlarks were located in the Western and Grand River natural sections. Other records indicated there are local small populations scattered throughout the remainder of the state.

Rangewide Distribution: Southwestern to south central Canada, western to west central United States & central Mexico

Abundance: Common & expanding northeast

Breeding Habitat: Drier fields & pastures with shorter vegetation than Eastern Meadowlark

Nest: Domed canopy of coarse grass, lined with finer grass & hair, with side opening in depression on ground

Eggs: 5 white with brown or purple marks

Incubation: 13–15 days

Fledging: 12 days

Spectacular songsters, Western Meadowlarks can best be identified by their song and call notes. The call notes are hollow "chucks," and the song is a gurgling, warbling jumble of flute-like notes. Distinguishing this species from the Eastern Meadowlark by sight is difficult but it can be accomplished based on a combination of field marks. Western Meadowlarks often associate with more arid, well-drained grasslands than do Eastern Meadowlarks. However, in Ontario there were no apparent habitat differences between the species. In the southwestern United States, Western Meadowlarks associate with damper sites (Lanyon

Abundance by Natural Division
Average Number of Birds / 100 Stops

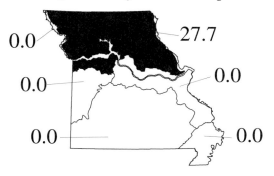

Breeding Phenology

EVIDENCE (# of Records)	MAR	APR	MAY	JUN	JUL	AUG	SEP
NE (1)			5/12	5/12			
NY (1)				6/04	6/04		
FY (13)			5/15			7/26	

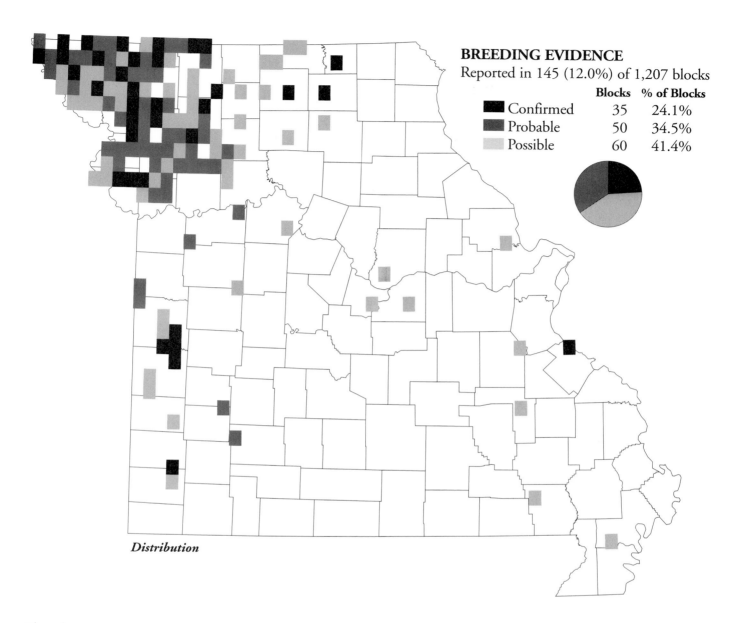

BREEDING EVIDENCE
Reported in 145 (12.0%) of 1,207 blocks

		Blocks	% of Blocks
■	Confirmed	35	24.1%
■	Probable	50	34.5%
■	Possible	60	41.4%

Distribution

Phenology

Nesting activity commenced in mid- to late April. Observations of fledglings on May 9 suggest meadowlarks likely laid eggs the last week of April. A late date of food being carried to young coincides with a date suggested by Lanyon (1994).

Notes

Neither of the two nests with eggs were observed to contain Brown-headed Cowbird eggs. Ehrlich et al. (1988) list them as an uncommon host.

Yellow-headed Blackbird
Xanthocephalus xanthocephalus

Rangewide Distribution: Southwestern Canada & western United States to northwestern Ohio
Abundance: Common to locally common
Breeding Habitat: Marshes, swamps with permanent water & emergent vegetation
Nest: Cup of woven vegetation, lined with dry grass over water in emergent vegetation
Eggs: 4 gray-white to pale green-white with brown or gray marks
Incubation: 11–13 days
Fledging: 9–12 days

Yellow-headed Blackbirds are frequently associated with emergent marshes. They take up residence in the deepest water areas, rather than the shallower areas associated with Red-winged Blackbirds. Their harsh song can be heard for great distances across the marsh. Although Missourians do not usually see this species, one or two Yellow-headed Blackbirds are occasionally found in Red-winged Blackbird flocks during spring and fall migrations.

Code Frequency

Only two observations of Yellow-headed Blackbirds were made during the Atlas Project. Ten to 20 other marshes that did not fall within the randomly-selected Atlas blocks were known to harbor this species at the same time this project was being conducted.

Distribution

Based on the results of the Atlas Project, Yellow-headed Blackbirds were distributed only in the Western Glaciated Plains Natural Section and the adjacent portion of the Big Rivers Natural Division. However, several restored marshes that were surveyed during the Atlas Project have since become new nesting sites for this species. Missouri is on the southern edge of this species' breeding range that extends eastward through northern Illinois (Terres 1987). Early in the century, nests were found over an area west of Clark County in the northeast to Saline and Jasper counties in central and southwestern Missouri. (Widmann 1907).

Abundance

Although no abundance information was collected on this species during the Atlas Project, Yellow-headed Blackbirds can be locally abundant. Squaw Creek National Wildlife Refuge in Holt County typically supports the largest population in the state. Twenty-three nests were located there in 1992.

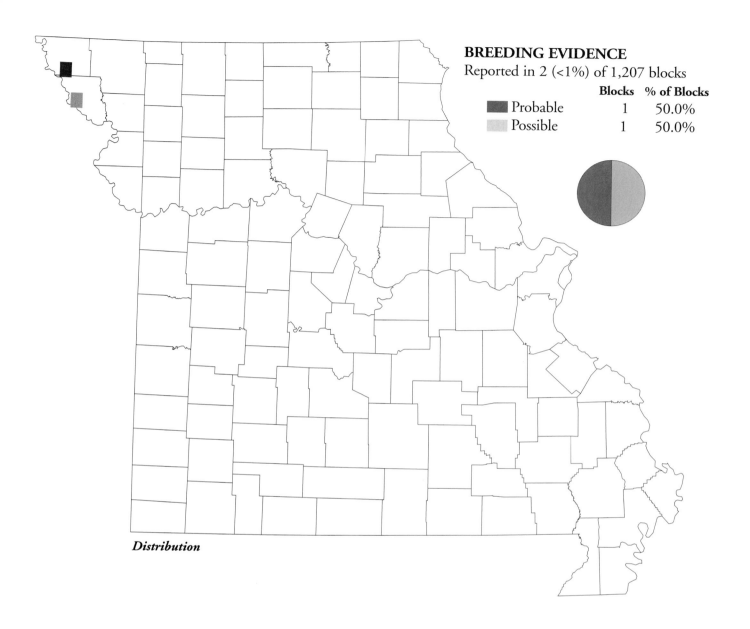

BREEDING EVIDENCE

Reported in 2 (<1%) of 1,207 blocks

		Blocks	% of Blocks
■	Probable	1	50.0%
■	Possible	1	50.0%

Distribution

Great-tailed Grackle
Quiscalus mexicanus

> **Rangewide Distribution:** South central United States to South America
> **Abundance:** Common, rapidly expanding northward
> **Breeding Habitat:** Open areas with scattered trees, cattail marsh & cultivated areas around human habitation
> **Nest:** Bulky cup of twigs, forbs, sedge & mud, lined with fine grass, rootlets & trash in tree near water
> **Eggs:** 3–4 greenish-blue with dark colored marks
> **Incubation:** 13–14 days
> **Fledging:** 20–23(?) days

This large grackle has shown a dramatic population increase and range expansion throughout the western United States, particularly in the Great Plains. The first documented evidence of Great-tailed Grackles in Missouri was a specimen collected at Bigelow Marsh in Holt County in 1976 (Robbins 1977). The first documented breeding record was at a cattail marsh next to Big Lake State Park in Holt County in 1979 (Robbins and Easterla 1986). Great-tailed Grackles breed in cattails (*Typha* spp.), other emergent vegetation and willow thickets in wetlands, especially if adjacent to agricultural fields or human habitation (Ehrlich et al. 1988). Farm ponds and large wetlands such as Squaw Creek National Wildlife Refuge also provide appropriate habitat in Missouri.

Code Frequency

Although Great-tailed Grackles are easily detected and identified, they are colonial in appropriate marsh habitat and could have been missed during the Atlas Project. Once located, however, Great-tailed Grackles were easily confirmed to breed. Therefore, the blocks in which possible breeding was recorded may not represent actual breeding sites but may involve vagrants or unmated individuals.

Distribution

The Great-tailed Grackle's Missouri range centers in the Osage Plains. Atlasers also detected them along the state's western border in Buchanan, Holt and Newton counties. Observations in Harrison, Johnson, Lafayette and Macon counties provided additional evidence for the continued eastward expansion of this species in Missouri. At most of these locations they bred in wetland vegetation, most commonly cattails. This species has recently been documented to breed as far east as Boone County (McKenzie 1996).

Phenology

Great-tailed Grackles begin to reappear in Missouri at the end of February and their numbers increase and peak in late March and early April (Robbins and Easterla 1992). Nests detected with young and fledglings on June 14 and July 14, respectively, may have represented second broods. Ehrlich et al. (1988) reported that Great-tailed Grackles produce two and, occasionally, three broods a season.

Breeding Phenology

EVIDENCE (# of Records)	MAR	APR	MAY	JUN	JUL	AUG	SEP
NY (2)		5/01		6/14			
FY (9)			5/20		7/13		

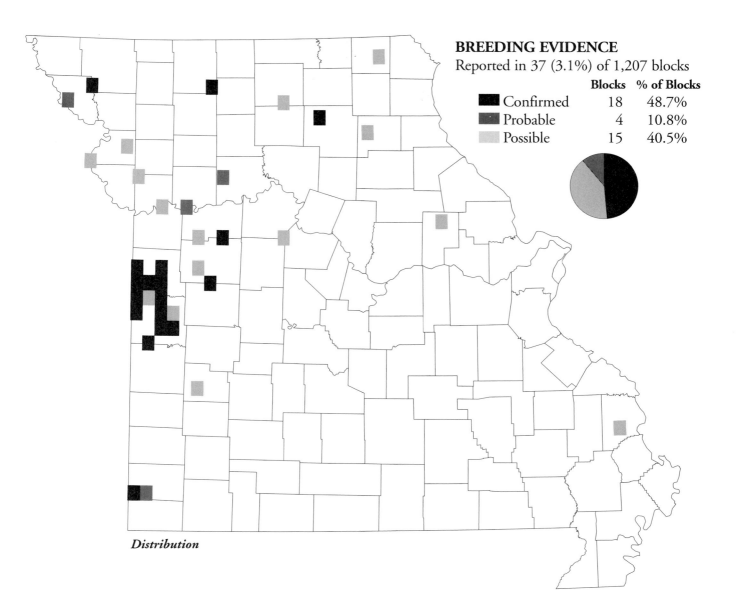

Reported in 37 (3.1%) of 1,207 blocks

	Blocks	% of Blocks
Confirmed	18	48.7%
Probable	4	10.8%
Possible	15	40.5%

Distribution

Common Grackle
Quiscalus quiscula

> **Rangewide Distribution:** Central through southeast-
> ern Canada & central through eastern United States
> **Abundance:** Common to abundant
> **Breeding Habitat:** Open woods, scattered trees &
> human habitat
> **Nest:** Grass, forbs, twigs, rushes, sedges & mud, lined
> with fine material & trash near water in shrub, tree,
> cavity or structure
> **Eggs:** 4–5 green-white to light brown, with brown or
> purple marks
> **Incubation:** 13–14 days
> **Fledging:** 16–20 days

Common Grackles are seen daily by most Missourians. They apparently benefit from close association with humans and they are common in suburbs, towns, parks and agrarian landscapes. Gregarious, noisy and numerous, grackles are easily detected because of their tendency to forage in open situations such as lawns and shorelines. They nest in small, loose colonies in overgrown fence rows, damp thickets, farm groves and residential trees and shrubs. They frequently select conifers (Peterjohn and Rice 1991) or other dense foliage to conceal their nests.

Code Frequency

Because Common Grackles are widespread, associate with human-altered habitats and are easy to detect, they would be expected to be recorded in most blocks. Although nests are well concealed, other evidence of breeding was observed frequently due to the abundant, conspicuous nature of the species. Common Grackles were easy for Atlasers to confirm in most blocks by observations of food being delivered to young or young out of the nest. Atlasers who were unable to locate nest sites or other breeding con-firmations may have been misled by wandering, foraging flocks or post-breeding individuals.

Distribution

Atlas Project findings indicate that Common Grackles range throughout the state. The species may have nested less frequently in the most-forested portions of the Ozarks, or perhaps it was simply more difficult to confirm there. Robbins and Easterla (1992) suggest that the species has increased in the Ozarks since Widmann's day. Fewer confir-mations in north central, central and extreme northwestern Missouri are difficult to explain. Fewer observations in some areas might be attributed to fewer foraging opportunities which force birds to travel greater distances, making them more likely to be missed by Atlasers.

Abundance

Although fewer confirmations were recorded in northern Missouri, this region exhibited the greatest abundance dur-ing the Atlas Project. Areas of abundance were highly vari-able among the natural divisions and produced no definite pattern. Robbins and Easterla (1992) reported that the Mississippi Lowlands had more than twice as many Common Grackles as the Ozark and Glaciated Plains natur-al divisions based on Breeding Bird Survey data from 1967 through 1989.

Abundance by Natural Division
Average Number of Birds / 100 Stops

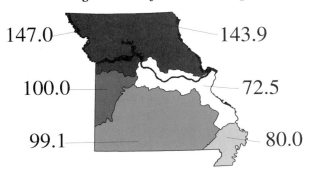

147.0   143.9

100.0   72.5

99.1   80.0

Breeding Phenology

EVIDENCE (# of Records)	MAR	APR	MAY	JUN	JUL	AUG	SEP
NB (44)	4/09				6/26		
NE (12)		4/20		6/03			
NY (42)		4/18				7/28	
FY (257)			5/01				8/07

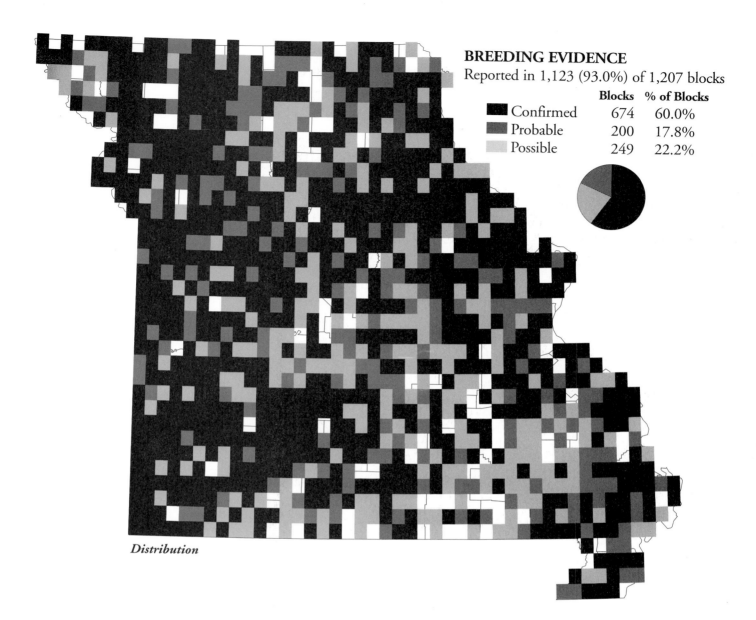

BREEDING EVIDENCE

Reported in 1,123 (93.0%) of 1,207 blocks

		Blocks	% of Blocks
■	Confirmed	674	60.0%
■	Probable	200	17.8%
■	Possible	249	22.2%

Distribution

Phenology

A total of 627 confirmed breeding reports provide a complete portrayal of Common Grackle breeding phenology. Migration into Missouri commences in late February and peaks in late March to early April (Robbins and Easterla

1992). Breeding is apparently initiated soon after arrival. Because Harrison (1975) stated that Common Grackles produce only one brood, late dates for nest building may represent renesting attempts.

Brown-headed Cowbird
Molothrus ater

Rangewide Distribution: Southern Canada & entire United States to central Mexico
Abundance: Common & expanding its range
Breeding Habitat: Fields, woods, edges, pasture, parks & residential
Nest: None; lays eggs in other birds' nests
Eggs: Cowbirds in captivity lay up to 77 eggs per season
Incubation: 10–12 days
Fledging: 11 days by host

Brown-headed Cowbirds, North America's best known brood parasites, are easily sighted. They often perch in full view presumably to court mates and select host nests. Because of the glossy blackness of their plumage and their brown heads, adult males are easily recognized. Female cowbirds are smaller, dusty brown and drab, and less easily identified. Brown-headed Cowbirds emit a variety of easily-recognized calls, from liquid "glugs" to squeaks and whistles.

Code Frequency

Due to their conspicuous habits, Brown-headed Cowbirds were easy to detect wherever they occurred. Certain codes such as nest building and food for young are inappropriate for obligate brood parasites. Atlasers accomplished confirmation of breeding only by the observation of a cowbird egg or young in another bird's nest, by the observation of a female cowbird visiting a nest, or by the observation of a fledgling cowbird being fed by a host species. Except for host species feeding of Brown-headed Cowbird young, confirmation of breeding was difficult for this species. Additionally, Brown-headed Cowbirds will commute up to seven kilometers between foraging and nesting territories (Robinson et al. 1993), and commuting birds may have accounted for some records in which breeding was not confirmed.

Distribution

As expected, Brown-headed Cowbirds were distributed essentially statewide, although they were not recorded at all in many adjacent blocks in Reynolds and Shannon counties. This may have been due to the lack of crop fields and pastures in this region. Thompson (1994) postulated that a lack of forest edge limits a cowbird's ability to find hosts. Also, Brown-headed Cowbirds are likely less numerous in areas that lack foraging habitat. Therefore, cowbirds may have avoided the more solidly-forested regions. They were less frequent in the Mississippi Lowlands, which has fewer domestic livestock and grazed habitats.

Abundance

There was an unexpected lack of variation in abundance statewide as determined by Miniroute and Breeding Bird Surveys. Slightly greater numbers were recorded in the Big Rivers and Ozark Border natural divisions than in the remaining four natural divisions.

Abundance by Natural Division
Average Number of Birds / 100 Stops

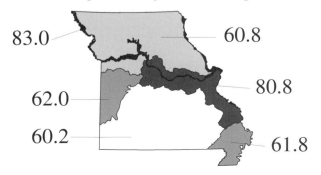

83.0
60.8
62.0
80.8
60.2
61.8

Breeding Phenology

EVIDENCE (# of Records)	MAR	APR	MAY	JUN	JUL	AUG	SEP
NE (78)		4/24				8/04	
NY (11)			5/15		7/16		
FY (13)			5/20			8/04	

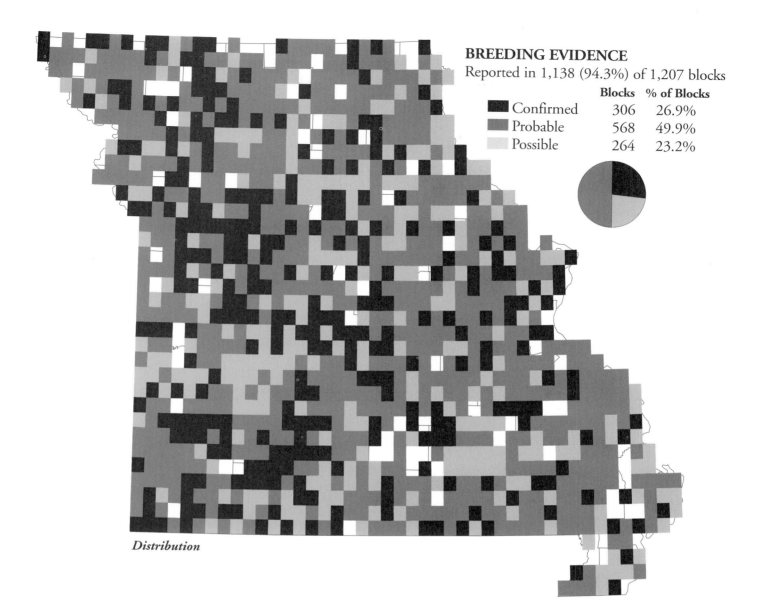

BREEDING EVIDENCE

Reported in 1,138 (94.3%) of 1,207 blocks

		Blocks	% of Blocks
■	Confirmed	306	26.9%
■	Probable	568	49.9%
■	Possible	264	23.2%

Distribution

Phenology

In the wild, Brown-headed Cowbird females ovulate and lay up to 40 eggs during their two-month breeding season (Friedmann and Kiff 1985). Atlasers observed nests with cowbird eggs from April 24 through August 4. The FY records in the phenology chart are observations of adults of several host species feeding young cowbirds.

Notes

Brown-headed Cowbird parasitism was recorded in 46 host species (Appendix E). Lowther (1993) listed 144 species that have reared cowbird young. He reported that the 15 most parasitized species in North America, beginning with the most frequently parasitized, were Yellow Warbler, Song Sparrow, Red-eyed Vireo, Chipping Sparrow, Eastern Phoebe, Eastern Towhee, Ovenbird, Common Yellowthroat, American Redstart, Indigo Bunting, Yellow-breasted Chat, Red-winged Blackbird, Kentucky Warbler, Willow Flycatcher and Bell's Vireo.

Orchard Oriole
Icterus spurius

Rangewide Distribution: Central to eastern United
   States & Mexico
Abundance: Widespread & locally common
Breeding Habitat: Lawns, parks, brushy areas & open
   areas with thickets
Nest: Woven green grass blades, lined with fine grass
   & plant down, suspended in shrub
Eggs: 3–5 pale bluish-white with brown, purple or
   gray marks
Incubation: 12 days
Fledging: 11–14 days

Chestnut and black with a beautiful warbling song, Orchard Orioles are familiar to many Missourians. They are smaller, less-known cousins of the brilliant orange and black Baltimore Orioles. Orchard Orioles are found throughout the state in shrubby and brushy areas of grasslands or other edge-type habitats. In addition to stream-side forest-edge habitats, they frequent trees in backyards and parks throughout the state. Farther west, or in open prairies, this species forages for insects in grasslands as well as trees. Family groups wander in late summer to seek fruit-bearing trees and shrubs (Widmann 1907). They sometimes nest in large loose colonies. For

example, one hundred and fourteen nests were found on seven acres in the state of Louisiana (Terres 1987).

Code Frequency

Atlasers used all codes to document this species, reporting them in 70 percent of all blocks statewide. Most were possible breeding records, although they may have represented actual breeding sites. Of all confirmed records, 44 percent were observations of nests at various stages of development.

Distribution

The statewide distribution map for Orchard Orioles is likely a true reflection of their range in Missouri. Parts of the Glaciated Plains, Mississippi Lowlands and Osage Plains natural divisions show several small areas where this species was absent. These areas may reflect a lack of suitable habitat or reflect differences in Atlasers' abilities to detect this species' song.

Abundance

The greatest abundance was recorded in the Ozark, Ozark Border and Big Rivers natural divisions where there is an abundance of forests and streams. These natural divisions averaged about 8.2 birds/100 stops. Natural divisions which have more open land averaged about one-third the relative abundance at 3.0 birds/100 stops.

Phenology

Most birds arrive in late April but an early Atlas Project record of April 6 was only one day later than the early date for the state (Robbins and Easterla 1992). Only one nest with young was recorded later than June, and it may have been a renesting attempt. Nesting activity occurred during May, with one bird reported on a nest by May 8 and fledg-

Abundance by Natural Division
Average Number of Birds / 100 Stops

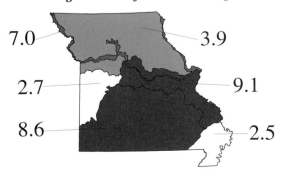

7.0   3.9
2.7   9.1
8.6   2.5

Breeding Phenology

EVIDENCE (# of Records)	MAR	APR	MAY	JUN	JUL	AUG	SEP
NB (27)			5/16		6/30		
NE (2)			6/11		6/24		
NY (26)			6/03		7/25		
FY (63)			5/25			8/18	

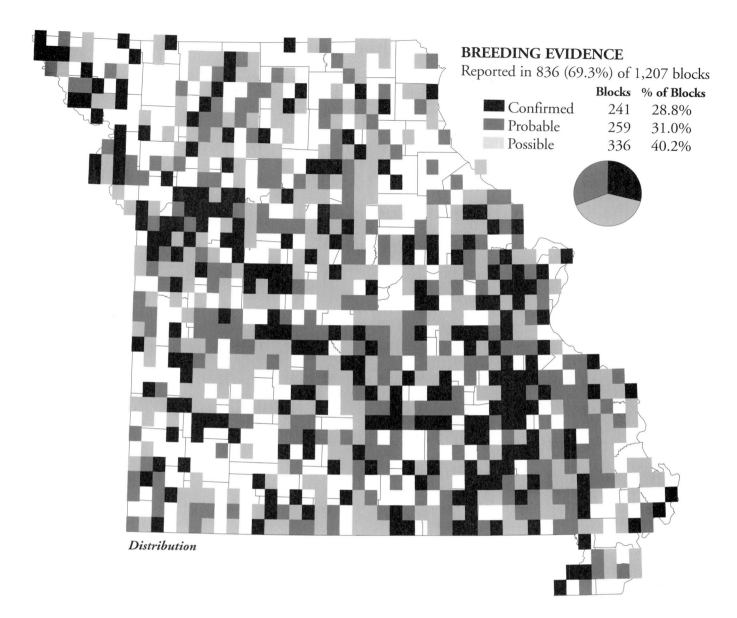

BREEDING EVIDENCE

Reported in 836 (69.3%) of 1,207 blocks

		Blocks	% of Blocks
■	Confirmed	241	28.8%
■	Probable	259	31.0%
■	Possible	336	40.2%

Distribution

lings observed by May 31. Most Orchard Orioles depart in late August and early September (Robbins and Easterla 1992).

Notes

There were three records of Brown-headed Cowbirds parasitizing Orchard Orioles. Two reports were of fledglings being feed by adult Orchard Orioles and one report involved a squabble between a cowbird and an Orchard Oriole undertaking a second nesting attempt. Ehrlich et al. (1988) listed the Orchard Oriole as a common cowbird host.

Baltimore Oriole
Icterus galbula

Rangewide Distribution: Southern Canada & entire United States to northern Mexico
Abundance: Common, widespread in the west
Breeding Habitat: Deciduous trees, lawns & open woodlands
Nest: Plant fiber, lined with fine grass, plant down & hair on drooping branch of deciduous tree
Eggs: 4–5 pale grayish- to bluish-white marked with dark colors
Incubation: 12–14 days
Fledging: 12–14 days

This colorful species was known as the Northern Oriole during the Atlas Project. In 1995, however, the name was officially changed to Baltimore Oriole (AOU 1995). The beautiful, bright orange and black male is usually seen as a flash of colors in the top of a shade tree. The swinging, sac-like nests hang from small drooping outer branches of willows, elms, maples and sycamores, and are most easily detected in winter.

Code Frequency

Baltimore Orioles are brightly colored, beautiful songsters that frequently nest in trees in yards. Atlasers easily recorded their presence during the breeding season. Their characteristic swinging nests made them easy to confirm especially during winter. Atlasers recorded Baltimore Orioles in 66 percent of all blocks. The breeding evidence most frequently recorded by Atlasers was nest building, nest visitation and food deliveries to young and fledglings.

Distribution

Baltimore Orioles were found statewide. This species was recorded in 93 percent of the blocks in the Glaciated Plains and 73 percent of the Osage Plains. They were less common in the Mississippi Lowlands and Ozark natural divisions where they were reported in only 52 and 33 percent of the blocks, respectively. The forested Ozark Natural Division appears to have a very small breeding population. Most breeding records of the Ozark Natural Division are in towns and open forests near the adjacent section to the east and west. The sparseness of population in upland forests in the Ozark Natural Division is mirrored in upland forests of Arkansas (James and Neal, 1986). The low number of records in the Mississippi Lowlands Natural Division presumably reflects the extensive treeless condition of this agricultural region. In the Mississippi Lowlands, potential nesting areas with large shade trees are restricted to urban areas.

Abundance

Abundance increased with latitude, with 7–13 birds/100 stops north of the Ozark Natural Division and fewer than one bird/100 stops within it.

Phenology

Most summer residents arrive by mid-April and depart by late August with migrants present in September and early

Abundance by Natural Division
Average Number of Birds / 100 Stops

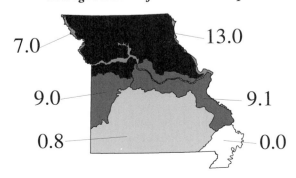

7.0
13.0
9.0
9.1
0.8
0.0

Breeding Phenology

EVIDENCE (# of Records)	MAR	APR	MAY	JUN	JUL	AUG	SEP
NB (41)		5/05			7/20		
NE (4)		5/17			7/02		
NY (29)		5/01			7/20		
FY (75)			5/20			8/09	

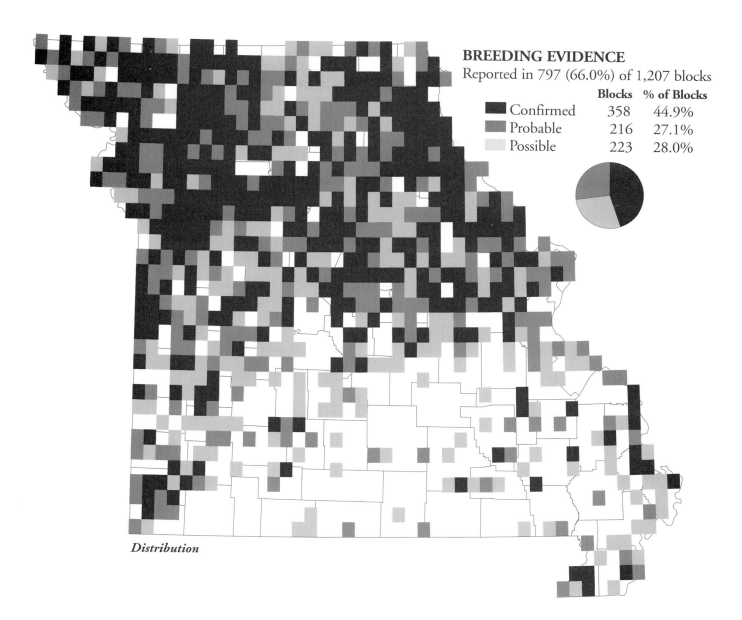

BREEDING EVIDENCE

Reported in 797 (66.0%) of 1,207 blocks

		Blocks	% of Blocks
■	Confirmed	358	44.9%
▨	Probable	216	27.1%
░	Possible	223	28.0%

Distribution

October (Robbins and Easterla 1992). Records of fledglings extended through August 5. Some late-season Atlasers located 49 used nests between February 2 and November 28.

Notes

There was one report of a fledgling Brown-headed Cowbird being fed by a Baltimore Oriole in the eastern Glaciated Plains Natural Section. This species was considered an uncommon host species by Ehrlich et al. (1988). In large forests of the Ozarks, Atlasers noted Baltimore Orioles which likely were in migrating groups and not breeders. Widmann (1907) also reported such individuals on dry wooded ridges in May in the Ozarks.

House Finch

Carpodacus mexicanus

Rangewide Distribution: Southern Canada, western & eastern (not central) United States

Abundance: Abundant

Breeding Habitat: Areas with scattered trees, bushes & thickets

Nest: Twigs, grass, debris, leaves, roots & hair in tree or building

Eggs: 4–5 blue-white or blue-green with brown or black marks & wreathed

Incubation: 12–14 days

Fledging: 11–19 days

This recent addition to Missouri's avifauna is seldom found far from human habitation. It is associated with urban, suburban and rural areas that have suitable nesting sites and where bird feeders are maintained. House Finches select nest sites with overhead protection including hanging flower pots, under awnings and extremely dense cover such as evergreen trees, shrubs and hedges. Natives of the Rocky Mountains westward, House Finches were introduced into the eastern United States during the 1940s (Elliott and Arbib 1953). Initially restricted to the vicinity of New York City, their numbers increased substantially during the 1960s as they expanded their breeding range across the United States (Hill 1993). They were first detected in Missouri in the St. Louis area in 1974 and then again in 1977 (Jones 1978). The first nest was reported in St. Louis in 1983 (Wilson 1984).

Code Frequency

House Finches were still expanding their range westward across Missouri and increasing in numbers when the Atlas Project was initiated in 1986. House Finches are considered a relatively easy species to detect and confirm due to their loud, often-repeated song and their association with human habitation.

Distribution

House Finches were sparsely scattered to the western border of Missouri indicating their westward range expansion across the state was complete by the conclusion of the Atlas Project. Detected in only 100 blocks, House Finches were rarely recorded in the western Ozarks and were most frequently recorded in the Glaciated Plains, Osage Plains, Mississippi Lowlands and eastern Ozark natural divisions.

Phenology

House Finches are permanent residents (Robbins and Easterla 1992) that initiate nesting in spring when conditions are favorable. The nest with young discovered on May 11 indicates that nesting was actually initiated in mid- to late April. This may be typical of the first brood. House Finches often raise two or three broods (Hill 1993). Atlas observations of nest building and the late observation of a nest with young likely represent second broods.

Breeding Phenology

EVIDENCE (# of Records)	MAR	APR	MAY	JUN	JUL	AUG	SEP
NB (2)				6/10	6/20		
NE (2)			5/12	6/07			
NY (3)			5/11	6/17			
FY (3)			5/19		7/16		

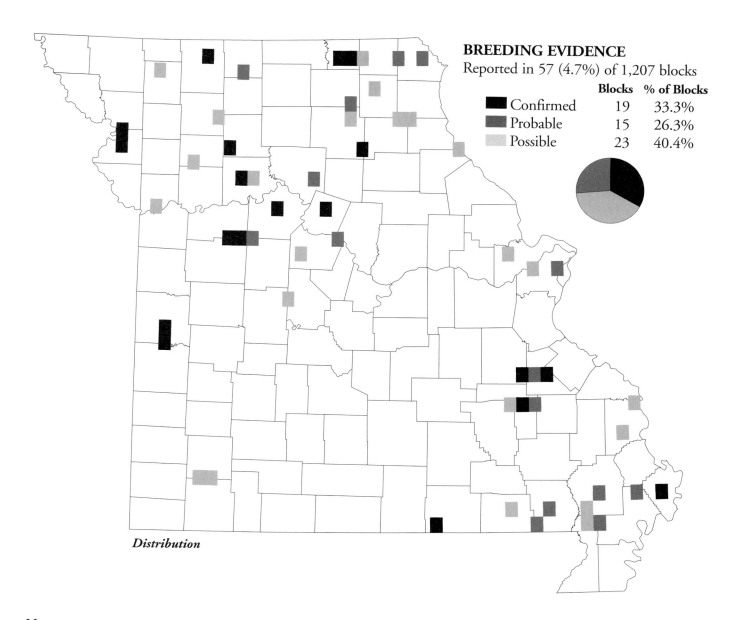

BREEDING EVIDENCE
Reported in 57 (4.7%) of 1,207 blocks

	Blocks	% of Blocks
Confirmed	19	33.3%
Probable	15	26.3%
Possible	23	40.4%

Distribution

Notes

Brown-headed Cowbirds generally do not parasitize the nests of House Finches (Hill 1993). No brood parasitism was observed during the Atlas Project.

Pine Siskin
Carduelis pinus

Rangewide Distribution: Southern Canada, western to
west central & northeastern corner of United States
Abundance: Common
Breeding Habitat: Coniferous & mixed forests, parks
& residential areas
Nest: Twigs, rootlets & grass, lined with fine rootlets,
moss, fur & feathers in deciduous tree
Eggs: 3–4 pale greenish-blue, wreathed with browns or
black spots
Incubation: 13 days
Fledging: 14–15 days

Irregular winter residents and migrants in Missouri, Pine
Siskins will occasionally remain into the summer following
years of sizable winter populations. Such individuals and
pairs are usually discovered at bird feeders, and although
nesting is suspected, confirmations of breeding are rare.

Code Frequency

Bird-bander Marquerite Solomon captured and banded a
Pine Siskin within a Camden County block.

Phenology

The female banded in Camden County had a brood
patch on May 30, suggesting that the bird was incubating a
clutch of eggs nearby.

Notes

Four other nesting attempts have been recorded for
Missouri since 1960, two in Jackson and two in Cass counties (Robbins and Easterla 1992).

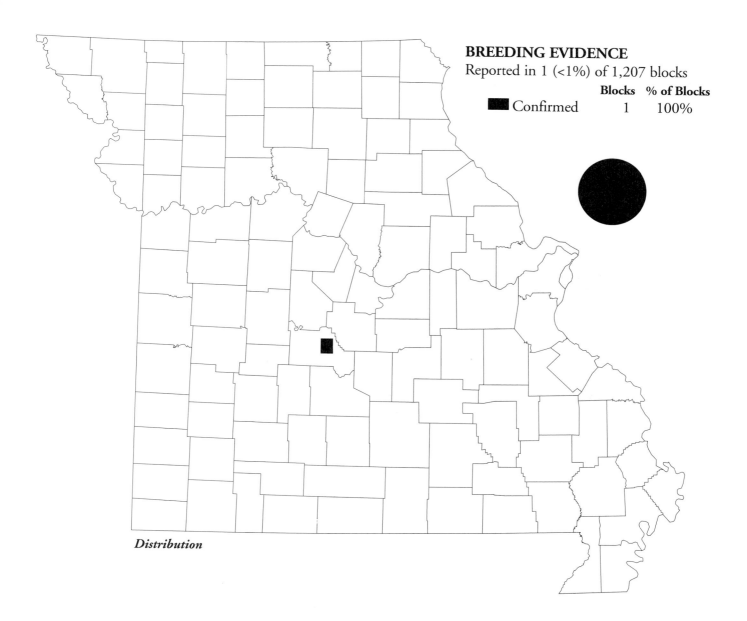

Reported in 1 (<1%) of 1,207 blocks

	Blocks	% of Blocks
Confirmed	1	100%

Distribution

American Goldfinch

Carduelis tristis

Rangewide Distribution: South central & southeastern Canada, northern, central & southeastern United States

Abundance: Common & widespread

Breeding Habitat: Edges of deciduous forest & open, weedy & cultivated areas

Nest: Tightly woven forbs & vegetation, lined with plant down, in trees

Eggs: 4–6 pale blue or bluish-white, unmarked

Incubation: 10–12 days

Fledging: 11–17 days

During the breeding season, these widespread finches occupy rural areas, especially fallow, weedy fields interspersed with brushy thickets. They are prevalent in weedy margins of wetlands and brushy borders of farmsteads and forests. Small parcels such as fence rows, roadside ditches and even small forest openings are suitable if brushy cover is present. Goldfinches construct a thick-walled cup nest in an upright fork of a shrub or small sapling at a height of 1–3 meters above the ground (Peck and James 1987).

American Goldfinches typically initiate breeding in July or August, the latest of any North American passerine (Peterjohn and Rice 1991). The comparatively late nesting of this species appears to be correlated with the maturation of thistles, which provide seeds for food and down for nesting material (Harrison 1975).

Code Frequency

American Goldfinches are easy to detect and, therefore, where not located in a block, they presumably occurred in extremely low numbers or were not present. Their distinctive twittering usually attracts attention first, and they are often then seen passing overhead in characteristic undulating flight. Because flying birds are often observed in pairs, Atlasers were able to easily elevate them to the probable breeding level.

Breeding confirmations, however, were difficult to observe. Apparently Atlasers were rarely successful in their attempts to follow the movements of goldfinches suspected of nesting. Perhaps the bulk of the surveying effort was conducted too early in the season to confirm this unusually late-nesting species. Considering these limitations, American Goldfinches presumably bred in most blocks in which they were found.

Distribution

American Goldfinches were recorded statewide. Their only region of scarcity was in the Mississippi Lowlands where extensively tilled lands and lack of brushy, weedy fields offer few potential nest sites. Atlasers obtained only possible breeding evidence in many adjacent blocks in the lower Ozarks, perhaps because American Goldfinches rarely breed in its extensive forests which have little brush and edge habitat. Another region of fewer probable and confirmed breeding records was centered on Shelby County in northeastern Missouri, which has much favorable habitat. Reasons for the lower detection rate are unknown. Perhaps the cursory coverage that region received via Block-Busting and other single-weekend surveys, combined with late-season nesting of the species, resulted in reduced detections.

Abundance

American Goldfinches were recorded as most abundant

Abundance by Natural Division
Average Number of Birds / 100 Stops

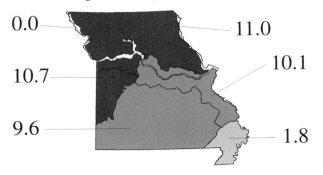

0.0

11.0

10.7

10.1

9.6

1.8

Breeding Phenology

EVIDENCE (# of Records)	MAR	APR	MAY	JUN	JUL	AUG	SEP
NB (30)			5/27				8/26
NE (2)				7/10			8/29
NY (8)			6/08				8/31
FY (8)				6/28			9/03

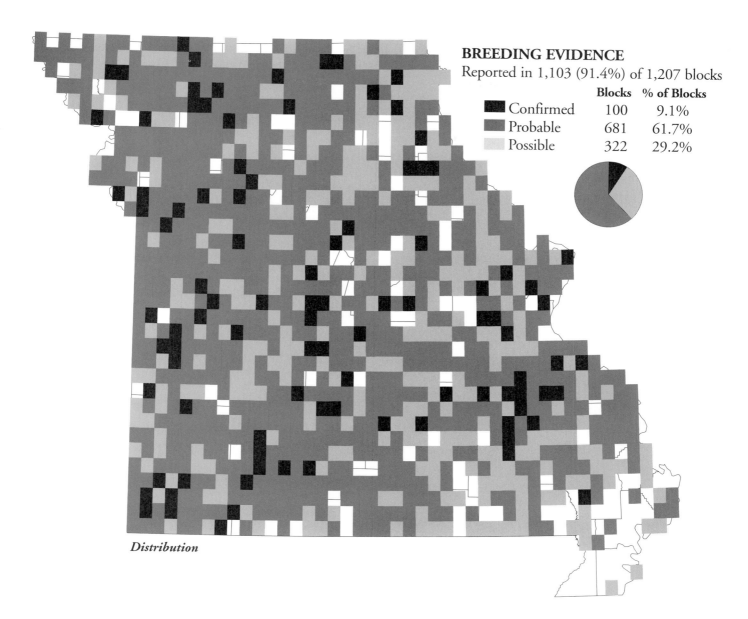

BREEDING EVIDENCE
Reported in 1,103 (91.4%) of 1,207 blocks

	Blocks	% of Blocks
Confirmed	100	9.1%
Probable	681	61.7%
Possible	322	29.2%

Distribution

in the Glaciated Plains and Osage Plains natural divisions. This concurs with Robbins and Easterla (1992) who reported on BBS data. Goldfinches also appeared nearly as abundant in the Ozark Border and Ozark natural divisions. As expected from the distributional map, they were least abundant in the Mississippi Lowlands.

Phenology

In Ohio, pair formation continues into June but most goldfinches do not initiate nesting until after July 15 (Nice 1939). Atlasers observed confirmed breeding evidence unexpectedly early in comparison. A nest building record on May 27 was three days earlier than the earliest record reported in Ohio. A nest containing hatched young, recorded on June 8, was even more remarkable. Middleton (1993)

reported a few early nests with eggs in May and early June; however, nesting peaks for this species in the second half of July and continues into September. Paul McKenzie, United States Fish and Wildlife Service, suggested in editorial review that earlier nesting may be a response to earlier-maturing, exotic thistles that are now more abundant and may provide an early source of nesting material. The Atlas Project's latest nest with eggs was observed on August 29. Nesting events likely continued but were not reported as Atlasers were asked to conclude their field survey by September 15.

Notes

Although American Goldfinches are common Brown-headed Cowbird hosts (Ehrlich et al. 1988), Atlasers did not document any parasitism during the project.

House Sparrow

Passer domesticus

Rangewide Distribution: Southern Canada, United
    States & Mexico
Abundance: Abundant
Breeding Habitat: Open areas with scattered trees &
    bushes
Nest: Grass & forbs, lined with feathers & hair, occa-
    sionally a ball-shaped nest, usually in tree cavities
Eggs: 4–6 white or greenish-bluish, with gray or
    brown marks
Incubation: 10–13 days
Fledging: 14–17 days

These Old World natives have become adaptable residents of our cities and farms. They were introduced into the
United States in 1850 at Brooklyn, New York and subsequently at other locations (Lowther and Cink 1992). They
were first recorded in Missouri in 1870 at St. Louis
(Widmann 1907) and may have reached their greatest numbers here during the early 1900s before mechanization
replaced horse farming. The Missouri Breeding Bird Survey
indicates that this species has declined at an average annual
rate of 2.6 percent from 1967 through 1989 (Wilson 1990).
House Sparrows are often unwanted because of their high

population densities and because their straggling and loosely
constructed nests are typically placed on or in buildings.
They have adversely affected our native avifauna, including
Eastern Bluebirds, Tufted Titmice, Cliff Swallows and various chickadees (Peterjohn and Rice 1991).

Code Frequency

House Sparrows are one of the easiest species to detect
and to confirm as breeders. They had the highest percentage
of confirmation of any species. It is likely that they bred in
all blocks where located even where not confirmed.
Similarly, they were probably absent in the few blocks where
they were not recorded.

Distribution

House Sparrows were absent from the most densely-
forested parts of the state. House Sparrows were recorded in
fewer than half the blocks in Carter, Reynolds and Shannon
counties. Additionally, they were absent or rarely recorded in
adjacent counties and in certain blocks in Stone, Douglas,
Ozark and Pulaski counties, perhaps due to the lack of suitable nesting and foraging habitat. Otherwise, House
Sparrows bred solidly across Missouri. They likely also
occurred in all blocks in the Glaciated Plains and Osage
Plains natural divisions.

Abundance

The relative abundance of House Sparrows was greatest
in the Big River, Glaciated Plains and Osage Plains natural
divisions, perhaps due to the availability of nesting and foraging habitat.

Phenology

House Sparrows are permanent residents and initiate
breeding early in spring. Nest building observed in mid-

Abundance by Natural Division
Average Number of Birds / 100 Stops

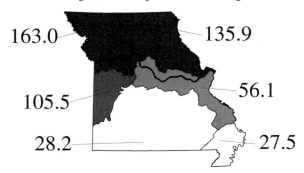

163.0   135.9

105.5   56.1

28.2   27.5

Breeding Phenology

EVIDENCE (# of Records)	MAR	APR	MAY	JUN	JUL	AUG	SEP
NB (135)	3/11					8/04	
NE (30)		4/12			7/11		
NY (138)		4/13				8/05	
FY (90)		4/20					8/26

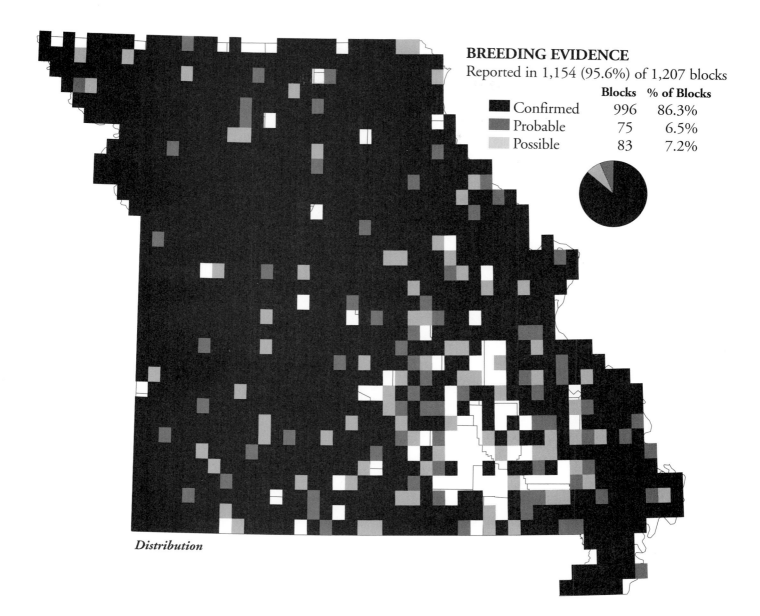

BREEDING EVIDENCE
Reported in 1,154 (95.6%) of 1,207 blocks

		Blocks	% of Blocks
■	Confirmed	996	86.3%
■	Probable	75	6.5%
■	Possible	83	7.2%

Distribution

March is likely typical of the season's first nesting. House Sparrows can rear multiple broods and often nest through early September (Lowther and Cink 1992). Late observations of nest building and food being carried to young are evidence of a protracted nesting season in Missouri.

Notes

House Sparrows are apparently a rare host of Brown-headed Cowbirds (Friedmann 1963). Three instances of parasitism were recorded during the Atlas Project.

Eurasian Tree Sparrow
Passer montanus

> **Rangewide Distribution:** Native to Old World, eastern to central Missouri & western Illinois
> **Abundance:** Introduced & now locally common
> **Breeding Habitat:** Open areas with scattered trees & bushes & agricultural areas
> **Nest:** Grass & forbs, lined with feathers in artificial or tree cavity
> **Eggs:** 4–6 white to pale gray, with brown marks
> **Incubation:** 13–14 days
> **Fledging:** 12–14 days

Introduced to St. Louis in 1870, this close relative of the House Sparrow has not expanded its range much beyond the St. Louis vicinity (Porter 1989). Nonetheless, it is now a permanent resident in Hannibal, Missouri and Quincy, Illinois (Robbins and Easterla 1992). It often shares its city, town and farm habitat with the House Sparrow. The two species are similar in appearance except that the Eurasian Tree Sparrow has a black ear patch on a white cheek and, unlike the House Sparrow, the coloration of the female's plumage is similar to the male's.

Code Frequency

Although Atlasers were cognizant of Eurasian Tree Sparrows in the St. Louis and Hannibal areas, it is likely that many Atlasers in outlying blocks misidentified them as House Sparrows. Therefore, Eurasian Tree Sparrows may have had a wider distribution than suggested by the map. Breeding confirmations were apparently easy to obtain with breeding confirmed in 12 of the 13 blocks in which this species was found.

Distribution

Eurasian Tree Sparrows bred primarily in St. Louis and St. Charles counties. There were three additional blocks scattered in areas adjacent to the Mississippi River north to the Hannibal area. The observations remote from St. Louis were predictable as Musselman (1950) reported them breeding at Hannibal in 1946 and Dinsmore et al. (1984) reported them as far north as Davenport, Iowa. It is surprising that they were not recorded farther west and south of St. Louis. In 1986, six birds were recorded outside of Atlas blocks in Montgomery County (Barksdale 1987) and single birds were seen as far south as Farmington during the mid-1970s (Robbins and Easterla 1992). If they did occur within blocks beyond the range indicated on the map, they may have been missed by Atlasers or misidentified as House Sparrows.

Phenology

According to the limited number of observations obtained during the Atlas Project, Eurasian Tree Sparrows bred from May through July.

Breeding Phenology

EVIDENCE (# of Records)	MAR	APR	MAY	JUN	JUL	AUG	SEP
NY (3)				6/24	7/25		
FY (2)			5/18	6/02			

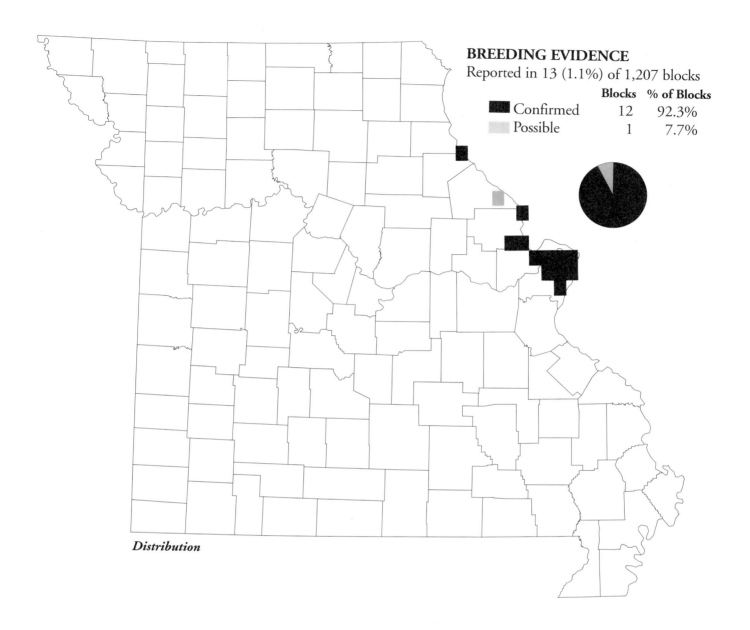

Appendix A
Atlas Block Descriptions and Summaries of Species, Volunteers and Survey Hours

Quadrangle (1)	Number(2)	Lat.(3)	Long.(4)	Spec.(5)	Hrs.(6)	Vol.(7)	Block-busted (8)
Acorn Ridge	1086E	365230	900000	50	15.5	1	
Adrian	577 A	382230	941500	58	8.0	2	
Advance	1043C	370000	895230	53	10.0	1	
Agency	233 F	393730	943730	54	9.5	3	
Alanthus Grove	79 D	401500	943000	53	34.5	4	
Albany North	81 E	401500	941500	61	12.0	1	
Albany South	114 F	400730	941500	45	6.0	1	
Aldrich	846 C	373000	933000	69	18.0	1	
Allbright	957 D	371500	900730	46	4.0	1	B
Allendale	46 E	402230	941500	52	7.0	1	
Alley Spring	989 E	370730	912230	62	18.0	3	B
Alma	406 D	390000	933000	56	12.0	1	
Altamont	179 D	395230	940000	65	22.0	1	
Altenburg	837 C	373730	893000	76	33.0	1	
Alton	1165D	363730	912230	61	7.0	1	
Amazonia	172 C	395230	945230	54	4.0	1	
Americus	490 C	384500	913000	76	15.0	4	
Amity	205 B	394500	942230	60	16.0	1	
Amoret	611 D	381500	943000	61	16.0	2	
Anderson	1141F	363730	942230	38	54.0	1	
Annada	362 F	391500	904500	80	14.0	1	
Anniston	1137E	364500	891500	37	0.0	1	
Anson	33 A	403000	914500	31	3.0	1	B
Anthonies Mill	709 C	380000	910000	42	47.0	1	
Anutt	820 B	373730	913730	66	7.5	2	B
Appleton City	650 A	380730	940000	66	12.0	2	
Arbela	64 E	402230	920000	55	6.5	3	B
Arbyrd	1247A	360000	900730	36	7.0	2	
Argo	672 F	380730	911500	63	4.5	1	
Argyle	631 E	381500	920000	61	26.0	1	
Arley	300 E	392230	942230	41	16.0	1	
Armstrong	347 A	391500	923730	63	28.5	1	
Arnica	764 C	374500	933730	64	11.0	1	
Arrow Rock	411 E	390000	925230	57	29.0	1	
Asbury	922 B	371500	943000	72	41.0	1	
Ash Grove	930 F	371500	933000	60	22.0	2	
Ashburn	292 D	393000	910730	54	0.0	1	
Ashland	484 A	384500	921500	78	22.0	1	
Atchison East	261 C	393000	950000	40	3.5	1	
Atlanta	192 F	395230	922230	53	10.0	3	B
Auburn	395 F	390730	905230	47	17.0	2	
Aurora	1057A	365230	933730	56	5.0	1	
Austin	541 A	383000	941500	58	9.5	2	
Auxvasse	419 B	390000	915230	52	6.5	1	
Ava	1065B	365230	923730	56	7.0	4	B

Note: Column headings are as follows: (1) Name of quadrangle where survey block is located; (2) Sequentially assigned block number; (3) Latitude; (4) Longitude; (5) Total number of species observed in block; (6) Total hours of observation; (7) Number of volunteers who surveyed within block; and (8) "B" indicates survey coverage by block-busting.

Quadrangle (1)	Number(2)	Lat.(3)	Long.(4)	Spec.(5)	Hrs.(6)	Vol.(7)	Block-busted (8)
Avalon	243 B	393730	932230	51	4.0	2	B
Avilla	967 B	370730	940730	54	11.0	1	
Axtell	221 E	394500	922230	64	16.0	4	
Azen	31 B	403000	920000	48	6.0	2	B
Bado Se	942 F	371500	920000	56	21.0	1	
Bagnell	662 C	380730	923000	76	25.0	1	
Bahner	550 C	383000	930730	60	7.0	2	B
Bakersfield	1203C	363000	920730	58	13.0	1	
Bancroft	150 F	400000	934500	64	32.0	2	
Banner	827 A	373730	904500	68	40.0	2	
Bardley	1168E	363730	910000	58	22.0	1	
Barnard	110 F	400730	944500	62	10.5	1	
Barnesville	191 F	395230	923000	64	16.0	3	B
Barnett	590 C	382230	923730	50	37.0	2	
Barnumton	694 E	380000	925230	70	7.5	1	
Bartlett	1031E	370000	912230	61	14.0	2	B
Bassville	933 B	371500	930730	58	8.0	1	
Bates City	402 C	390000	940000	59	9.0	1	
Bayouville	1182F	363730	891500	61	5.0	2	
Beach	893 E	372230	925230	68	13.5	1	
Beaman	477 E	384500	930730	58	5.0	2	B
Bearcreek	804 A	373730	933730	62	9.5	2	
Belew Creek	643 F	381500	903000	78	89.0	4	
Belgique	795 F	374500	894500	68	5.0	1	
Belgrade	787 B	374500	904500	62	32.0	1	
Bell City	1044C	370000	894500	49	14.0	4	B
Bellamy	800 F	373730	940730	70	21.0	1	
Belle	634 B	381500	913730	53	7.0	1	
Bellflower North	424 C	390000	911500	53	20.5	1	
Bellflower South	456 C	385230	911500	66	12.0	1	
Belton	466 A	384500	943000	66	17.0	4	
Benjamin	135 E	400730	913730	47	5.5	3	B
Bennett Spring	811 E	373730	924500	49	0.0	1	
Berger	529 F	383730	911500	51	16.0	1	
Berlin	146 E	400000	941500	58	15.5	1	
Bernie	1177E	363730	895230	33	7.0	1	
Berryman	746 C	375230	910000	56	27.0	2	
Bertrand	1091B	365230	892230	45	2.5	3	B
Bethany	83 B	401500	940000	66	8.0	3	
Bethel	195 A	395230	920000	51	14.0	3	B
Bethpage	1143A	363730	940730	35	35.0	1	
Beulah	859 F	373000	915230	80	21.5	1	
Bevier North	220 D	394500	923000	48	20.0	4	B
Bevier South	250 E	393730	923000	76	19.0	3	
Bible Grove	97 A	401500	921500	54	5.0	2	B
Big Bend	701 F	380000	920000	66	22.0	1	
Big Lake	138 B	400000	951500	66	6.0	1	B
Big Piney	817 E	373730	920000	82	44.0	1	
Big Spring	1079A	365230	905230	64	11.0	1	B
Billings	1014A	370000	933000	43	4.5	1	

Quadrangle (1)	Number(2)	Lat.(3)	Long.(4)	Spec.(5)	Hrs.(6)	Vol.(7)	Block-busted (8)
Billingsville	444 C	385230	924500	59	13.0	2	
Billmore	1211B	363000	910730	63	12.0	1	
Birch Tree	1075E	365230	912230	58	21.0	1	
Blackburn	407 F	390000	932230	64	10.0	1	
Blairstown	544 E	383000	935230	59	22.0	1	
Blanchard	6 E	403000	950730	54	17.0	1	
Block	12 C	403000	942230	57	7.5	1	
Blockton	13 E	403000	941500	60	6.0	1	
Blockton Se	816 D	373730	920730	65	17.0	3	B
Bloodland	1087B	365230	895230	62	14.5	3	B
Bloomfield	716 E	380000	900730	60	5.0	1	
Bloomsdale	400 F	390000	941500	57	19.5	1	
Blue Springs	1231A	362230	894500	40	3.0	2	B
Boekerton	307 B	392230	933000	64	8.0	2	B
Bogard	142 F	400000	944500	65	12.0	1	
Bolckow	109 B	400730	945230	56	10.5	1	
Bolckow Nw	847 B	373000	932230	51	8.5	1	
Bolivar	659 B	380730	925230	42	12.0	1	
Bollinger Creek	845 D	373000	933730	62	45.0	2	
Bona	1226B	363000	891500	58	8.0	2	
Bondurant	750 B	375230	903000	83	30.0	1	
Bonne Terre	445 B	385230	923730	58	10.5	1	
Boonville	309 A	392230	931500	50	13.5	1	
Bosworth	359 F	391500	910730	63	8.0	1	B
Bowling Green	623 A	381500	930000	69	8.5	3	B
Boylers Mill	1108D	364500	925230	70	21.0	1	
Bradleyville	1236F	361500	895230	38	10.0	4	B
Bragg City	730 C	375230	930000	70	8.0	1	
Branch	1163A	363730	913730	59	10.5	1	
Brandsville	1151D	363730	930730	76	23.0	1	
Branson	129 C	400730	922230	49	8.5	10	
Brashear	271 B	393000	934500	56	30.0	1	
Braymer	700 C	380000	920730	56	6.0	1	
Brays	593 C	382230	921500	77	36.0	2	
Brazito	210 E	394500	934500	57	6.0	4	B
Breckenridge	1169A	363730	905230	61	21.0	1	
Briar	119 B	400730	933730	58	12.0	2	B
Brimson	666 C	380730	920000	52	5.0	1	
Brinktown	30 F	403000	920730	59	13.0	2	B
Brock	798 F	373730	942230	85	25.0	2	
Bronaugh	216 D	394500	930000	51	8.0	2	B
Brookfield	973 B	370730	932230	46	20.5	1	
Brookline	48 F	402230	940000	67	8.0	2	
Brooklyn	1175E	363730	900730	44	21.5	1	
Broseley	1109B	364500	924500	69	22.0	2	
Brown Branch	815 D	373730	921500	69	9.5	1	B
Brownfield	155 E	400000	930730	52	4.0	2	B
Browning	416 D	390000	921500	63	15.5	1	
Browns	698 B	380000	922230	70	20.0	1	
Brumley	1019F	370000	925230	58	21.0	1	

Appendix A *Continued*
Atlas Block Descriptions and Summaries of Species, Volunteers and Survey Hours

Quadrangle (1)	Number(2)	Lat.(3)	Long.(4)	Spec.(5)	Hrs.(6)	Vol.(7)	Block-busted (8)
Bruner	954 F	371500	903000	77	99.5	1	
Brunot	311 B	392230	930000	56	20.0	1	
Brunswick East	310 B	392230	930730	71	19.0	1	
Brunswick West	853 C	373000	923730	64	0.0	4	B
Brush Creek	1067F	365230	922230	51	7.0	2	B
Brushyknob	217 D	394500	925230	47	6.0	1	
Bucklin	188 E	395230	925230	49	4.0	2	B
Bucklin Nw	369 B	390730	940730	52	0.0	2	
Buckner	423 E	390000	912230	44	31.0	2	
Buell	809 B	373730	930000	57	12.5	1	
Buffalo	808 E	373730	930730	50	17.5	1	
Buffalo Nw	480 A	384500	924500	65	27.0	1	
Bunceton	907 B	372230	910730	61	10.5	1	
Bunker	28 C	403000	922230	58	8.5	3	B
Bunker Hill	960 C	371500	894500	61	17.5	2	B
Burfordville	225 F	394500	915230	50	4.5	3	B
Burksville	40 B	402230	950000	63	8.5	1	
Burlington Junction	382 A	390730	923000	69	39.0	1	
Burton	511 A	383730	933000	80	23.0	3	
Burtville	613 B	381500	941500	54	16.0	2	
Butler	648 A	380730	941500	64	14.0	2	
Butler and Vicinity	279 A	393000	924500	56	20.5	1	
Bynumville	984 C	370730	920000	73	16.0	1	
Cabool Ne	983 B	370730	920730	71	17.0	1	
Cabool Nw	1026D	370000	920000	54	7.0	1	
Cabool Se	1025B	370000	920730	46	40.0	1	
Cabool Sw	50 C	402230	934500	59	7.0	1	
Cainsville	583 F	382230	933000	72	8.0	1	
Calhoun East	582 D	382230	933730	79	6.5	1	
Calhoun West	519 B	383730	923000	54	11.0	1	
California North	555 B	383000	923000	64	14.0	1	
California South	452 C	385230	914500	63	15.0	4	
Calwood	345 A	391500	925230	59	16.5	1	
Cambridge	370 F	390730	940000	72	21.0	1	
Camden	298 D	392230	943730	72	20.0	1	
Camden Point	696 A	380000	923730	72	60.0	2	
Camdenton	237 E	393730	940730	64	21.5	2	
Cameron East	236 E	393730	941500	59	31.0	4	
Cameron West	1229F	362230	900000	43	16.0	2	
Campbell	136 D	400730	913000	50	7.0	2	
Canton	1148E	363730	933000	71	15.5	1	
Cape Fair	962 B	371500	893000	76	10.0	1	
Cape Girardeau	920 C	372230	893000	59	13.0	1	
Cape Girardeau Ne	763 A	374500	934500	82	24.0	2	
Caplinger Mills	1246B	360000	901500	53	12.5	2	
Cardwell	964 A	370730	943000	72	11.5	2	
Carl Junction	341 E	391500	932230	62	9.0	1	
Carrollton East	340 E	391500	933000	53	8.0	1	
Carrollton West	966 D	370730	941500	70	16.0	1	
Carthage	1244C	360730	893730	39	15.5	1	

Atlas Block Descriptions and Summaries of Species, Volunteers and Survey Hours

Quadrangle (1)	Number(2)	Lat.(3)	Long.(4)	Spec.(5)	Hrs.(6)	Vol.(7)	Block-busted (8)
Caruthersville	956 E	371500	901500	52	12.0	2	B
Cascade	1146A	363730	934500	70	19.0	1	
Cassville	1223C	363000	893730	37	2.5	2	B
Catron	1204B	363000	920000	68	11.0	1	
Caulfield	1021E	370000	923730	51	8.0	5	B
Cedar Gap	642 B	381500	903730	48	29.0	2	
Cedar Hill	848 C	373000	931500	60	11.5	1	
Cedar Vista	904 C	372230	913000	61	0.0	1	
Cedargrove	843 D	373000	935230	51	20.0	1	
Cedarville	289 B	393000	913000	48	16.5	1	
Center	520 D	383730	922230	73	11.0	1	
Centertown Nw	509 D	383730	934500	63	0.0	1	
Centerview	909 B	372230	905230	53	21.0	1	
Centerville	385 D	390730	920730	55	20.0	1	
Centralia	386 E	390730	920000	56	7.0	1	
Centralia Ne	1062B	365230	930000	60	19.0	2	
Chadwick	1003B	370730	893730	33	5.5	1	
Chaffee	434 D	385230	940000	67	17.0	1	
Chapel Hill	850 C	373000	930000	68	18.0	1	
Charity	1092A	365230	891500	57	6.0	1	B
Charleston	1179F	363730	893730	25	2.0	4	B
Charter Oak	914 A	372230	901500	63	16.0	2	
Cherokee Pass	783 F	374500	911500	51	15.0	2	
Cherryville	1013C	370000	933730	54	15.5	1	
Chesapeake	535 A	383730	903000	62	37.5	2	
Chesterfield	545 B	383000	934500	67	13.0	2	
Chilhowee	212 F	394500	933000	57	12.0	4	B
Chillicothe	184 F	395230	932230	45	6.0	2	B
Chula	252 B	393730	921500	54	6.0	4	B
Clarence	350 F	391500	921500	69	10.5	3	
Clark	518 D	383730	923730	53	8.5	1	
Clarksburg	204 A	394500	943000	67	29.0	1	
Clarksdale	361 C	391500	905230	58	7.5	1	B
Clarksville	537 F	383730	901500	54	73.0	2	
Clayton	1028C	370000	914500	55	6.0	1	
Clear Springs	986 B	370730	914500	64	21.0	1	
Clear Springs Ne	7 D	403000	950000	52	7.5	1	
Clearmont	994 D	370730	904500	66	44.0	3	B
Clearwater Dam	478 A	384500	930000	57	6.0	4	B
Clifton City	314 C	392230	923730	53	19.0	1	
Clifton Hill	693 B	380000	930000	56	6.5	1	
Climax Springs	1088A	365230	894500	46	6.0	1	
Clines Island	581 D	382230	934500	45	99.5	2	
Clinton North	617 D	381500	934500	79	33.5	3	
Clinton South	806 A	373730	932230	57	5.0	2	
Cliquot	26 E	403000	923730	64	16.0	2	
Coatsville	148 A	400000	940000	78	71.0	2	
Coffey	792 C	374500	900730	73	18.0	1	
Coffman	955 F	371500	902230	60	11.5	2	B
Coldwater	586 D	382230	930730	66	5.5	2	B

Quadrangle (1)	Number(2)	Lat.(3)	Long.(4)	Spec.(5)	Hrs.(6)	Vol.(7)	Block-busted (8)
Cole Camp	281 C	393000	923000	68	80.0	2	
College Mound	273 C	393000	933000	55	12.5	1	
Coloma	100 E	401500	915230	64	4.5	4	B
Colony	448 E	385230	921500	84	99.5	4	
Columbia	501 E	384500	900730	61	96.0	1	
Columbia Bottom	897 F	372230	922230	61	10.5	1	
Competition	438 B	385230	933000	48	26.0	1	
Concordia	94 C	401500	923730	56	11.0	1	
Connelsville	735 C	375230	922230	76	36.0	1	
Conns Creek	782 F	374500	912230	56	17.5	2	
Cook Station	598 D	382230	913730	69	11.5	1	
Cooper Hill	510 A	383730	933730	65	24.0	1	
Cornelia	105 D	400730	952230	42	18.5	1	
Corning	908 A	372230	910000	71	11.0	1	
Corridon	950 F	371500	910000	55	4.5	1	B
Corridon Se	203 C	394500	943730	52	22.0	1	
Cosby	1209D	363000	912230	75	15.5	1	
Couch	785 D	374500	910000	40	40.5	1	
Courtois	270 A	393000	935230	63	9.0	3	
Cowgill	106 B	400730	951500	42	14.0	3	
Craig	1058A	365230	933000	55	24.0	2	
Crane	579 C	382230	940000	62	11.0	2	
Creighton	536 A	383730	902230	72	15.0	3	
Creve Coeur	844 A	373000	934500	63	15.0	1	
Crisp	736 B	375230	921500	69	19.0	5	B
Crocker	587 C	382230	930000	58	6.0	2	B
Crockerville	692 D	380000	930730	70	17.0	1	
Cross Timbers	836 E	373730	893730	48	4.5	1	
Crosstown	34 A	403000	913730	52	10.0	1	B
Croton	706 F	380000	912230	56	49.0	2	
Cuba	1159E	363730	920730	65	30.0	1	
Cureall Nw	358 D	391500	911500	54	6.5	3	B
Curryville	675 E	380730	905230	79	27.5	1	
Cyclone Hollow	360 D	391500	910000	58	8.0	1	B
Cyrene	887 F	372230	933730	65	13.5	4	
Dadeville	715 B	380000	901500	58	7.0	1	
Danby	862 E	373000	913000	59	16.0	1	
Darien	113 D	400730	942230	46	5.5	1	
Darlington	784 E	374500	910730	60	16.0	2	
Davisville	940 C	371500	921500	67	11.5	1	
Dawson	1106B	364500	930730	66	28.0	2	
Day	578 D	382230	940730	71	22.5	2	
Dayton	262 D	393000	945230	47	23.0	1	
De Kalb	678 A	380730	903000	59	46.0	2	
De Soto	263 A	393000	944500	57	32.5	1	
Dearborn	733 B	375230	923730	57	3.5	3	B
Decaturville	760 F	374500	940730	64	14.0	2	
Dederick	133 D	400730	915230	54	6.0	1	
Deer Ridge	757 B	374500	943000	46	13.0	2	
Deerfield	1242F	360730	895230	36	12.5	1	

Atlas Block Descriptions and Summaries of Species, Volunteers and Survey Hours

Quadrangle (1)	Number(2)	Lat.(3)	Long.(4)	Spec.(5)	Hrs.(6)	Vol.(7)	Block-busted (8)
Deering	533 A	383730	904500	67	26.5	3	
Defiance	1249B	360000	895230	49	21.0	2	
Denton	953 F	371500	903730	59	25.0	2	
Des Arc	912 F	372230	903000	57	23.0	4	
Des Arc Ne	777 F	374500	920000	62	32.0	1	
Devils Elbow	1132B	364500	895230	71	99.5	2	
Dexter	741 A	375230	913730	53	9.0	3	
Dillon	565 B	383000	911500	64	15.0	1	
Dissen	738 E	375230	920000	70	6.0	3	B
Dixon	1020C	370000	924500	59	28.0	2	
Dogwood	1001B	370730	895230	62	11.0	3	B
Dongola	1170F	363730	904500	63	8.5	2	B
Doniphan North	1214F	363000	904500	53	8.0	1	
Doniphan South	1114C	364500	920730	66	17.0	1	
Dora	863 E	373000	912230	66	18.5	1	
Doss	74 F	401500	950730	62	28.5	2	
Dotham	373 A	390730	933730	50	15.5	2	
Dover	62 D	402230	921500	54	9.0	5	B
Downing	61 C	402230	922230	52	11.5	3	B
Downing Nw	855 B	373000	922230	62	10.0	1	
Drew	575 C	382230	943000	53	14.5	2	
Drexel	814 A	373730	922230	78	13.0	1	B
Drynob	926 F	371500	940000	57	24.0	1	
Dudenville	1131B	364500	900000	51	32.0	2	
Dudley	937 D	371500	923730	60	6.5	1	
Duncan	283 B	393000	921500	59	25.0	1	
Duncans Bridge	198 A	395230	913730	66	28.0	5	B
Durham	1070F	365230	920000	58	12.0	1	
Dyestone Mountain	1190A	363000	934500	71	10.0	1	
Eagle Rock	49 A	402230	935230	61	8.0	3	
Eagleville	506 B	383730	940730	65	8.0	1	
East Lynne	1136C	364500	892230	21	2.5	1	
East Prairie	932 F	371500	931500	58	5.0	1	
Ebenezer	710 D	380000	905230	60	49.5	2	
Ebo	819 B	373730	914500	61	4.5	3	B
Edgar Springs	867 B	373000	905230	74	20.0	1	
Edgehill	264 C	393000	943730	57	21.0	1	
Edgerton	131 F	400730	920730	59	48.0	5	
Edina	164 C	400000	920000	57	6.0	2	B
Edina Se	657 C	380730	930730	59	6.0	1	
Edwards	627 C	381500	923000	58	13.0	2	
Eldon	722 F	375200	940000	68	11.0	2	
Eldorado Springs North	761 A	374500	940000	61	13.5	2	
Eldorado Springs South	772 C	374500	923730	63	6.0	3	B
Eldridge East	771 F	374500	924500	64	18.0	4	B
Eldridge West	985 F	370730	915230	63	34.0	3	
Elk Creek	892 F	372230	930000	66	10.5	1	
Elkland	766 F	374500	932230	78	18.0	1	
Elkton	993 E	365230	903730	67	10.0	1	
Ellington	1036B	370000	904500	67	6.0	1	B

Quadrangle (1)	Number(2)	Lat.(3)	Long.(4)	Spec.(5)	Hrs.(6)	Vol.(7)	Block-busted (8)
Ellington Se	1081B	365230	903730	66	5.5	1	B
Ellsinore	470 C	384500	940000	70	32.0	3	
Elm	190 E	395230	923730	51	14.0	1	
Elmer	268 C	393000	940730	76	67.5	2	
Elmira	396 E	390730	904500	57	14.0	1	
Elsberry	1103E	364500	933000	70	41.0	2	
Elsey	226 D	394500	914500	57	9.5	5	B
Emden	990 B	370730	911500	51	4.0	2	B
Eminence	592 D	382230	922230	85	53.0	1	
Enon	394 E	390730	910000	67	30.0	1	
Eolia	1133F	364500	894500	29	2.5	2	B
Essex	628 C	381500	922230	71	52.5	3	
Eugene Sw	570 A	383000	903730	68	4.0	1	
Eureka	540 B	383000	942230	66	19.5	2	
Everett	185 E	395230	931500	47	12.0	3	B
Eversonville	929 C	371500	933730	58	11.0	2	
Everton	335 B	391500	940730	69	54.0	4	
Excelsior Springs	992 E	370730	910000	55	11.0	3	B
Exchange	1145F	363730	935230	67	14.5	1	
Exeter	891 C	372230	930730	74	20.0	1	
Fair Grove	805 D	373730	933000	50	15.5	3	
Fair Play	1172A	363730	903000	79	32.0	1	
Fairdealing	72 F	401500	952230	48	8.5	1	
Fairfax	356 D	391500	913000	47	10.0	2	
Farber	4 C	403000	952230	57	15.0	2	
Farmers City	183 D	395230	933000	54	17.5	5	B
Farmersville	790 F	374500	902230	70	20.0	1	
Farmington	381 F	390730	923730	66	24.0	1	
Fayette	436 D	385230	934500	57	49.5	2	
Fayetteville	331 D	391500	943730	48	15.0	1	
Ferrelview	679 D	380730	902230	48	17.5	2	
Festus	1008A	370000	941500	66	17.0	1	
Fidelity	762 F	374500	935230	64	17.0	2	
Filley	141 C	400000	945230	54	19.0	1	
Fillmore	1130D	364500	900730	52	26.0	1	
Fisk	818 F	373730	915230	64	4.5	2	B
Flat	240 F	393730	934500	54	6.0	2	B
Flat Creek	789 D	374500	903000	60	25.5	4	
Flat River	1171B	363730	903730	60	14.0	1	
Flatwoods	677 B	380730	903730	78	24.0	1	
Fletcher	552 C	383000	925230	64	7.5	3	B
Florence	321 A	392230	914500	61	20.0	1	
Florida	500 E	384500	901500	63	14.0	1	
Florissant	429 E	390000	903730	62	11.0	1	
Foley	171 F	395230	950000	55	19.5	1	
Forbes	145 C	400000	942230	54	14.0	1	
Ford City	206 B	394500	941500	65	8.0	1	
Fordham	977 E	370730	925230	54	19.5	2	
Fordland	346 E	391500	924500	51	28.0	2	
Forest Green	495 B	384500	905230	78	27.0	2	

Atlas Block Descriptions and Summaries of Species, Volunteers and Survey Hours

Quadrangle (1)	Number(2)	Lat.(3)	Long.(4)	Spec.(5)	Hrs.(6)	Vol.(7)	Block-busted (8)
Foristell	1152A	363730	930000	79	25.0	1	
Forsyth	553 C	383000	924500	51	2.5	3	B
Fortuna	244 D	393730	931500	69	65.0	1	
Fountain Grove	325 C	392230	911500	70	6.0	1	B
Frankford	412 B	390000	924500	59	28.5	1	
Franklin	562 E	383000	913730	66	13.0	1	
Fredericksburg	872 B	373000	901500	58	15.0	1	
Fredericktown	632 D	381500	915230	55	7.5	2	
Freeburg	539 A	383000	943000	59	16.0	1	
Freeman	1077F	365230	910730	59	21.5	1	
Fremont	751 E	375230	902230	62	25.0	1	
French Village	876 C	373000	894500	52	4.5	1	
Friedheim	691 C	380000	931500	73	15.5	1	
Fristoe	487 E	384500	915230	70	19.5	2	
Fulton	939 C	371500	922230	60	14.5	1	
Fuson	618 C	381500	933730	80	27.5	3	
Gaines	1201F	393000	922230	57	15.0	1	
Gainesville	1157D	363000	922230	61	18.0	1	
Gainesville Nw	1104A	364500	932230	64	36.5	1	
Galena	180 D	395230	935230	75	19.0	3	
Gallatin	975 F	370730	930730	63	19.0	1	
Galloway	121 E	400730	932230	62	12.5	2	
Galt	1150A	363730	931500	53	36.0	1	
Garber	542 E	383000	940730	56	16.0	2	
Garden City	84 C	401500	935230	50	7.0	1	
Gardner	797 D	373730	943000	39	7.0	1	
Garland	1107C	364500	930000	60	0.0	1	
Garrison	1035A	370000	905230	57	9.5	2	B
Garwood	527 C	383730	913000	69	7.0	4	
Gasconade	1212D	363000	910000	63	8.5	2	B
Gatewood	80 C	401500	942230	58	26.0	4	
Gentry	1113A	364500	921500	73	16.5	1	
Gentryville	601 D	382230	911500	59	8.5	2	
Gerald	1230E	362230	895230	35	7.0	1	
Gideon	159 A	400000	923730	73	17.0	1	B
Gifford	118 D	400730	934500	94	99.5	2	
Gilman City East	117 D	400730	935230	66	7.0	2	
Gilman City West	999 A	370730	900730	64	19.0	2	B
Gipsy	905 E	372230	912230	62	29.5	1	
Gladden	380 C	390730	924500	70	29.5	1	
Glascow	958 B	371500	900000	46	15.0	2	B
Glenallen	1219A	363000	900730	40	23.0	1	
Glennonville	911 B	372230	903730	69	53.0	4	
Glover	599 F	382230	913000	64	19.0	1	
Goerlisch Ridge	1191C	363000	933730	52	0.0	1	
Golden	884 A	372230	940000	58	6.5	1	
Golden City	1064A	365230	924500	58	10.5	1	B
Goodhope	961 C	371500	893730	56	10.0	1	B
Gordonville	51 E	402230	933730	58	7.0	1	
Goshen	286 A	393000	915230	70	34.0	1	

Quadrangle (1)	Number(2)	Lat.(3)	Long.(4)	Spec.(5)	Hrs.(6)	Vol.(7)	Block-busted (8)
Goss	265 C	393000	943000	44	7.0	1	
Gower	463 D	385230	902230	72	28.0	2	
Grafton	1052A	365230	941500	71	42.0	3	
Granby	375 F	390730	932230	56	10.5	1	
Grand Pass	1125A	364500	904500	70	6.0	2	B
Grandin	1124D	364500	905230	63	9.0	2	
Grandin Sw	430 D	385230	943000	41	13.0	2	
Grandview	538 C	383730	900730	47	15.0	1	
Granite City	828 F	373730	903730	93	99.5	2	
Graniteville	45 C	402230	942230	56	6.0	1	
Grant City	285 C	393000	920000	54	12.0	4	B
Granville	625 C	381500	924500	64	39.5	2	
Gravois Mills	605 B	382230	904500	72	43.5	2	
Gray Summit	58 A	402230	924500	47	6.0	1	
Graysville	865 D	373000	910730	50	14.0	1	
Greeley	695 A	380000	924500	59	44.5	2	
Green Bay Terrace	92 A	401500	925230	42	6.0	1	
Green City	512 B	383730	932230	65	11.5	1	
Green Ridge North	548 D	383000	932230	63	7.0	1	
Green Ridge South	886 E	372230	934500	54	5.0	1	
Greenfield	98 D	401500	920730	52	4.0	3	
Greensburg	95 B	401500	923000	54	16.0	10	
Greentop	997 D	370730	902230	56	11.0	2	
Greenville Sw	1039E	370000	902230	72	30.5	1	
Greer	1121D	364500	911500	71	27.0	1	
Grovespring	896 B	372230	923000	72	7.5	2	B
Guilford	111 F	400730	943730	65	22.0	3	
Guthrie	486 B	384500	920000	57	8.0	1	
Hagers Grove	223 A	394500	920730	52	4.0	3	B
Hahatonka	732 B	375230	924500	71	33.0	1	
Hale	275 D	393000	931500	58	27.0	1	
Half Rock	88 E	401500	932230	52	5.0	4	B
Half Way	849 A	373000	930730	64	11.0	1	
Halifax	714 C	380000	902230	64	12.0	2	
Halls	231 C	393730	945230	45	13.0	1	
Hallsville	417 C	390000	920730	72	18.5	2	
Halltown	971 C	370730	933730	61	45.0	6	
Halltown, Ne	972 D	370730	933000	39	15.5	1	
Hamburg, IL	397 E	390703	903730	65	8.5	1	
Hamburg, Iowa	2 C	403000	953730	52	25.5	1	
Hamilton East	239 E	393730	935230	71	9.5	3	
Hamilton West	238 E	393730	940000	64	13.0	2	B
Hancock	737 B	375230	920730	59	5.5	3	B
Handy	1123B	364500	910000	61	21.0	1	
Hanleyville	1174B	363730	901500	41	16.5	1	
Hannibal East	260 C	393730	911500	65	20.0	4	B
Hannibal Se	291 D	393000	911500	54	18.0	3	
Hannibal West	259 B	393730	912230	59	13.0	1	
Hardin	338 F	391500	934500	44	13.0	2	
Harris	89 A	401500	931500	58	3.0	19	B

Appendix A *Continued*
Atlas Block Descriptions and Summaries of Species, Volunteers and Survey Hours

Quadrangle (1)	Number(2)	Lat.(3)	Long.(4)	Spec.(5)	Hrs.(6)	Vol.(7)	Block-busted (8)
Harrisburg	383 D	390730	922230	56	9.0	1	
Harrisonville	505 A	383730	941500	62	15.5	2	
Hartsburg	521 B	383730	921500	54	0.0	1	
Hartshorn	945 C	371500	913730	60	39.5	1	
Hartville	938 B	371500	923000	64	11.5	1	
Hartwell	580 A	382230	935230	65	14.5	2	
Harviell	1173B	363730	902230	52	6.0	3	
Harwood	721 A	375230	940730	83	29.5	2	
Hatfield	14 D	403000	940730	68	16.0	2	
Hatton	418 B	390000	920000	70	8.0	1	
Hawk Point	457 E	385230	910730	64	30.0	2	
Hayti Heights	1243B	360730	894500	35	9.0	2	B
Helena	174 B	395230	943730	51	20.0	2	
Hemple	234 E	393730	943000	44	31.5	3	
Henderson Mound	1181D	363730	892230	53	12.0	3	B
Hendrickson	1083C	365230	902230	62	44.0	1	
Herculaneum	644 F	381500	902230	53	56.5	3	
Hermann	528 B	383730	912230	68	14.0	1	
Hermitage	728 D	375230	931500	80	20.0	1	
Higbee	348 B	391500	923000	57	24.0	1	
Higdon	873 D	373000	900730	65	37.0	1	
Higginsville	405 D	390000	933730	54	12.0	1	
High Gate	669 B	380730	913730	55	28.0	1	
High Prairie	936 B	371500	924500	75	15.0	1	
Highlandville	1060A	365230	931500	83	79.0	2	
Hilda	1153B	363730	925230	68	21.5	1	
Hill Store	1178E	363730	894500	29	4.0	2	B
Hilldale	414 E	390000	923000	62	27.0	1	
Hogan Hollow	1126A	364500	903730	76	29.5	1	
Holden	508 D	383730	935230	67	15.0	2	
Hollister	1195E	363000	930730	59	0.0	1	
Holt	301 F	392230	941500	73	40.0	2	
Hopkins	9 E	403000	944500	57	6.5	1	
Hopkins Sw	8 E	403000	945230	57	20.5	1	
Hornersville	1248E	360000	900000	67	35.5	2	
Horton	720 A	375230	941500	63	28.0	1	
House Springs	607 D	382230	903000	71	30.0	4	
Houston	943 F	371500	915230	72	28.0	1	
Houstonia	440 E	385230	931500	47	0.0	2	
Howes Mill Spring	823 F	373730	911500	66	10.5	1	B
Hubbard Lake	1225C	363000	892230	55	9.0	1	
Huggins	941 F	371500	920730	60	12.5	1	
Hughesville	476 E	384500	931500	63	12.0	1	
Humansville	765 D	374500	933000	59	7.0	1	
Hume	681 E	380000	943000	69	13.0	1	
Hunnewell	256 D	393730	914500	58	21.5	1	
Hunter	1080E	365230	904500	60	6.0	1	B
Huntsdale	447 E	385230	922230	57	7.0	1	
Huntsville	315 A	392230	923000	66	41.0	1	
Hurdland	130 E	400730	921500	47	4.0	2	

Quadrangle (1)	Number(2)	Lat.(3)	Long.(4)	Spec.(5)	Hrs.(6)	Vol.(7)	Block-busted (8)
Hurley	1059A	365230	932230	59	64.0	3	
Hurricane	916 E	372230	900000	65	7.0	1	B
Hutchison	355 B	391500	913730	26	3.0	1	
Huzzah	745 B	375230	910730	56	36.0	2	
Iantha	839 E	373000	942230	62	9.0	1	
Iberia	699 F	380000	921500	61	6.0	4	B
Iconium	689 D	380000	933000	60	20.5	2	
Independence	399 B	390000	942230	56	15.0	1	
Indian Grove	277 F	393000	930000	54	5.0	2	
Indian Spring	743 C	375230	912230	55	4.5	1	B
Ionia	549 E	383000	931500	61	8.0	1	
Iron Mountain Lake	829 B	373730	903000	71	24.0	1	
Irondale	788 D	374500	903730	63	20.0	1	
Ironton	869 B	373000	903730	96	90.0	2	
Isabella	1200E	363000	923000	63	23.0	1	
Jacket	1188F	363000	940000	72	15.0	2	
Jackson	919 A	372230	893730	60	23.5	2	B
Jacksonville	282 B	393000	922230	71	42.0	1	
Jam up Cave	1030D	370000	913100	69	4.0	1	
Jameson	149 A	400000	935230	90	99.5	2	
Jamesport	181 D	395230	934500	65	0.0	1	
Jamestown	483 A	384500	922230	53	13.5	2	
Jane	1186C	363000	941500	70	19.0	2	
Jasper	924 D	371500	941500	51	6.0	1	
Jefferson City	558 C	383000	920730	51	16.0	2	
Jefferson City Nw	522 F	383730	920730	81	15.5	2	
Jenkins	1102B	364500	933730	37	9.0	2	
Jerico Springs	842 C	373000	940000	57	10.5	1	
Joanna	288 D	393000	913730	67	24.5	1	
Johnson Mountain	826 A	373730	905230	53	17.0	2	
Johnson Shut-ins	868 F	373000	904500	76	28.0	4	
Johnstown	615 F	381500	940000	72	14.0	2	
Jonesburg	492 E	384500	912230	64	25.5	1	
Joplin East	1007E	370000	942230	67	22.5	3	
Joplin West	1006F	370000	943000	69	25.0	1	
Kahoka	67 E	402230	913730	48	2.5	1	B
Kahoka Se	103 D	401500	913000	46	5.5	1	
Kaintuck Hollow	778 F	374500	915230	62	31.5	2	
Kampville	498 E	384500	903000	58	39.0	3	
Kansas City	398 D	390000	943000	56	20.0	1	
Kaskaskia	755 C	375230	895230	62	12.0	2	
Kearney	334 F	391500	941500	65	58.0	4	
Kearney Sw	333 A	391500	942230	59	35.0	1	
Keltner	1063D	365230	925230	48	9.5	2	
Kennett South	1241E	360730	900000	30	2.5	2	
Kenoma	883 B	372230	940730	59	18.0	2	
Kewanee	1180A	363730	893000	30	6.0	2	B
Keytesville	312 A	392230	925230	57	21.0	1	
Kidder	208 A	394500	940000	70	10.0	1	
Kilwinning	29 C	403000	921500	51	11.5	5	B

Quadrangle (1)	Number(2)	Lat.(3)	Long.(4)	Spec.(5)	Hrs.(6)	Vol.(7)	Block-busted (8)
Kimsey Creek	139 D	400000	950730	62	33.5	3	
King City	144 D	400000	943000	51	13.5	1	
Kingdom City	451 F	385230	915230	54	9.0	1	
Kings Point	927 E	371500	935230	65	14.5	1	
Kingsville	507 D	383730	940000	55	16.0	2	
Kirksville	128 B	400730	923000	43	8.0	1	
Kirkwood	572 E	383000	902230	48	13.5	1	
Knob Lick	831 A	373730	901500	70	18.5	1	
Knob Noster	474 B	384500	933000	55	20.5	3	
Knob Noster Nw	437 E	385230	933730	63	11.0	2	
Knobby	658 D	380730	930000	73	13.5	1	
Knox City	132 D	400730	920000	72	4.5	2	B
Knoxville	303 F	392230	940000	49	39.0	1	
Koshkonong	1207D	363000	913730	50	7.0	1	
La Belle	165 F	400000	915230	40	18.5	1	
La Grange	168 C	400000	913000	57	31.0	1	
La Monte	475 C	384500	932230	57	10.0	1	
La Plata	161 B	400000	922230	51	21.0	5	B
La Russell	968 C	370730	940000	59	16.5	1	
Labadie	569 F	383000	904500	70	14.5	1	
Laclede	215 E	394500	930730	70	0.0	1	
Laddonia	389 E	390730	913730	37	11.5	2	
Lagonda	249 A	393730	923730	69	12.0	1	
Lake Jacomo	432 E	385230	941500	72	71.0	1	
Lake Killarney	870 C	373000	903000	80	31.0	1	
Lake Ozark	661 D	380730	923730	50	29.0	1	
Lake Thunderhead	23 F	403000	930000	53	7.0	1	
Lakenan	255 C	393730	915230	52	5.0	2	B
Lakeview Heights	622 F	381500	930730	59	7.5	4	B
Lamar North	840 E	373000	941500	45	8.0	1	
Lamar South	882 B	372230	941500	52	8.0	1	
Lamoni South	16 E	403000	935230	56	8.0	4	
Lampe	1193E	363000	932230	67	20.0	2	
Langdon	71 F	401500	953000	57	13.0	1	
Lanton	1206D	363000	914500	60	14.0	2	
Laredo	153 D	400000	932230	59	15.0	2	
Latham	554 D	383000	923730	60	10.5	1	
Lathrop	267 E	393000	941500	62	23.0	2	
Lawrenceton	752 C	375230	901500	35	4.0	2	
Lawson	302 A	392230	940730	55	27.5	2	
Leadmine	770 D	374500	925230	71	11.5	3	B
Leasburg	707 D	380000	911500	62	5.5	1	
Lebanon	812 A	373730	923730	56	9.5	1	
Lecoma	780 F	374500	913730	67	4.0	2	B
Lees Summit	431 C	385230	942230	64	33.0	4	
Leesville	619 B	381500	933000	65	8.0	1	
Leeton	546 B	383000	933730	70	8.0	1	
Lentner	253 F	393730	920730	53	13.0	4	B
Leonard	194 C	395230	920730	52	22.5	3	B
Leslie	602 D	382230	910730	62	11.0	2	

Appendix A *Continued*
Atlas Block Descriptions and Summaries of Species, Volunteers and Survey Hours

Quadrangle (1)	Number(2)	Lat.(3)	Long.(4)	Spec.(5)	Hrs.(6)	Vol.(7)	Block-busted (8)
Lesterville	910 C	372230	904500	85	20.0	1	
Lesterville Se	952 A	371500	904500	73	37.5	2	
Lewis Hollow	946 E	371500	913000	75	30.0	2	
Lewistown	166 F	400000	914500	53	8.0	4	B
Lexington East	372 C	390730	934500	68	32.0	2	
Lexington West	371 D	390730	935230	66	0.0	2	
Liberal	838 E	373000	943000	81	20.0	1	
Liberty	367 A	390730	942230	68	22.5	1	
Licking	902 B	372230	914500	79	38.0	1	
Lincoln	585 B	382230	931500	48	4.0	2	B
Lincoln Nw	584 E	382230	932230	68	7.0	1	
Lincoln Se	621 B	381500	931500	59	6.0	4	B
Lindley	154 F	400000	931500	37	6.0	2	B
Linn	597 F	382230	914500	78	24.5	1	
Linneus	186 A	395230	930730	49	3.5	2	B
Lithium	794 C	374500	895230	54	5.0	1	
Livonia	59 D	402230	923730	63	5.5	2	
Lockwood	885 E	372230	935230	59	11.5	1	
Locust Hill	162 A	400000	921500	58	7.5	2	B
Loggers Lake	906 E	372230	911500	68	24.0	1	
Lohman	557 E	383000	921500	59	25.0	1	
Lone Elm	481 E	384500	923730	63	13.5	2	
Lonedell	641 A	381500	904500	78	45.0	4	
Long Lane	851 C	373000	925230	80	13.0	2	B
Longwood	441 C	385230	930730	53	21.5	1	
Loose Creek	560 B	383000	915230	56	16.0	1	
Louisiana	327 F	392230	910000	52	12.5	1	
Louisville	393 C	390730	910730	67	26.0	1	
Low Wassie	1076D	365230	911500	59	20.0	1	
Lowndes	998 F	370730	901500	61	16.0	3	B
Lowry City	653 A	380730	933730	71	8.0	1	
Lucerne	54 A	402230	931500	47	6.0	1	
Luckett Ridge	428 D	390000	904500	77	11.0	1	
Luystown	561 B	383000	914500	65	7.0	1	
Macks Creek	731 E	375230	925230	68	20.5	1	
Macon	251 C	393730	922230	61	24.5	1	
Madison	318 F	392230	920730	64	32.0	1	
Main City	576 C	382230	942230	82	33.5	3	
Maitland	108 B	400730	950000	50	19.0	1	
Malden	1221A	363000	895230	38	14.5	1	
Malta Bend	376 B	390730	931500	57	10.0	2	
Manchester	571 F	383000	903000	73	99.5	1	
Manes	898 C	372230	921500	54	9.5	1	
Mansfield	1022C	370000	923000	57	10.5	1	
Mansfield Ne	980 F	370730	923000	51	20.5	3	B
Mansfield Nw	979 E	370730	923730	65	12.0	1	
Many Springs	1166A	363730	911500	60	23.0	1	
Maple Grove	925 B	371500	940730	54	8.5	2	
Maples	860 C	373000	914500	75	24.0	1	
Maramec Spring	742 A	375230	913000	67	14.0	1	

Atlas Block Descriptions and Summaries of Species, Volunteers and Survey Hours

Quadrangle (1)	Number(2)	Lat.(3)	Long.(4)	Spec.(5)	Hrs.(6)	Vol.(7)	Block-busted (8)
Marble Hill	959 B	371500	895230	57	5.0	2	B
Marceline	247 B	393730	925230	54	15.0	1	
Marquand	915 C	372230	900730	57	8.5	1	B
Marshall North	377 E	390730	930730	51	11.5	1	
Marshall South	409 E	390000	930730	52	10.0	2	
Marshfield	935 E	371500	925230	68	18.0	1	
Marthasville	531 C	383730	910000	63	13.0	1	
Martinsburg	421 D	390000	913730	73	16.0	3	
Maryknoll	460 E	385230	904500	56	28.0	3	
Maryville East	77 C	401500	944500	57	13.5	1	
Maryville West	76 B	401500	945230	54	13.5	1	
Matkins	115 E	400730	940730	53	7.0	1	
Maxville	608 C	382230	902230	62	13.0	1	
Maysville	177 E	395230	941500	47	6.5	1	
Mayview	404 A	390000	934500	62	15.0	1	
Maywood	199 B	395230	913000	52	48.5	2	
Mcdowell	1101A	364500	934500	75	13.5	3	
Mcgee	1041C	370000	900730	66	22.0	2	
Mcnatt	1142E	363730	941500	66	25.0	1	
Meadville	214 C	394500	931500	68	6.0	4	B
Medill	66 D	402230	914500	34	2.5	1	B
Memphis	63 D	402230	920730	45	7.0	1	
Mendon	276 D	393000	930730	67	29.0	1	
Mendota	24 C	403000	925230	57	7.0	1	
Meramec State Park	674 A	380730	910000	86	99.5	2	
Meta Ne	595 E	382230	920000	72	41.0	2	
Meta Nw	594 F	382230	920730	73	38.0	1	
Meta Sw	630 D	381500	920730	64	18.0	1	
Metz	719 A	375230	942230	63	22.5	1	
Mexico East	388 E	390730	914500	63	8.5	1	
Mexico Se	420 E	390000	914500	56	25.0	1	
Mexico West	387 D	390730	915230	68	7.5	1	
Miami	343 A	391500	930730	45	16.0	1	
Miami Station	342 A	391500	931500	58	8.0	1	
Middle Grove	317 A	392230	921500	78	50.0	1	
Middletown	391 A	390730	912230	51	4.0	1	
Midridge	949 D	371500	910730	61	7.2	1	B
Mike	278 C	393000	925230	51	4.0	2	
Milan East	124 D	400730	930000	35	17.0	3	B
Milan Se	156 C	400000	930000	69	27.0	1	
Milan West	123 C	400730	930730	63	10.5	3	
Milford	841 A	373000	940730	47	9.5	1	
Mill Grove	87 A	401500	933000	62	6.0	2	
Mill Spring	1037F	370000	903730	73	9.0	1	
Millard	160 D	400000	923000	51	9.0	1	
Miller	970 B	370730	934500	61	14.0	1	
Millersburg	449 D	385230	920730	74	20.5	3	
Millersburg Ne	450 C	385230	920000	69	12.0	1	
Millersburg sw	485 D	384500	920730	55	19.0	2	
Millersville	918 B	372230	894500	68	42.0	1	

Quadrangle (1)	Number(2)	Lat.(3)	Long.(4)	Spec.(5)	Hrs.(6)	Vol.(7)	Block-busted (8)
Millville	304 B	392230	935230	65	20.0	1	
Mincy	1196C	363000	930000	71	30.0	1	
Mindenmines	880 D	372230	943000	63	6.5	1	
Mineral Point	749 B	375230	903730	66	13.5	1	
Minnith	793 F	374500	900000	60	4.5	1	
Missouri City	368 F	390730	941500	72	83.0	3	
Mitchellville	116 D	400730	940000	69	47.0	2	
Moberly	316 F	392230	922230	69	27.0	1	
Modena	86 F	401500	933730	62	12.0	3	B
Mokane East	525 C	383730	914500	58	10.5	1	
Mokane West	524 D	383730	915230	55	40.0	1	
Molino	353 C	391500	915230	47	17.5	1	
Monegaw Springs	687 B	380000	934500	83	17.0	2	
Monegaw Springs Nw	651 D	380730	935230	59	4.0	1	
Monett	1055D	365230	935230	72	20.5	3	
Monroe City	257 A	393730	913730	47	12.0	1	
Montauk	903 D	372230	913730	63	21.0	1	
Montevallo	801 B	373730	940000	57	17.5	1	
Montgomery City	454 E	385230	913000	55	20.0	1	
Monticello	167 D	400000	913730	62	26.0	1	
Montier	1074E	365230	910000	54	14.5	1	
Montreal	734 F	375230	923000	68	55.0	5	B
Montrose	616 F	381500	935230	66	18.0	2	
Moody	1205E	363000	915230	56	10.0	1	
Mooring	1239A	361500	893000	47	10.0	2	B
Morehouse	1134D	364500	893730	27	3.5	3	B
Morley	1046F	370000	893000	52	5.5	4	B
Morrison	526 A	383730	913730	68	9.0	4	
Morrisville	889 E	372230	932230	47	25.0	2	
Moselle	604 C	382230	905230	66	45.0	2	
Mound City	107 D	400730	950730	47	7.0	1	
Moundville	758 C	374500	942230	42	22.0	2	
Mount Moriah	85 F	401500	934500	51	8.5	1	
Mount Sterling	32 F	403000	915230	37	4.0	2	B
Mountain Grove North	982 B	370730	921500	57	10.5	1	
Mountain Grove South	1024F	370000	921500	49	4.5	2	B
Mountain View	1073F	365230	913730	57	11.5	1	
Mt Vernon	1012A	370000	934500	53	61.0	1	
Myrtle	1210D	363000	911500	46	8.0	2	
Mystic	125 B	400730	925230	52	7.0	1	
Nagogami Lodge	702 C	380000	915230	71	7.0	2	B
Napton	410 B	390000	930000	50	28.5	2	
Nashua	332 C	391500	943000	70	21.0	2	
Nashville	881 B	372230	942230	55	11.0	1	
Naylor	1216C	363000	903000	63	37.0	2	
Neck City	923 C	371500	942230	70	21.5	2	
Neelys Landing	878 A	373000	893000	70	29.0	1	B
Neeper	101 D	401500	914500	48	8.0	2	B
Nelson	442 A	385230	930000	60	10.0	1	
Nelsonville	197 F	395230	914500	55	10.0	5	B

Quadrangle (1)	Number(2)	Lat.(3)	Long.(4)	Spec.(5)	Hrs.(6)	Vol.(7)	Block-busted (8)
Neosho East	1097F	364500	941500	66	30.0	1	
Neosho West	1096A	364500	942230	62	37.0	1	
Nettleton	209 B	394500	935230	68	17.0	3	
Nevada	759 A	374500	941500	66	14.0	2	
New Bloomfield	523 B	383730	920000	66	99.5	2	
New Boston	189 A	395230	924500	42	6.0	1	
New Cambria East	219 E	394500	923730	46	6.0	2	B
New Cambria West	218 F	394500	924500	47	5.0	2	B
New Florence	455 C	385230	912230	51	0.0	1	
New Frankfort	344 A	391500	930000	51	14.0	1	
New Franklin	413 E	390000	923730	69	15.0	2	
New Hampton	82 A	401500	940730	49	7.5	1	
New Hartford	392 F	390730	911500	44	8.0	1	
New Haven	566 F	383000	910730	60	58.5	1	
New Home	647 C	380730	942230	56	12.0	2	
New London	290 D	393000	912230	62	23.0	1	
New Madrid	1224A	363000	893000	38	9.5	4	B
New Melle	532 C	383730	905230	75	15.0	1	
New Point	140 D	400000	950000	61	10.0	1	
Newark	196 E	395230	915230	52	10.5	2	
Newburg	739 B	375230	915230	59	6.0	3	B
Newtonia	1053E	365230	940730	62	10.0	1	
Niangua	894 C	372230	924500	71	14.0	1	
Nichols Knob	1069A	365230	920730	65	9.5	1	
Nind	158 E	400000	924500	55	9.0	1	
Nixa	1016A	370000	931500	50	8.0	1	
Noel	1185B	363000	942230	80	27.5	2	
Norborne	339 E	391500	933730	47	8.0	1	
North Kansas City	366 E	390730	943000	44	30.0	1	
Norwood	1023F	370000	922230	60	9.0	1	
Novelty	163 B	400000	920730	61	7.5	2	B
Novinger	127 E	400730	923730	51	3.0	1	
O'fallon	497 F	384500	903730	54	10.0	2	
Oak Grove	401 C	390000	940730	80	40.0	1	
Oak Grove Heights	976 F	370730	930000	65	99.5	1	
Oak Hill	671 E	380730	912230	64	6.5	1	
Oak Mills	295 B	392230	950000	85	0.0	3	
Oak Ridge	877 B	373000	893730	55	4.0	1	
Oakland	813 A	373730	923000	56	26.5	5	B
Oates	866 A	373700	910000	62	13.5	1	
Odessa North	403 C	390000	935230	55	19.0	1	
Odessa South	435 D	385230	935230	73	23.5	2	
Oglesville	1218B	363000	901500	51	22.0	1	
Ohio	652 F	380730	934500	63	21.0	2	
Okete	427 F	390000	905230	83	0.0	1	
Old Mines	711 F	380000	904500	68	16.0	3	
Olean	591 A	382230	923000	59	12.5	1	
Omaha	25 F	403000	924500	58	6.0	1	
Onondaga Cave	708 D	380000	910730	66	2.5	2	
Oran	1045C	370000	893730	44	35.0	1	

Quadrangle (1)	Number(2)	Lat.(3)	Long.(4)	Spec.(5)	Hrs.(6)	Vol.(7)	Block-busted (8)
Oregon	170 B	395230	950730	50	53.0	1	
Osage City	559 E	383000	920000	57	19.5	1	
Osceola	688 C	380000	933730	63	17.0	2	
Osgood	122 B	400730	931500	54	6.5	1	B
Otterville East	516 A	383730	925230	51	35.5	1	
Otterville West	515 B	383730	930000	71	11.5	1	B
Owens	981 D	370730	922230	54	4.0	2	B
Owensville East	636 C	381500	912230	76	15.0	2	
Owensville West	635 E	381500	913000	63	18.0	1	
Oxly	1215E	363000	903730	54	6.0	1	
Ozark	1017B	370000	930730	58	17.0	1	
Ozark Springs	775 D	374500	921500	52	4.0	1	
Pacific	606 D	382230	903730	67	23.0	2	
Palmer	786 D	374500	905230	57	23.5	3	
Palmyra	228 D	394500	913000	65	7.5	5	B
Papinsville	684 D	380000	940730	63	41.5	2	
Paris East	320 A	392230	915230	52	9.0	1	
Paris West	319 B	392230	920000	63	22.0	1	
Parker Lake	833 D	373730	900000	67	7.5	1	
Parkville	365 D	390730	943730	70	15.0	1	
Parma	1222C	363000	894500	32	11.0	2	
Parnell East	44 C	402230	943000	57	7.0	1	
Parnell West	43 B	402230	943730	53	7.0	1	
Patterson	996 E	370730	903000	53	6.5	1	B
Patton	874 B	373000	900000	55	0.0	1	
Pattonsburg	147 A	400000	940730	61	25.0	1	
Pawnee	15 C	403000	940000	64	16.5	2	
Paydown	668 A	380730	914500	62	5.0	2	B
Peace Valley	1118B	364500	913730	59	8.0	1	
Peculiar	504 D	383730	942230	58	17.0	1	
Perrin	235 C	393730	942230	55	17.5	1	
Perry	322 B	392230	913730	57	15.0	1	
Perry Ne	323 E	392230	913000	47	5.5	4	B
Perryville East	835 F	373730	894500	49	4.0	1	
Perryville West	834 C	373730	895230	64	14.0	1	
Pershing	563 E	383000	913000	68	22.0	1	
Peru	35 D	402230	953730	53	22.5	1	
Philadelphia	227 B	394500	913730	56	21.0	3	
Phillipsburg	852 E	373000	924500	51	5.0	2	B
Pickering	42 F	402230	944500	56	6.5	1	
Piedmont	995 D	370730	903730	66	9.0	1	
Piedmont Hollow	1120B	364500	912230	59	21.0	1	
Piedmont Se	1038C	370000	903000	65	11.0	1	B
Pierce City	1054D	365230	940000	47	20.0	2	
Pilot Grove North	443 C	385230	925230	57	12.0	1	
Pilot Grove South	479 B	384500	925230	73	26.0	1	
Pine Crest	1029D	370000	913730	63	99.5	2	
Pinnacle Lake	491 F	384500	912230	66	21.5	1	
Pittsville	471 C	384500	935230	61	10.0	1	
Platte City	330 E	391500	944500	50	35.5	1	

Appendix A *Continued*
Atlas Block Descriptions and Summaries of Species, Volunteers and Survey Hours

Quadrangle (1)	Number(2)	Lat.(3)	Long.(4)	Spec.(5)	Hrs.(6)	Vol.(7)	Block-busted (8)
Plattsburg	266 F	393000	942230	67	27.0	2	
Pleasant Gap	649 D	380730	940730	86	22.0	2	
Pleasant Hill	468 C	384500	941500	55	34.5	1	
Pleasant Hope	890 A	372230	931500	60	16.0	1	
Pleasanton	18 F	403000	933730	62	7.0	1	
Plymouth	272 D	393000	933730	62	27.5	3	
Polk	807 D	373730	931500	57	9.0	1	
Pollock	91 F	401500	930000	55	25.5	2	
Pollock Nw	55 E	402230	930730	46	4.5	1	
Pollock Sw	90 E	401500	930730	66	16.0	1	
Polo	269 A	393000	940000	57	26.0	1	
Pomona	1116B	364500	915230	54	20.0	1	
Poplar Bluff	1128C	364500	902230	66	0.0	1	
Portageville	1232C	362230	893730	45	8.0	1	
Potosi	748 C	375230	904500	62	14.5	1	
Pottersville	1160D	363730	920000	65	14.5	1	
Powder Mill Ferry	991 E	370730	910730	63	12.0	1	
Powe	1176B	363730	900000	50	7.0	2	B
Powell	1187B	363000	940730	73	10.0	1	
Poynor	1213D	363000	905230	66	31.0	2	
Prairie du Rocher	717 E	380000	900000	47	5.5	1	
Prairie Hill	280 E	393000	923730	57	31.5	2	
Prairie Home	482 A	384500	923000	57	8.5	1	
Prescott	901 E	372230	915230	78	23.0	2	
Preston	729 D	375230	930730	56	4.0	1	
Princeton	52 C	402230	933000	60	11.0	1	
Proctor Creek	624 A	381500	925230	58	10.5	1	B
Protem	1198E	363000	924500	51	10.5	2	
Protem Ne	1154B	363730	924500	71	23.0	1	
Protem Sw	1197F	363000	925230	52	22.5	2	
Purdy	1100D	364500	935230	70	15.0	3	
Pure Air	126 C	400730	924500	38	12.0	1	
Puxico	1085B	365230	900730	80	30.0	1	
Pyrmont	551 A	383000	930000	86	63.0	2	
Queen City	60 E	402230	923000	57	36.0	2	
Quick City	543 E	383000	940000	64	15.5	2	
Quincy	690 E	380000	932230	70	11.0	1	
Quincy Sw	229 D	394500	912230	50	26.0	1	
Quincy West	200 A	395230	912230	52	31.0	2	
Racine	1050A	365230	943000	65	19.0	1	
Rader	895 F	372230	923730	52	7.5	1	
Ravanna	53 F	402230	932230	48	9.0	1	
Ravenwood	78 B	401500	943730	56	22.0	5	
Raymondville	944 C	371500	914500	62	23.0	1	
Raymore	467 D	384500	942230	61	27.0	1	
Rayville	336 D	391500	940000	59	7.0	1	
Readsville	489 C	384500	913730	64	10.0	1	
Redbird	670 A	380730	913000	63	14.0	1	
Redford	951 A	371500	905230	42	7.5	1	
Reeds	1009D	370000	940730	47	0.0	1	

Quadrangle (1)	Number(2)	Lat.(3)	Long.(4)	Spec.(5)	Hrs.(6)	Vol.(7)	Block-busted (8)
Reeds Spring	1149A	363730	932230	47	13.5	1	
Reform	488 E	384500	914500	66	18.0	1	
Renick	349 A	391500	922230	56	23.5	1	
Rensselaer	258 C	393730	913000	53	3.0	4	B
Republic	1015E	370000	932230	55	9.0	3	
Rescue	969 F	370730	935230	57	13.0	1	
Rhodes Mountain	871 C	373000	902230	56	34.5	4	
Rhyse	861 B	373000	913730	64	30.0	4	
Rich Hill	683 D	380000	941500	74	20.0	2	
Richards	718 A	375230	943000	66	12.0	1	
Richland	774 F	374500	922230	62	17.0	2	B
Richmond	337 A	391500	935230	71	4.5	1	
Richwoods	676 B	380730	904500	68	23.0	2	
Riverton	1167A	363730	910730	76	11.0	1	
Roads	306 E	392230	933730	55	8.0	1	
Roby	857 D	373000	920730	74	19.5	1	
Rocheport	446 D	385230	923000	86	13.0	2	
Rock Pile Mountain	913 D	372230	902230	55	6.5	1	B
Rock Port	36 F	402230	953000	51	13.0	1	
Rockbridge	1112A	364500	922230	63	16.0	1	
Rockport	293 E	393000	910000	65	0.0	1	
Rockville	685 B	380000	940000	53	12.5	2	
Rockwood	796 C	374500	893730	56	7.0	1	
Rocky Comfort	1144E	363730	940000	63	7.5	1	
Rocky Mount	626 D	381500	923730	57	17.0	2	
Rogersville	1018F	370000	930000	73	13.0	2	
Rolla	740 E	375230	914500	57	47.0	2	
Rombauer	1129D	364500	901500	63	15.5	2	B
Rosati	705 B	380000	913000	51	2.0	2	B
Roscoe	724 C	375230	934500	74	14.0	1	
Rosebud	600 D	382230	912230	65	30.0	2	
Rothville	246 F	393730	930000	42	5.5.0	2	B
Roubidoux	899 F	372230	920730	61	9.0	2	
Round Spring	947 C	371500	912230	66	30.0	2	
Rover	1164D	363730	913000	57	15.0	2	B
Rowena	352 E	391500	920000	60	0.0	1	
Russ	854 A	373000	923000	60	4.5	3	B
Russellville	556 D	383000	922230	49	9.0	1	
Rutledge	99 F	401500	920000	44	4.0	2	
Safe	704 D	380000	913730	49	10.5	2	B
Salem	821 A	373730	913000	62	8.0	2	B
Saline City	379 C	390730	925230	60	19.5	2	
Salisbury	313 D	392230	924500	56	21.0	1	
Sampsel	211 E	394500	933730	64	34.5	3	
Santa Fe	354 F	391500	914500	55	17.5	1	
Sarcoxie	1010A	370000	940000	70	28.0	1	
Savannah	173 C	395230	944500	47	25.0	2	
Scopus	917 E	372230	895230	55	4.0	1	B
Scott City	1004E	370730	893000	53	9.0	1	
Seaton	781 E	374500	913000	51	3.5	2	B

Quadrangle (1)	Number(2)	Lat.(3)	Long.(4)	Spec.(5)	Hrs.(6)	Vol.(7)	Block-busted (8)
Sedalia East	514 A	383730	930730	55	28.0	6	B
Sedalia West	513 C	383730	931500	58	4.0	2	B
Sedgewickville	875 F	373000	895230	43	6.5	2	B
Seligman	1189A	363000	935230	69	15.0	1	
Selmore	1061D	365230	930730	71	27.0	4	
Senath	1240D	360730	900730	53	21.5	2	
Seneca	1095D	364500	943000	64	20.0	2	
Sentinel	767 B	374500	931500	60	11.5	1	
Seymour	978 A	370730	924500	65	12.0	2	
Shackleford	408 C	390000	931500	73	14.0	1	
Shawnee Bend	620 C	381500	932230	60	6.5	1	
Shearwood	182 B	395230	933730	69	16.0	4	B
Shelbina	254 C	393730	920000	37	6.0	4	B
Shelby	187 B	395230	930000	43	6.0	1	B
Shelbyville	224 F	394500	920000	56	12.0	4	B
Sheldon	799 A	373730	941500	70	28.0	1	
Shell Knob	1147C	363730	933730	63	12.5	1	
Sheridan	11 F	403000	943000	62	11.0	1	
Shirley	747 E	375230	905230	52	17.0	2	
Shook	1040A	370000	901500	63	30.0	5	B
Short Bend	822 B	373730	912230	55	34.5	1	
Sikeston North	1090A	365230	893000	31	5.0	1	B
Sikeston South	1135D	364500	893000	43	6.0	3	B
Silex	426 A	390000	910000	55	16.0	1	
Siloam Springs	1115E	364500	920000	70	15.0	1	
Skidmore	75 E	401500	950000	58	10.0	1	
Skidmore Nw	39 D	402230	950730	54	4.0	1	
Slabtown Spring	858 A	373000	920000	72	30.5	1	
Slater	378 D	390730	930000	54	18.5	1	
Smallett	1110D	364500	923730	72	17.5	1	
Smithville	299 C	392230	943000	67	22.0	3	
South Fork	1161D	363730	915230	58	12.0	1	
South Greenfield	928 F	371500	934500	62	17.0	1	
Southwest City	1184C	363000	943000	68	24.0	2	
Spencerburg	324 B	392230	912230	55	35.0	1	
Spikard	120 F	400730	933000	48	5.0	2	B
Spokane	1105E	364500	931500	50	17.0	1	
Sprague	682 C	380000	942230	61	28.0	2	
Spring Bluff	638 F	381500	910730	26	14.0	1	
Springfield	974 A	370730	931500	39	18.0	1	
Sprott	791 D	374500	901500	70	20.0	2	
Spruce	614 D	381500	940730	54	19.0	3	
St. Anthony	664 F	380730	921500	72	10.5	1	
St. Charles	499 A	384500	902230	51	21.5	2	
St. Clair	640 D	381500	905230	76	23.0	5	
St. Elizabeth	629 D	381500	921500	51	6.0	4	
St. John	22 E	403000	930730	45	5.0	1	
St. Joseph North	202 E	394500	944500	41	99.5	1	
St. Joseph South	232 B	393730	944500	71	50.0	2	
St. Patrick	102 C	401500	913730	54	19.0	1	

Quadrangle (1)	Number(2)	Lat.(3)	Long.(4)	Spec.(5)	Hrs.(6)	Vol.(7)	Block-busted (8)
Stafford	93 C	401500	924500	17	17.0	2	
Stahl	112 F	400730	943000	47	6.5	1	
Stanberry	308 C	392230	932230	49	21.0	2	
Standish	1238C	361500	893730	40	1.5	2	B
Stanley	639 A	381500	910000	63	7.0	1	
Stanton	754 F	375230	900000	69	10.5	1	
Ste Genevieve	1250D	360000	894500	46	4.0	2	B
Steele	744 A	375230	911500	59	38.0	2	
Steelville	1033C	370000	910730	57	15.5	5	
Stegall Mountain	1098C	364500	940730	56	51.0	1	
Stella	305 B	392230	934500	63	26.0	1	
Stet	803 C	373730	934500	45	22.0	2	
Stockton	864 D	373000	911500	61	5.0	2	B
Stone Hill	1011F	370000	935230	42	14.0	2	
Stotts City	773 D	374500	923000	54	1.5	4	B
Stoutland	287 C	393000	914500	54	41.5	2	
Stoutsville	588 E	382230	925230	60	6.0	1	
Stover	934 A	371500	930000	62	30.0	1	
Strain	637 E	381500	911500	64	24.0	2	
Strasburg	469 E	384500	940730	65	34.0	2	
Stringtown	1127B	364500	903000	75	48.0	2	B
Sturdivant	1042B	370000	900000	76	0.0	2	
Sturgeon	384 F	390730	921500	77	35.0	2	
Sturgeon Sw	415 F	390730	921500	81	36.5	4	
Success	900 C	372230	920000	79	36.5	2	
Sue City	193 B	395230	921500	44	4.0	4	B
Sullivan	673 C	380730	910730	60	16.0	2	
Summerfield	633 E	381500	914500	64	7.5	2	B
Summersville	987 A	370730	913730	78	23.0	1	
Summerville Ne	988 E	370730	913000	68	3.5	2	B
Sumner	245 D	393730	930730	42	5.5	2	B
Sunrise Beach	660 D	380730	924500	65	51.0	2	
Sweden	1066B	365230	923000	56	4.0	4	B
Sweet Springs	439 B	385230	932230	60	20.0	2	
Swiss	564 E	383000	912230	55	11.5	2	
Sycamore	1158B	363730	921500	80	11.0	1	
Taberville	686 C	380000	935230	59	24.5	4	
Table Rock Dam	1194D	363000	931500	70	14.5	4	
Tarkio East	38 F	402230	951500	47	5.0	1	
Tarkio Se	73 D	401500	951500	46	7.0	1	
Tarkio West	37 C	402230	952230	57	18.0	1	
Tarsney Lakes	433 B	385230	940730	79	99.5	2	
Tennemo	1245E	360730	893000	49	9.0	2	B
Thayer	1208F	363000	913000	61	6.0	1	
The Sinks	948 C	371500	911500	66	25.0	1	
Thebes	1005A	370730	892230	55	7.0	3	B
Thebes Sw	1047F	370000	892230	69	7.5	3	B
Theodosia	1199C	363000	923730	60	11.5	1	
Thomasville	1119A	364500	913000	58	13.0	1	
Thornfield	1155E	363730	923730	65	17.0	1	

Atlas Block Descriptions and Summaries of Species, Volunteers and Survey Hours

Quadrangle (1)	Number(2)	Lat.(3)	Long.(4)	Spec.(5)	Hrs.(6)	Vol.(7)	Block-busted (8)
Tiff	712 C	380000	903730	58	24.0	1	
Tiff City	1140E	363730	943000	63	12.5	1	
Tiffin	723 F	375230	935230	64	11.0	2	
Tina	274 C	393000	932230	50	22.0	1	
Tipton	517 D	383730	924500	60	26.0	1	
Tipton Ford	1051D	365230	942230	70	19.5	1	
Toronto	697 A	380000	923000	70	23.0	1	
Tracy	297 A	392230	944500	57	36.0	4	
Trask	1072D	365230	914500	63	14.5	3	
Treloar	530 C	383730	910730	63	15.0	1	
Trenton East	152 B	400000	933000	54	5.0	2	B
Trenton West	151 B	400000	933730	69	18.0	1	
Troy	459 B	385230	905230	66	9.5	2	
Truxton	425 A	390000	910730	55	10.5	1	
Tulip	351 B	391500	920730	65	20.0	1	
Tunas	769 A	374500	930000	65	21.5	1	
Tuscumbia	663 A	380730	922230	74	6.0	1	
Udall	1202D	363000	921500	65	22.0	1	
Union	603 A	382230	910000	67	17.0	2	
Union Star	175 C	395230	943000	46	16.0	2	
Unionville East	57 C	402230	925230	58	15.0	1	
Unionville West	56 A	402230	930000	53	7.0	1	
Urbana	768 A	374500	930730	68	8.0	1	
Utica East	242 C	393730	933000	52	6.5	1	B
Utica West	241 D	393730	933730	55	13.5	6	B
Valhalla	654 A	380730	933000	73	25.0	2	
Valley Ridge	1220C	363000	900000	47	5.0	2	B
Van Buren North	1034F	370000	910000	65	10.5	4	B
Van Buren South	1078A	365230	910000	80	0.0	3	
Van Cleve	665 D	380730	920730	60	3.5	2	
Vandalia	357 C	391500	912230	49	10.0	3	
Vandalia Lake	390 B	390730	913000	51	7.0	2	
Vanduser	1089C	365230	893730	38	10.0	3	B
Vanzant	1068A	365230	921500	54	6.0	2	B
Vastus	1217F	363000	902230	59	40.0	1	
Vera	326 C	392230	910730	49	7.5	1	
Verona	1056D	365230	934500	79	22.0	3	
Versailles	589 E	382230	924500	61	41.0	1	
Viburnum East	825 F	373730	910000	82	39.0	2	
Viburnum West	824 E	373730	910730	47	30.5	1	
Vichy	703 B	380000	914500	59	44.0	3	
Vienna	667 D	380730	915230	67	6.5	2	B
Vineland	713 E	380000	903000	50	0.0	2	
Viola	1192F	363000	933000	58	7.5	2	
Virginia	612 D	381500	942230	48	16.0	2	
Vista	725 F	375230	933730	64	13.0	1	
Wachita Mountain	830 D	373730	902230	85	17.0	1	
Wagoner	802 A	373730	935230	56	17.5	1	
Walnut Grove	888 B	372230	933000	58	8.0	4	
Wappapello	1084C	365230	901500	61	33.5	2	

Quadrangle (1)	Number(2)	Lat.(3)	Long.(4)	Spec.(5)	Hrs.(6)	Vol.(7)	Block-busted (8)
Wardell	1237D	361500	894500	36	13.5	1	
Warrensburg East	473 F	384500	933730	61	12.0	1	
Warrensburg West	472 D	384500	934500	69	13.0	1	
Warrenton	493 E	384500	910730	64	5.5	1	
Warrenton Ne	458 D	385230	910000	68	14.0	2	
Warsaw	104 C	401500	912230	50	12.0	1	
Warsaw East	656 A	380730	931500	71	6.0	2	
Warsaw West	655 D	380730	932230	68	13.0	1	
Washington Center	47 E	402230	940730	63	24.5	3	
Washington East	568 D	383000	905230	53	16.0	3	
Washington West	567 A	383000	910000	58	20.5	1	
Wasola	1111C	364500	923000	43	3.0	2	B
Waverly	374 E	390730	933000	61	24.5	3	
Wayland	68 A	402230	913730	68	4.5	1	B
Waynesville	776 A	374500	920730	54	22.5	2	
Weatherby	178 C	395230	940730	59	13.0	2	
Weaubleau	726 B	375230	933000	62	11.0	1	
Webb City	965 F	370730	942230	70	19.0	1	
Webster Groves	573 A	383000	901500	52	22.0	1	
Weingarten	753 F	375230	900730	67	14.5	1	
Weldon Spring	534 A	383730	903730	54	16.0	2	
Wellsville	422 B	390000	913000	72	32.5	3	
Wentzville	496 C	384500	904500	63	16.5	2	
West Line	503 D	383730	943000	71	25.0	1	
West Plains	1162B	363730	914500	59	12.0	1	
Weston	296 F	392230	945230	51	13.5	1	
Westphalia East	596 D	382230	915230	74	20.0	1	
Wheatland	727 F	375230	932230	72	14.0	1	
Wheaton	1099D	364500	940000	75	22.5	3	
Wheeling	213 E	394500	932230	50	8.0	4	B
Whitesville	143 B	400000	943730	56	14.0	2	
Whitewater	1002A	370730	894500	56	12.0	2	B
Wickliffe Sw	1138E	364500	890730	33	5.5	1	B
Wien	248 E	393730	924500	43	6.5	1	
Wilcox	41 F	402230	945230	58	34.5	2	
Wilderness	1122A	364500	910730	52	20.0	1	
Willard	931 B	371500	932230	62	21.0	1	
Willhoit	1156F	363730	923000	54	14.5	1	
Williamsburg	453 C	385230	913730	80	99.5	1	
Williamstown	134 F	400730	914500	60	5.5	3	B
Williamsville	1082B	365230	903000	66	26.5	1	
Willmathsville	96 D	401500	922230	61	14.0	1	
Willow Springs	1117A	364500	914500	57	20.0	3	
Willow Springs North	1027C	370000	915230	63	26.5	4	
Willow Springs South	1071A	365230	915230	54	5.0	1	
Windsor	547 F	383000	933000	52	20.5	1	
Windyville	810 B	373730	925230	65	0.0	1	
Winfield	461 F	385230	903730	73	17.5	1	
Winigan	157 E	400000	925230	50	5.0	1	B
Winnipeg	856 B	373000	921500	53	7.0	3	B

Atlas Block Descriptions and Summaries of Species, Volunteers and Survey Hours

Quadrangle (1)	Number(2)	Lat.(3)	Long.(4)	Spec.(5)	Hrs.(6)	Vol.(7)	Block-busted (8)
Winona	1032F	370000	911500	60	13.5	1	
Winston	207 C	394500	940730	63	35.0	3	
Wolf Island	1183C	363730	890730	41	22.0	1	
Womack	832 E	373730	900730	53	11.0	1	
Wood	176 C	395230	942230	45	5.0	1	
Woodlawn	284 F	393000	920730	68	24.5	1	
Worland	646 F	380730	943000	60	19.0	2	
Wright City	494 B	384500	910000	50	14.0	1	
Wyaconda	65 D	402230	915230	45	8.5	3	B
Wyatt	1093E	365230	890730	38	6.5	1	
Yancy Mills	779 F	374500	914500	64	22.5	2	
Zalma	1000B	370730	900000	64	11.5	3	B

Appendix B
Abundance Data Routes and Data Gathering Procedure

Route Number	Natural Division	Quadname (1)	Source of Data MR or BBS (2)
1	Mississippi Lowlands	Charleston	BBS
2	Ozarks	Poynor	BBS
3	Ozarks	Peace Valley	BBS
4	Ozarks	Sweden	BBS
5	Ozarks	Cassville	BBS
6	Ozarks	Jane	BBS
7	Ozark Border	Cape Girardeau	BBS
8	Ozarks	Cascade	BBS
10	Ozarks	Cedargrove	BBS
11	Ozarks	Cook Station	BBS
12	Ozarks	Dixon	BBS
13	Ozarks	Bennett Spring	BBS
14	Ozarks	Morrisville	BBS
15	Osage Plains	Filley	BBS
16	Ozarks	Reeds	BBS
18	Ozarks	Paydown	BBS
19	Ozark Border	Montgomery City	BBS
20	Ozarks	Boylers Mill	BBS
21	Ozark Border	Centertown Nw	BBS
22	Ozark Border	Cole Camp	BBS
23	Osage Plains	Ohio	BBS
24	Glaciated Plains	Tarsney Lakes	BBS
25	Glaciated Plains	Rensselaer	BBS
27	Glaciated Plains	Madison	BBS
28	Glaciated Plains	Huntsville	BBS
29	Glaciated Plains	Richmond	BBS
30	Glaciated Plains	Stet	BBS
32	Glaciated Plains	Agency	BBS
33	Glaciated Plains	Kahoka	BBS
34	Glaciated Plains	Novinger	BBS
35	Glaciated Plains	Trenton East	BBS
36	Glaciated Plains	Bolckow	BBS
37	Glaciated Plains	Skidmore Nw	BBS
109	Ozarks	Graniteville	BBS
117	Ozark Border	Moselle	BBS
122	Glaciated Plains	Sedalia East	BBS
124	Osage Plains	Kingsville	BBS
126	Glaciated Plains	Frankford	BBS
131	Glaciated Plains	Cameron East	BBS
137	Glaciated Plains	Tarkio East	BBS
217	Ozark Border	Union	BBS
11 F	Glaciated Plains	Sheridan	MR
26 E	Glaciated Plains	Coatsville	MR
37 C	Glaciated Plains	Tarkio West	MR
53 F	Glaciated Plains	Ravanna	MR
85 F	Glaciated Plains	Mount Moriah	MR
107 D	Glaciated Plains	Mound City	MR

Note: (1) Quadrangle Name; and (2) Source of data either Atlas Project Miniroute (MR) or United States Fish and Wildlife Service Breeding Bird Survey (BBS).

Route Number	Natural Division	Quadname (1)	Source of Data MR or BBS (2)
114 F	Glaciated Plains	Albany South	MR
123 C	Glaciated Plains	Milan West	MR
133 D	Glaciated Plains	Deer Ridge	MR
179 D	Glaciated Plains	Altamont	MR
204 A	Glaciated Plains	Clarksdale	MR
242 C	Glaciated Plains	Utica East	MR
249 A	Glaciated Plains	Lagonda	MR
297 A	Glaciated Plains	Tracy	MR
368 F	Glaciated Plains	Missouri City	MR
376 B	Big Rivers	Malta Bend	MR
386 E	Glaciated Plains	Centralia Ne	MR
486 B	Glaciated Plains	Guthrie	MR
511 A	Osage Plains	Burtville	MR
518 D	Ozark Border	Clarksburg	MR
561 B	Ozark Border	Luystown	MR
576 C	Osage Plains	Main City	MR
580 A	Osage Plains	Hartwell	MR
584 E	Osage Plains	Lincoln Nw	MR
593 C	Ozark Border	Brazito	MR
636 C	Ozark Border	Owensville East	MR
692 D	Ozarks	Cross Timbers	MR
710 D	Ozarks	Ebo	MR
714 C	Ozark Border	Halifax	MR
719 A	Osage Plains	Metz	MR
753 F	Ozark Border	Weingarten	MR
788 D	Ozark Border	Irondale	MR
807 D	Ozarks	Polk	MR
814 A	Ozarks	Drynob	MR
820 B	Ozarks	Anutt	MR
880 D	Osage Plains	Mindenmines	MR
908 A	Ozarks	Corridon	MR
928 F	Ozarks	South Greenfield	MR
933 B	Ozarks	Bassville	MR
940 C	Ozarks	Dawson	MR
943 F	Ozarks	Houston	MR
995 D	Ozarks	Piedmont	MR
1034F	Ozarks	Van Buren North	MR
1042B	Ozark Border	Sturdivant	MR
1053E	Ozarks	Newtonia	MR
1057A	Ozarks	Aurora	MR
1060A	Ozarks	Highlandville	MR
1069A	Ozarks	Nichols Knob	MR
1083C	Ozarks	Hendrickson	MR
1132B	Mississippi Lowlands	Dexter	MR
1147C	Ozarks	Shell Knob	MR
1174B	Mississippi Lowlands	Hanleyville	MR
1188F	Ozarks	Jacket	MR
1193E	Ozarks	Lampe	MR
1198E	Ozarks	Protem	MR
1205E	Ozarks	Moody	MR
1209D	Ozarks	Couch	MR

Abundance Data Routes and Data Gathering Procedure

Route Number	Natural Division	Quadname (1)	Source of Data MR or BBS (2)
1214F	Ozarks	Doniphan South	MR
1222C	Mississippi Lowlands	Parma	MR
1249B	Mississippi Lowlands	Denton	MR

Appendix C
Relative Abundance of Breeding Bird Species by Natural Division, 1986–1992

| Species | Glaciated Plains Route Number | | | | | | | | |
| | 24 | 25 | 27 | 28 | 29 | 30 | 32 | 33 | 34 |
	Average Number of Birds/100 Stops								
Great Blue Heron	10.7	0.9	6.5	3.1	3.1	4.7	2.5	4.0	0.6
Green Heron	4.7	0.0	0.5	1.4	0.9	1.3	0.0	1.0	1.7
Canada Goose	8.0	0.0	0.5	7.4	1.1	0.0	6.0	0.0	5.4
Wood Duck	0.0	0.0	1.5	0.0	0.0	0.0	0.0	0.0	0.0
Turkey Vulture	8.7	1.1	1.0	3.4	3.1	0.3	1.0	8.0	2.3
Red-shouldered Hawk	0.0	0.0	0.0	0.0	0.3	0.0	0.0	0.0	0.0
Red-tailed Hawk	2.7	1.1	2.0	4.3	5.4	1.3	7.0	4.0	4.6
Ring-necked Pheasant	0.0	0.0	36.0	0.0	0.0	0.3	4.0	18.0	0.6
Wild Turkey	0.0	0.0	1.5	4.6	0.6	0.0	0.0	1.0	10.6
Northern Bobwhite	76.0	69.4	198.0	90.9	124.0	105.0	175.0	87.0	103.0
Killdeer	10.7	11.1	23.5	10.9	19.4	34.3	195.0	14.0	12.9
Upland Sandpiper	0.0	2.6	1.5	0.9	0.9	7.0	17.0	0.0	0.9
Rock Dove	13.3	5.1	11.0	2.3	9.1	27.3	0.5	0.0	6.6
Mourning Dove	60.7	58.3	134.0	80.0	103.0	106.0	127.0	91.0	70.0
Yellow-billed Cuckoo	18.0	3.7	10.5	11.1	41.4	17.7	21.0	7.0	8.0
Chimney Swift	52.0	22.3	13.5	11.4	4.9	14.7	5.0	6.0	6.3
Ruby-throated Hummingbird	0.7	0.0	0.0	0.3	0.6	0.0	0.0	2.0	0.6
Belted Kingfisher	0.0	0.6	0.5	0.0	0.0	0.0	1.0	0.0	0.6
Red-headed Woodpecker	7.3	13.1	78.5	16.0	27.1	11.3	16.0	37.0	24.9
Red-bellied Woodpecker	10.7	15.4	24.0	26.3	22.3	11.3	22.0	12.0	13.7
Downy Woodpecker	2.7	2.0	10.0	6.3	18.9	7.0	1.5	6.0	3.7
Hairy Woodpecker	0.0	0.6	2.5	0.6	3.4	0.3	1.0	0.0	1.4
Northern Flicker	6.7	4.3	10.0	8.6	18.9	7.0	11.0	16.0	15.7
Pileated Woodpecker	0.0	0.6	3.5	1.1	0.3	0.0	0.0	1.0	0.3
Eastern Wood-Pewee	12.0	0.6	8.0	13.7	21.4	2.7	5.0	3.0	9.1
Acadian Flycatcher	0.0	0.0	0.0	0.0	2.6	0.0	0.0	0.0	0.0
Eastern Phoebe	8.0	1.7	2.5	11.7	8.0	2.3	3.0	11.0	2.6
Great Crested Flycatcher	6.7	2.0	6.5	15.1	17.4	7.0	8.0	3.0	18.3
Eastern Kingbird	8.7	10.3	64.5	36.0	19.1	13.3	30.0	13.0	32.6
Scissor-tailed Flycatcher	0.0	0.0	0.0	0.3	0.3	0.0	0.0	0.0	0.0
Horned Lark	16.0	33.4	135.0	10.6	16.0	73.0	1.0	21.0	22.0
Purple Martin	10.0	9.1	5.0	4.0	1.7	2.7	0.0	3.0	0.6
No Rough-winged Swallow	5.3	2.9	0.5	1.7	3.4	7.7	0.0	4.0	8.6
Bank Swallow	0.0	2.3	0.0	0.0	0.0	0.3	0.0	4.0	3.4
Barn Swallow	48.0	35.4	66.5	49.4	43.1	58.7	37.5	56.0	57.4
Blue Jay	26.0	20.6	21.5	54.3	53.4	16.7	31.5	31.0	37.7
American Crow	50.0	26.0	40.5	24.6	54.0	18.0	28.5	39.0	50.3
Black-capped Chickadee	29.3	2.9	6.5	20.3	40.9	10.0	14.5	9.0	20.0
Carolina Chickadee	0.0	0.0	0.0	0.0	0.0	0.0	0.0	0.0	0.0
Tufted Titmouse	22.0	10.3	25.0	37.1	41.1	9.3	7.5	6.0	22.3
White-breasted Nuthatch	1.3	3.1	1.0	6.3	12.6	2.3	3.5	6.0	8.0
Carolina Wren	1.3	1.1	2.0	1.1	6.9	1.3	0.5	2.0	0.0
House Wren	4.0	5.4	35.0	11.1	12.9	17.0	16.5	26.0	44.0
Bewick's Wren	0.0	0.0	0.0	0.9	0.3	0.0	0.0	0.0	0.6
Blue-gray Gnatcatcher	2.7	0.0	3.5	2.0	10.3	0.7	0.5	3.0	0.6
Eastern Bluebird	24.7	8.3	16.5	36.9	19.4	7.3	17.0	18.0	34.0
Wood Thrush	0.7	0.0	0.0	1.7	3.4	0.7	0.0	0.0	0.0
American Robin	107.0	97.7	160.0	117.0	122.0	88.3	79.5	106.0	129.0

Relative Abundance of Breeding Bird Species by Natural Division, 1986–1992

Species	24	25	27	Glaciated Plains Route Number 28	29	30	32	33	34
				Average Number of Birds/100 Stops					
Gray Catbird	2.0	1.1	18.5	11.1	6.3	7.0	3.5	26.0	12.0
Northern Mockingbird	15.3	11.4	3.5	10.0	16.9	7.3	4.5	1.0	2.9
Brown Thrasher	12.0	16.9	37.0	38.0	16.6	26.7	26.5	22.0	53.4
Cedar Waxwing	0.3	0.0	1.0	2.3	0.9	3.0	0.0	0.0	0.9
Loggerhead Shrike	6.0	3.4	6.5	6.9	2.9	4.3	9.0	1.0	0.3
European Starling	90.0	69.1	165.0	62.6	26.6	45.0	13.0	363.0	54.0
White-eyed Vireo	0.0	0.0	0.0	0.3	0.0	0.0	0.5	2.0	0.3
Bell's Vireo	0.7	0.0	4.0	0.9	0.0	2.3	0.0	4.0	0.0
Yellow-throated Vireo	1.3	0.3	0.0	2.0	2.9	0.0	0.0	2.0	3.1
Warbling Vireo	4.7	2.3	6.0	1.7	9.7	4.3	3.0	4.0	2.9
Red-eyed Vireo	0.7	0.0	2.0	1.4	4.0	0.0	0.0	3.0	0.0
Blue-winged Warbler	0.0	0.0	0.0	2.9	0.0	0.0	0.0	0.0	0.0
Northern Parula	2.7	0.0	1.0	2.6	3.4	0.3	0.0	0.0	0.3
Yellow Warbler	0.0	0.0	0.0	0.0	0.0	5.7	0.0	0.0	0.0
Yellow-throated Warbler	0.0	0.0	0.0	0.0	0.3	0.0	0.0	0.0	0.0
Pine Warbler	0.0	0.0	0.0	0.0	0.0	0.0	0.0	0.0	0.0
Prairie Warbler	0.0	0.0	0.0	0.0	0.0	0.0	0.0	1.0	0.0
Black-and-white Warbler	0.7	0.0	0.0	0.0	0.0	0.0	0.0	0.0	0.0
Worm-eating Warbler	0.0	0.0	0.0	0.3	0.0	0.0	0.0	1.0	0.0
Louisiana Waterthrush	0.0	0.0	0.0	0.6	2.0	0.0	0.0	0.0	0.0
Kentucky Warbler	0.0	0.0	1.5	3.1	4.6	0.0	0.0	2.0	0.3
Common Yellowthroat	12.7	4.3	68.5	28.6	30.6	53.3	12.0	41.0	21.7
Yellow-breasted Chat	0.0	0.9	1.0	5.7	1.7	0.7	0.0	11.0	1.1
Summer Tanager	0.7	0.0	1.0	5.4	2.0	0.0	0.0	2.0	0.9
Scarlet Tanager	0.0	0.0	0.0	2.0	1.4	0.0	0.0	2.0	0.3
Northern Cardinal	0.0	37.1	43.0	61.1	105.0	38.3	42.0	36.0	42.3
Rose-breasted Grosbeak	0.7	3.4	20.0	17.7	11.1	5.3	0.5	12.0	29.1
Blue Grosbeak	0.7	1.1	14.0	8.0	7.7	1.7	0.0	0.0	3.4
Indigo Bunting	11.3	9.7	50.5	60.6	128.0	53.0	26.5	55.0	35.1
Dickcissel	73.3	30.9	178.0	51.4	252.0	352.0	117.0	56.0	61.1
Eastern Towhee	0.7	0.0	3.0	29.7	2.6	0.3	0.5	4.0	12.3
Chipping Sparrow	0.0	0.0	1.0	7.4	1.7	0.0	0.0	14.0	7.1
Field Sparrow	16.7	4.3	18.0	70.6	42.9	2.3	16.0	15.0	54.3
Vesper Sparrow	0.0	0.0	0.0	0.3	1.4	0.0	0.0	0.0	1.4
Lark Sparrow	0.7	0.0	2.5	2.6	15.7	7.7	0.5	0.0	5.4
Grasshopper Sparrow	60.0	1.7	72.0	24.9	39.4	22.3	33.0	21.0	39.1
Song Sparrow	0.0	3.4	45.0	1.1	0.0	12.3	3.5	26.0	6.9
Bobolink	0.0	0.0	6.5	0.0	0.6	0.0	4.0	0.0	8.3
Red-winged Blackbird	121.0	406.0	828.0	206.0	193.0	443.0	146.0	258.0	294.0
Eastern Meadowlark	126.0	86.6	246.0	129.0	152.0	105.0	62.0	88.0	130.0
Western Meadowlark	0.0	0.0	0.0	0.0	0.9	2.3	51.5	1.0	18.3
Common Grackle	187.0	209.0	255.0	92.0	71.1	123.0	102.0	420.0	82.3
Brown-headed Cowbird	71.3	33.4	52.5	73.1	52.0	57.7	84.5	48.0	95.4
Orchard Oriole	1.3	0.9	8.5	9.1	1.0	5.3	0.0	1.0	4.6
Baltimore Oriole	9.3	7.1	10.5	16.6	18.9	21.0	14.5	10.0	25.1
House Finch	0.0	0.0	0.0	0.0	0.0	0.0	0.0	1.0	0.3
American Goldfinch	6.0	3.4	15.5	16.3	30.3	22.0	9.5	36.0	16.0
House Sparrow	278.0	257.0	817.0	82.0	159.0	147.0	174.0	153.0	140.0

Relative Abundance of Breeding Bird Species by Natural Division, 1986–1992

			Glaciated Plains Route Number					
Species	35	36	37	122	126	131	137	11F	26E
				Average Number of Birds/100 Stops					
Great Blue Heron	0.7	2.0	0.5	2.0	0.3	2.0	1.0	0.0	0.0
Green Heron	2.3	1.6	0.0	4.0	0.0	0.6	0.0	0.0	0.0
Canada Goose	0.7	0.0	0.0	0.0	0.0	10.0	0.0	0.0	0.0
Wood Duck	1.3	0.0	1.5	0.0	0.7	0.3	0.0	0.0	7.0
Turkey Vulture	2.3	8.8	14.0	2.0	8.0	1.1	25.0	0.0	3.0
Red-shouldered Hawk	0.0	0.0	0.0	0.0	0.0	0.0	0.0	0.0	0.0
Red-tailed Hawk	4.3	5.2	3.0	3.0	2.7	2.3	1.0	<0.1	0.0
Ring-necked Pheasant	27.7	23.6	27.5	0.0	0.0	0.6	8.0	53.0	7.0
Wild Turkey	8.3	0.0	0.0	1.0	2.0	1.4	0.0	<0.1	30.0
Northern Bobwhite	153.0	123.0	27.5	134.0	81.7	147.0	26.0	110.0	153.0
Killdeer	20.3	30.4	13.0	23.0	7.0	45.1	10.0	13.0	13.0
Upland Sandpiper	4.7	1.6	0.5	6.0	0.0	8.9	0.0	0.0	0.0
Rock Dove	2.3	10.0	4.5	6.0	2.7	35.1	9.0	13.0	0.0
Mourning Dove	134.0	86.0	115.0	140.0	103.0	86.0	163.0	120.0	127.0
Yellow-billed Cuckoo	8.3	10.0	11.0	19.0	7.3	9.1	10.0	7.0	0.0
Chimney Swift	13.0	17.6	21.0	40.0	5.3	70.0	10.0	20.0	0.0
Ruby-throated Hummingbird	0.3	0.0	0.0	3.0	0.0	0.9	0.0	0.0	0.0
Belted Kingfisher	1.0	0.8	0.5	1.0	0.0	0.9	0.0	3.0	0.0
Red-headed Woodpecker	33.3	21.2	15.5	8.0	10.7	16.0	11.0	17.0	30.0
Red-bellied Woodpecker	11.7	10.0	3.5	15.0	10.0	5.7	3.0	10.0	27.0
Downy Woodpecker	5.0	6.0	6.5	4.0	3.0	4.0	7.0	0.0	10.0
Hairy Woodpecker	0.3	0.8	0.0	0.0	0.6	0.0	0.0	0.0	0.0
Northern Flicker	10.7	5.2	6.5	7.0	3.3	6.6	9.0	40.0	10.0
Pileated Woodpecker	0.0	0.0	0.0	0.0	0.0	0.0	0.0	0.0	0.0
Eastern Wood-Pewee	9.7	10.4	1.5	9.0	2.0	4.0	1.0	<0.1	10.0
Acadian Flycatcher	0.3	0.8	0.0	0.0	0.0	0.0	0.0	0.0	0.0
Eastern Phoebe	4.0	3.6	1.5	12.0	3.0	6.3	4.0	3.0	3.0
Great Crested Flycatcher	13.7	8.4	3.5	11.0	4.7	1.1	5.0	7.0	23.0
Eastern Kingbird	39.0	22.0	16.0	20.0	10.7	61.4	26.0	43.0	30.0
Scissor-tailed Flycatcher	0.0	0.0	0.0	2.0	0.0	0.0	0.0	0.0	0.0
Horned Lark	28.3	8.8	5.0	30.0	57.3	55.4	1.0	7.0	3.0
Purple Martin	1.3	2.0	8.0	20.0	1.0	2.9	3.0	0.0	0.0
No Rough-winged Swallow	2.3	3.2	7.5	2.0	5.3	5.1	8.0	7.0	0.0
Bank Swallow	2.0	0.0	0.0	0.0	2.7	1.1	0.0	0.0	0.0
Barn Swallow	32.7	45.2	73.0	70.0	19.7	34.6	48.0	73.0	47.0
Blue Jay	33.7	27.2	18.0	48.0	35.3	27.4	16.0	70.0	37.0
American Crow	46.7	9.2	4.0	73.0	12.0	17.7	4.0	240.0	37.0
Black-capped Chickadee	8.7	10.8	4.5	7.0	12.3	16.6	13.0	23.0	30.0
Carolina Chickadee	0.0	0.0	0.0	0.0	0.3	0.0	0.0	0.0	0.0
Tufted Titmouse	11.0	4.4	2.5	3.3	10.7	12.9	3.0	7.0	23.0
White-breasted Nuthatch	2.3	4.8	1.0	2.0	0.3	2.3	2.0	10.0	3.0
Carolina Wren	1.3	0.0	0.0	1.0	4.0	0.6	1.0	0.0	<0.1
House Wren	43.7	17.6	11.0	18.0	3.0	28.3	26.0	47.0	73.0
Bewick's Wren	0.0	0.0	0.0	3.0	0.0	0.0	0.0	0.0	0.0
Blue-gray Gnatcatcher	1.7	0.0	1.5	2.0	2.7	0.9	0.0	0.0	0.0
Eastern Bluebird	11.3	9.2	11.5	25.0	15.7	12.0	4.0	20.0	23.0
Wood Thrush	1.0	2.0	0.0	0.0	1.3	0.0	1.0	13.0	3.0
American Robin	133.0	73.6	101.0	140.0	61.7	194.0	65.0	50.0	220.0

Species	35	36	37	Glaciated Plains Route Number 122	126	131	137	11F	26E
				Average Number of Birds/100 Stops					
Gray Catbird	16.3	6.8	7.0	12.0	10.3	7.4	10.0	20.0	30.0
Northern Mockingbird	2.3	5.6	1.5	14.0	2.3	3.7	2.0	0.0	0.0
Brown Thrasher	36.7	20.0	27.5	34.0	15.3	29.1	14.0	20.0	53.0
Cedar Waxwing	16.7	0.8	0.0	0.0	0.0	3.1	0.0	0.0	10.0
Loggerhead Shrike	5.0	10.4	5.0	5.0	0.0	5.1	6.0	0.0	0.0
European Starling	30.7	36.0	64.5	295.0	14.3	126.0	24.0	20.0	183.0
White-eyed Vireo	0.0	0.0	0.0	0.0	0.3	0.0	0.0	0.0	0.0
Bell's Vireo	4.0	2.0	2.0	2.0	0.3	2.0	0.0	0.0	0.0
Yellow-throated Vireo	0.3	0.4	1.0	0.0	0.3	0.3	1.0	0.0	0.0
Warbling Vireo	8.3	6.0	7.0	2.0	1.3	10.9	4.0	3.0	0.0
Red-eyed Vireo	1.7	0.4	0.5	1.0	0.0	0.0	0.0	0.0	0.0
Blue-winged Warbler	0.0	0.0	0.0	0.0	0.0	0.0	0.0	0.0	0.0
Northern Parula	0.7	0.8	0.0	2.0	1.3	0.0	0.0	0.0	3.0
Yellow Warbler	0.0	0.0	0.5	0.0	0.7	0.0	0.0	0.0	0.0
Yellow-throated Warbler	0.0	0.0	0.0	0.0	0.0	0.0	2.0	0.0	0.0
Pine Warbler	0.0	0.0	0.0	0.0	0.0	0.0	0.0	0.0	0.0
Prairie Warbler	0.0	0.0	0.0	0.0	0.0	0.0	0.0	0.0	0.0
Black-and-white Warbler	0.0	0.0	0.0	0.0	0.0	0.0	0.0	0.0	0.0
Worm-eating Warbler	0.0	0.0	0.0	0.0	0.0	0.0	0.0	0.0	0.0
Louisiana Waterthrush	0.0	0.0	0.0	0.0	0.0	0.0	0.0	0.0	0.0
Kentucky Warbler	1.0	0.0	0.0	0.0	0.3	0.0	0.0	0.0	7.0
Common Yellowthroat	45.7	25.6	16.0	27.0	19.7	33.1	11.0	47.0	57.0
Yellow-breasted Chat	0.0	0.0	0.0	0.0	0.3	1.4	0.0	0.0	0.0
Summer Tanager	0.3	0.0	0.0	2.0	0.3	0.0	0.0	0.0	0.0
Scarlet Tanager	0.0	0.0	0.0	0.0	0.0	0.3	0.0	0.0	3.0
Northern Cardinal	37.7	44.0	36.0	69.0	22.3	18.6	37.0	43.0	43.0
Rose-breasted Grosbeak	15.7	2.0	2.0	1.0	1.7	8.3	5.0	17.0	13.0
Blue Grosbeak	1.0	0.0	0.0	7.0	0.7	0.3	0.0	0.0	0.0
Indigo Bunting	32.3	27.6	13.0	29.0	37.7	26.3	7.0	33.0	20.0
Dickcissel	117.0	108.0	82.0	161.0	40.3	151.0	84.0	43.0	17.0
Eastern Towhee	5.3	1.6	2.5	7.0	3.3	6.3	0.0	23.0	23.0
Chipping Sparrow	3.3	3.6	2.0	21.0	3.0	5.1	2.0	0.3	3.0
Field Sparrow	30.3	10.0	2.5	48.0	49.7	21.1	2.0	17.0	60.0
Vesper Sparrow	3.0	0.4	1.0	0.0	0.0	0.0	1.0	3.0	7.0
Lark Sparrow	12.0	1.2	4.0	2.0	1.0	2.9	4.0	3.0	7.0
Grasshopper Sparrow	46.3	21.2	6.0	44.0	11.7	29.7	12.0	17.0	23.0
Song Sparrow	35.0	11.6	4.5	0.0	4.7	14.3	6.0	40.0	23.0
Bobolink	3.0	1.6	0.0	0.0	0.0	5.7	0.0	0.0	0.0
Red-winged Blackbird	279.0	232.0	152.0	240.0	458.0	331.0	170.0	300.0	207.0
Eastern Meadowlark	126.0	49.0	1.0	228.0	90.7	1.17	0.0	50.0	167.0
Western Meadowlark	0.7	43.2	91.5	0.0	0.0	66.3	116.0	17.0	0.0
Common Grackle	159.0	64.4	27.5	219.0	145.0	122.0	14.0	57.0	113.0
Brown-headed Cowbird	117.0	72.0	65.0	83.0	21.7	57.7	81.0	23.0	113.0
Orchard Oriole	3.7	0.0	3.5	4.0	2.3	0.6	1.0	3.0	3.0
Baltimore Oriole	23.3	12.4	23.0	6.0	5.0	10.6	30.0	13.0	17.0
House Finch	0.0	0.0	0.0	0.0	0.3	0.0	0.0	0.0	0.0
American Goldfinch	29.7	21.2	12.0	7.0	7.0	14.9	11.0	10.0	10.0
House Sparrow	77.0	165.0	377.0	140.0	67.3	209.0	195.0	170.0	117.0

Relative Abundance of Breeding Bird Species by Natural Division, 1986–1992

Species	37C	53F	Glaciated Plains Route Number 85F	107D	114F	123C	133D	179D
			Average Number of Birds/100 Stops					
Great Blue Heron	<0.1	0.0	0.0	3.0	3.0	0.0	0.0	0.0
Green Heron	0.0	0.0	0.0	0.0	3.0	<0.1	0.0	0.0
Canada Goose	0.0	<0.1	0.0	0.0	0.0	0.0	0.0	0.0
Wood Duck	7.0	0.0	0.0	0.0	0.0	0.0	0.0	0.0
Turkey Vulture	0.0	0.0	0.0	0.0	0.0	0.0	0.0	7.0
Red-shouldered Hawk	0.0	0.0	0.0	0.0	0.0	0.0	0.0	0.0
Red-tailed Hawk	<0.1	7.0	0.0	0.0	0.0	0.0	3.0	3.0
Ring-necked Pheasant	37.0	63.0	40.0	50.0	100.0	3.0	0.0	3.0
Wild Turkey	0.0	3.0	0.0	0.0	3.0	13.0	3.0	7.0
Northern Bobwhite	113.0	160.0	180.0	113.0	127.0	173.0	173.0	100.0
Killdeer	3.0	10.0	3.0	33.0	10.0	40.0	0.0	30.0
Upland Sandpiper	0.0	7.0	0.0	17.0	<0.1	0.0	0.0	0.0
Rock Dove	0.0	0.0	17.0	0.0	0.0	0.0	7.0	0.0
Mourning Dove	127.0	183.0	90.0	107.0	157.0	113.0	203.0	60.0
Yellow-billed Cuckoo	20.0	7.0	0.0	13.0	7.0	0.0	7.0	13.0
Chimney Swift	0.0	3.0	17.0	0.0	13.0	0.0	<0.1	3.0
Ruby-throated Hummingbird	0.0	0.0	0.0	0.0	0.0	0.0	0.0	0.0
Belted Kingfisher	0.0	0.0	0.0	0.0	0.0	7.0	0.0	0.0
Red-headed Woodpecker	13.0	47.0	37.0	33.0	17.0	37.0	17.0	33.0
Red-bellied Woodpecker	7.0	13.0	10.0	7.0	3.0	13.0	17.0	13.0
Downy Woodpecker	0.0	3.0	3.0	7.0	0.0	10.0	3.0	0.0
Hairy Woodpecker	0.0	0.0	0.0	0.0	0.0	0.0	0.0	0.0
Northern Flicker	3.0	0.0	20.0	3.0	10.0	23.0	10.0	7.0
Pileated Woodpecker	0.0	0.0	0.0	0.0	0.0	0.0	0.0	0.0
Eastern Wood-Pewee	0.0	10.0	7.0	3.0	3.0	7.0	0.0	3.0
Acadian Flycatcher	0.0	0.0	0.0	0.0	0.0	0.0	0.0	0.0
Eastern Phoebe	0.0	10.0	23.0	3.0	10.0	3.0	0.0	7.0
Great Crested Flycatcher	3.0	13.0	17.0	3.0	7.0	10.0	10.0	10.0
Eastern Kingbird	10.0	53.0	73.0	53.0	43.0	20.0	10.0	13.0
Scissor-tailed Flycatcher	0.0	0.0	0.0	0.0	0.0	0.0	0.0	0.0
Horned Lark	0.0	0.0	0.0	3.0	27.0	0.0	70.0	7.0
Purple Martin	0.0	0.0	0.0	0.0	0.0	0.0	0.0	17.0
No Rough-winged Swallow	7.0	7.0	7.0	0.0	0.0	0.0	13.0	0.0
Bank Swallow	0.0	0.0	0.0	0.0	3.0	0.0	0.0	0.0
Barn Swallow	90.0	30.0	60.0	57.0	103.0	73.0	23.0	37.0
Blue Jay	23.0	43.0	60.0	23.0	53.0	97.0	20.0	33.0
American Crow	10.0	43.0	60.0	0.0	57.0	113.0	60.0	33.0
Black-capped Chickadee	7.0	20.0	7.0	7.0	7.0	13.0	7.0	3.0
Carolina Chickadee	0.0	0.0	0.0	0.0	3.0	0.0	0.0	0.0
Tufted Titmouse	0.0	0.0	13.0	0.0	7.0	57.0	13.0	30.0
White-breasted Nuthatch	0.0	3.0	0.0	0.0	0.0	13.0	3.0	7.0
Carolina Wren	0.0	0.0	0.0	0.0	0.0	0.0	0.0	0.0
Bewick's Wren	0.0	0.0	0.0	0.0	0.0	0.0	0.0	0.0
House Wren	7.0	50.0	43.0	10.0	27.0	30.0	10.0	10.0
Blue-gray Gnatcatcher	0.0	0.0	0.0	0.0	0.0	0.0	0.0	0.0
Eastern Bluebird	7.0	23.0	13.0	13.0	17.0	40.0	20.0	7.0
Wood Thrush	0.0	0.0	0.0	0.0	0.0	0.0	0.0	0.0
American Robin	23.0	180.0	33.0	127.0	217.0	43.0	97.0	53.0

Relative Abundance of Breeding Bird Species by Natural Division, 1986–1992

Species	37C	53F	Glaciated Plains Route Number 85F	107D	114F	123C	133D	179D
			Average Number of Birds/100 Stops					
Gray Catbird	3.0	80.0	7.0	7.0	13.0	7.0	1.0	0.0
Northern Mockingbird	3.0	53.0	0.0	0.0	7.0	0.0	0.0	13.0
Brown Thrasher	27.0	20.0	17.0	37.0	60.0	13.0	37.0	27.0
Cedar Waxwing	0.0	0.0	0.0	3.0	0.0	0.0	0.0	0.0
Loggerhead Shrike	3.0	0.0	0.0	0.0	7.0	10.0	3.0	7.0
European Starling	37.0	293.0	13.0	23.0	50.0	67.0	0.0	<0.1
White-eyed Vireo	0.0	0.0	0.0	0.0	0.0	0.0	0.0	0.0
Bell's Vireo	0.0	7.0	3.0	0.0	0.0	3.0	0.0	10.0
Yellow-throated Vireo	3.0	0.0	0.0	0.0	3.0	3.0	0.0	3.0
Warbling Vireo	0.0	0.0	7.0	0.0	7.0	0.0	0.0	0.0
Red-eyed Vireo	0.0	0.0	0.0	0.0	0.0	0.0	0.0	0.0
Blue-winged Warbler	0.0	0.0	0.0	0.0	0.0	0.0	0.0	0.0
Northern Parula	0.0	0.0	0.0	0.0	0.0	0.0	0.0	0.0
Yellow Warbler	0.0	0.0	0.0	0.0	0.0	0.0	0.0	0.0
Yellow-throated Warbler	0.0	0.0	0.0	0.0	0.0	0.0	0.0	0.0
Pine Warbler	0.0	0.0	0.0	0.0	0.0	0.0	0.0	0.0
Prairie Warbler	0.0	0.0	0.0	0.0	0.0	0.0	0.0	0.0
Black-and-white Warbler	0.0	0.0	0.0	0.0	0.0	0.0	0.0	0.0
Worm-eating Warbler	0.0	0.0	0.0	0.0	0.0	0.0	0.0	0.0
Louisiana Waterthrush	0.0	0.0	0.0	0.0	0.0	0.0	0.0	0.0
Kentucky Warbler	0.0	0.0	0.0	0.0	0.0	3.0	0.0	0.0
Common Yellowthroat	10.0	87.0	57.0	10.0	27.0	43.0	27.0	3.0
Yellow-breasted Chat	0.0	0.0	0.0	3.0	0.0	3.0	3.0	0.0
Summer Tanager	0.0	0.0	0.0	0.0	0.0	0.0	0.0	0.0
Scarlet Tanager	0.0	0.0	0.0	0.0	0.0	0.0	0.0	0.0
Northern Cardinal	30.0	33.0	17.0	10.0	13.0	27.0	60.0	40.0
Rose-breasted Grosbeak	3.0	47.0	0.0	0.0	7.0	<0.1	7.0	20.0
Blue Grosbeak	0.0	0.0	0.0	0.0	0.0	0.0	0.0	0.0
Indigo Bunting	7.0	40.0	17.0	<0.1	20.0	30.0	133.0	13.0
Dickcissel	240.0	110.0	127.0	187.0	130.0	7.0	13.0	70.0
Eastern Towhee	0.0	<0.1	3.0	0.0	0.0	7.0	0.0	0.0
Chipping Sparrow	20.0	0.0	0.0	13.0	3.0	3.0	0.0	0.0
Field Sparrow	0.0	23.0	47.0	13.0	3.0	43.0	13.0	23.0
Vesper Sparrow	0.0	7.0	0.0	0.0	0.0	3.0	0.0	0.0
Lark Sparrow	3.0	0.0	0.0	7.0	3.0	0.0	3.0	3.0
Grasshopper Sparrow	50.0	43.0	77.0	50.0	117.0	20.0	0.0	17.0
Song Sparrow	0.0	57.0	13.0	7.0	13.0	20.0	0.0	7.0
Bobolink	0.0	17.0	17.0	0.0	10.0	3.0	0.0	0.0
Red-winged Blackbird	97.0	367.0	193.0	167.0	347.0	207.0	387.0	73.0
Eastern Meadowlark	13.0	97.0	147.0	57.0	67.0	57.0	37.0	87.0
Western Meadowlark	157.0	3.0	0.0	57.0	173.0	0.0	0.0	7.0
Common Grackle	97.0	83.0	43.0	7.0	337.0	87.0	303.0	167.0
Brown-headed Cowbird	37.0	177.0	90.0	30.0	87.0	70.0	3.0	40.0
Orchard Oriole	0.0	10.0	7.0	13.0	7.0	<0.1	0.0	3.0
Baltimore Oriole	7.0	20.0	7.0	7.0	13.0	3.0	37.0	10.0
House Finch	0.0	0.0	0.0	0.0	0.0	0.0	0.0	0.0
American Goldfinch	10.0	13.0	3.0	0.0	20.0	10.0	17.0	7.0
House Sparrow	373.0	67.0	57.0	57.0	273.0	27.0	3.0	323.0

| Species | Glaciated Plains Route Number | | | | | | |
| | 204A | 242C | 249A | 297A | 368F | 386E | 486B |
	Average Number of Birds/100 Stops						
Great Blue Heron	0.0	0.0	3.0	0.0	7.0	0.0	3.0
Green Heron	0.0	3.0	0.0	0.0	0.0	0.0	0.0
Canada Goose	0.0	0.0	0.0	0.0	0.0	3.0	0.0
Wood Duck	10.0	0.0	0.0	0.0	3.0	0.0	0.0
Turkey Vulture	0.0	3.0	0.0	0.0	0.0	0.0	0.0
Red-shouldered Hawk	0.0	0.0	0.0	0.0	0.0	0.0	0.0
Red-tailed Hawk	0.0	7.0	0.0	0.0	0.0	0.0	0.0
Ring-necked Pheasant	3.0	0.0	0.0	17.0	0.0	0.0	0.0
Wild Turkey	0.0	0.0	40.0	0.0	0.0	0.0	0.0
Northern Bobwhite	103.0	223.0	153.0	167.0	47.0	147.0	77.0
Killdeer	30.0	7.0	7.0	27.0	10.0	3.0	13.0
Upland Sandpiper	0.0	0.0	0.0	0.0	0.0	0.0	0.0
Rock Dove	17.0	13.0	3.0	0.0	10.0	0.0	13.0
Mourning Dove	113.0	180.0	87.0	70.0	87.0	187.0	130.0
Yellow-billed Cuckoo	7.0	0.0	3.0	27.0	37.0	0.0	3.0
Chimney Swift	0.0	0.0	0.0	13.0	53.0	17.0	7.0
Ruby-throated Hummingbird	0.0	0.0	0.0	0.0	<0.1	0.0	0.0
Belted Kingfisher	0.0	0.0	<0.1	0.0	0.0	0.0	0.0
Red-headed Woodpecker	30.0	30.0	37.0	13.0	23.0	27.0	10.0
Red-bellied Woodpecker	30.0	40.0	10.0	7.0	27.0	0.0	50.0
Downy Woodpecker	0.0	10.0	10.0	10.0	33.0	0.0	17.0
Hairy Woodpecker	<0.1	3.0	0.0	0.0	3.0	0.0	0.0
Northern Flicker	0.0	20.0	3.0	7.0	<0.1	3.0	7.0
Pileated Woodpecker	0.0	0.0	0.0	0.0	0.0	0.0	0.0
Eastern Wood-Pewee	<0.1	3.0	10.0	33.0	47.0	0.0	17.0
Acadian Flycatcher	0.0	7.0	0.0	0.0	3.0	0.0	0.0
Eastern Phoebe	10.0	<0.1	10.0	0.0	17.0	0.0	3.0
Great Crested Flycatcher	0.0	23.0	10.0	3.0	3.0	0.0	13.0
Eastern Kingbird	23.0	13.0	17.0	20.0	47.0	33.0	17.0
Scissor-tailed Flycatcher	0.0	0.0	0.0	0.0	0.0	0.0	0.0
Horned Lark	0.0	23.0	30.0	3.0	0.0	150.0	10.0
Purple Martin	0.0	0.0	3.0	0.0	13.0	<0.1	20.0
No Rough-winged Swallow	0.0	0.0	0.0	0.0	0.0	0.0	0.0
Bank Swallow	0.0	0.0	0.0	0.0	0.0	0.0	0.0
Barn Swallow	10.0	63.0	67.0	33.0	53.0	53.0	20.0
Blue Jay	20.0	77.0	10.0	27.0	37.0	17.0	63.0
American Crow	53.0	37.0	53.0	33.0	53.0	43.0	77.0
Black-capped Chickadee	7.0	7.0	20.0	27.0	47.0	7.0	3.0
Carolina Chickadee	0.0	0.0	0.0	0.0	0.0	0.0	0.0
Tufted Titmouse	7.0	20.0	30.0	50.0	40.0	20.0	37.0
White-breasted Nuthatch	0.0	7.0	23.0	10.0	7.0	0.0	10.0
Carolina Wren	0.0	3.0	0.0	0.0	40.0	0.0	0.0
Bewick's Wren	0.0	0.0	3.0	0.0	0.0	0.0	10.0
House Wren	3.0	3.0	30.0	0.0	3.0	0.0	33.0
Blue-gray Gnatcatcher	0.0	0.0	0.0	0.0	17.0	0.0	0.0
Eastern Bluebird	17.0	23.0	13.0	3.0	27.0	<0.1	53.0
Wood Thrush	0.0	0.0	0.0	7.0	10.0	0.0	0.0
American Robin	83.0	67.0	70.0	93.0	150.0	193.0	150.0

Relative Abundance of Breeding Bird Species by Natural Division, 1986–1992

Species	204A	242C	249A	297A	368F	386E	486B
			Glaciated Plains Route Number				
				Average Number of Birds/100 Stops			
Gray Catbird	7.0	13.0	13.0	3.0	0.0	7.0	<0.1
Northern Mockingbird	7.0	10.0	3.0	0.0	43.0	37.0	0.0
Brown Thrasher	23.0	43.0	20.0	27.0	13.0	37.0	17.0
Cedar Waxwing	0.0	0.0	0.0	0.0	0.0	7.0	0.0
Loggerhead Shrike	3.0	3.0	0.0	10.0	13.0	7.0	0.0
European Starling	40.0	30.0	40.0	40.0	103.0	37.0	147.0
White-eyed Vireo	0.0	0.0	0.0	0.0	0.0	0.0	0.0
Bell's Vireo	0.0	0.0	0.0	3.0	0.0	0.0	0.0
Yellow-throated Vireo	0.0	0.0	0.0	<0.1	0.0	0.0	0.0
Warbling Vireo	0.0	0.0	0.0	10.0	10.0	0.0	0.0
Red-eyed Vireo	0.0	0.0	0.0	3.0	7.0	0.0	10.0
Blue-winged Warbler	0.0	0.0	0.0	0.0	0.0	0.0	<0.1
Northern Parula	0.0	0.0	0.0	7.0	13.0	0.0	0.0
Yellow Warbler	0.0	0.0	3.0	0.0	0.0	0.0	0.0
Yellow-throated Warbler	0.0	0.0	0.0	0.0	0.0	0.0	0.0
Pine Warbler	0.0	0.0	0.0	0.0	0.0	0.0	0.0
Prairie Warbler	0.0	0.0	0.0	0.0	0.0	0.0	0.0
Black-and-white Warbler	0.0	0.0	0.0	0.0	0.0	0.0	0.0
Worm-eating Warbler	0.0	0.0	0.0	0.0	0.0	0.0	0.0
Louisiana Waterthrush	0.0	0.0	0.0	0.0	0.0	0.0	0.0
Kentucky Warbler	0.0	0.0	0.0	0.0	0.0	0.0	3.0
Common Yellowthroat	3.0	23.0	20.0	37.0	3.0	23.0	17.0
Yellow-breasted Chat	0.0	0.0	<0.1	0.0	0.0	0.0	7.0
Summer Tanager	0.0	<0.1	0.0	0.0	0.0	0.0	0.0
Scarlet Tanager	0.0	0.0	0.0	0.0	0.0	0.0	3.0
Northern Cardinal	40.0	57.0	87.0	33.0	133.0	20.0	87.0
Rose-breasted Grosbeak	3.0	3.0	0.0	13.0	7.0	10.0	7.0
Blue Grosbeak	0.0	0.0	7.0	0.0	0.0	0.0	13.0
Indigo Bunting	10.0	53.0	37.0	33.0	83.0	17.0	10.0
Dickcissel	120.0	220.0	133.0	120.0	40.0	213.0	53.0
Eastern Towhee	0.0	7.0	13.0	0.0	7.0	0.0	0.0
Chipping Sparrow	3.0	23.0	0.0	0.0	47.0	0.0	7.0
Field Sparrow	13.0	33.0	47.0	3.0	10.0	17.0	47.0
Vesper Sparrow	0.0	0.0	0.0	0.0	0.0	0.0	0.0
Lark Sparrow	7.0	3.0	0.0	0.0	0.0	3.0	7.0
Grasshopper Sparrow	0.0	37.0	53.0	3.0	0.0	17.0	37.0
Song Sparrow	0.0	17.0	0.0	0.0	0.0	13.0	<0.1
Bobolink	0.0	0.0	7.0	0.0	0.0	0.0	0.0
Red-winged Blackbird	133.0	197.0	273.0	223.0	133.0	503.0	327.0
Eastern Meadowlark	57.0	183.0	77.0	60.0	90.0	150.0	147.0
Western Meadowlark	20.0	23.0	0.0	13.0	0.0	0.0	0.0
Common Grackle	103.0	207.0	80.0	110.0	210.0	133.0	310.0
Brown-headed Cowbird	77.0	80.0	73.0	17.0	20.0	23.0	73.0
Orchard Oriole	3.0	0.0	0.0	7.0	3.0	7.0	0.0
Baltimore Oriole	30.0	10.0	7.0	17.0	23.0	0.0	0.0
House Finch	0.0	0.0	0.0	0.0	0.0	0.0	0.0
American Goldfinch	17.0	20.0	40.0	7.0	0.0	3.0	0.0
House Sparrow	60.0	287.0	90.0	37.0	143.0	123.0	102.0

Species	Big Rivers Route Number 376B	Ozark Border Route Number					
		7	19	21	22	117	217
		Average Number of Birds/100 Stops					
Great Blue Heron	0.0	0.3	2.7	2.9	0.8	1.0	5.3
Green Heron	3.0	1.7	2.3	2.9	1.2	4.0	4.7
Canada Goose	0.0	0.0	2.7	0.0	0.0	0.0	0.0
Wood Duck	0.0	0.0	0.0	1.1	0.0	3.0	0.7
Turkey Vulture	0.0	0.0	1.7	2.0	0.4	1.0	2.0
Red-shouldered Hawk	0.0	0.0	0.0	0.0	0.0	0.0	0.7
Red-tailed Hawk	0.0	2.3	0.3	2.0	1.2	0.0	2.7
Ring-necked Pheasant	0.0	0.0	0.0	0.0	0.0	0.0	0.0
Wild Turkey	0.0	0.0	0.3	2.6	0.4	2.0	1.3
Northern Bobwhite	123.0	63.0	162.0	86.9	106.0	60.0	30.7
Killdeer	57.0	25.7	127.0	11.1	19.6	0.0	5.3
Upland Sandpiper	0.0	0.0	0.3	0.0	5.6	0.0	0.0
Rock Dove	40.0	28.3	0.3	4.9	27.6	11.0	7.3
Mourning Dove	157.0	127.0	112.0	67.7	147.0	49.0	43.3
Yellow-billed Cuckoo	10.0	10.0	10.0	19.7	10.8	10.0	5.3
Chimney Swift	23.0	20.3	14.3	18.3	103.0	38.0	30.0
Ruby-throated Hummingbird	0.0	0.3	1.0	0.9	0.0	3.0	5.3
Belted Kingfisher	0.0	0.0	0.7	1.7	0.8	1.0	0.0
Red-headed Woodpecker	33.0	6.3	19.3	8.9	4.0	7.0	11.3
Red-bellied Woodpecker	10.0	5.7	18.3	24.3	5.2	39.0	29.3
Downy Woodpecker	3.0	5.3	2.7	5.7	2.8	16.0	10.0
Hairy Woodpecker	3.0	1.0	0.0	1.1	0.4	0.0	0.0
Northern Flicker	3.0	8.3	8.0	7.4	4.0	11.0	6.7
Pileated Woodpecker	0.0	0.0	1.0	1.7	0.0	3.0	6.0
Eastern Wood-Pewee	7.0	6.0	6.7	36.3	68.0	11.0	26.7
Acadian Flycatcher	0.0	0.7	0.0	1.1	0.0	4.0	0.0
Eastern Phoebe	0.0	1.7	3.3	12.3	2.0	11.0	18.7
Great Crested Flycatcher	0.0	3.7	5.0	24.9	4.8	21.0	24.0
Eastern Kingbird	7.0	8.3	13.0	22.3	20.8	9.0	18.0
Scissor-tailed Flycatcher	0.0	0.0	0.0	0.0	3.6	0.0	0.0
Horned Lark	33.0	9.0	61.0	7.4	10.4	0.0	0.7
Purple Martin	0.0	106.0	5.7	22.6	22.4	13.0	24.0
No Rough-winged Swallow	13.0	0.7	1.3	1.7	7.2	4.0	4.0
Bank Swallow	0.0	0.0	0.0	0.0	0.0	0.0	0.0
Barn Swallow	73.0	90.0	87.7	86.3	78.4	68.0	51.3
Blue Jay	27.0	26.0	55.3	71.4	38.8	69.0	65.3
American Crow	0.0	6.0	54.3	134.0	48.4	114.0	72.7
Black-capped Chickadee	0.0	0.0	3.0	4.9	4.4	4.0	8.0
Carolina Chickadee	0.0	4.7	0.0	0.3	0.4	5.0	4.0
Tufted Titmouse	0.0	24.0	18.0	45.4	6.8	73.0	63.3
White-breasted Nuthatch	0.0	1.7	10.3	16.9	1.2	16.0	14.7
Carolina Wren	0.0	7.7	3.3	3.1	2.0	5.0	23.3
Bewick's Wren	0.0	0.0	0.0	9.4	2.4	0.0	11.3
House Wren	13.0	3.0	13.3	2.6	9.2	15.0	8.0
Blue-gray Gnatcatcher	0.0	1.0	19.0	16.9	0.0	13.0	8.7
Eastern Bluebird	3.0	3.7	27.3	62.3	22.8	49.0	40.7
Wood Thrush	0.0	4.0	0.0	0.0	0.4	2.0	0.7
American Robin	200.0	97.0	114.0	63.4	136.0	95.0	83.3

Relative Abundance of Breeding Bird Species by Natural Division, 1986–1992

Species	Big Rivers Route Number 376B	Ozark Border Route Number					
		7	19	21	22	117	217
				Average Number of Birds/100 Stops			
Gray Catbird	13.0	1.3	5.0	2.9	2.4	2.0	1.3
Northern Mockingbird	7.0	31.3	5.0	29.7	27.2	37.0	25.3
Brown Thrasher	43.0	8.7	17.7	16.9	9.6	13.0	10.7
Cedar Waxwing	0.0	0.0	0.0	5.7	0.0	0.0	0.0
Loggerhead Shrike	0.0	3.7	2.7	1.4	2.4	2.0	0.7
European Starling	60.0	78.0	103.0	97.1	224.0	55.0	165.0
White-eyed Vireo	0.0	1.0	0.0	0.6	0.0	8.0	2.7
Bell's Vireo	0.0	0.0	0.0	0.0	3.6	0.0	0.0
Yellow-throated Vireo	0.0	0.0	0.0	0.6	0.0	0.0	2.0
Warbling Vireo	0.0	1.7	1.3	1.7	0.0	2.0	5.3
Red-eyed Vireo	0.0	1.0	1.3	2.6	0.4	6.0	2.7
Blue-winged Warbler	0.0	0.0	0.0	0.3	0.0	0.0	0.0
Northern Parula	0.0	0.7	0.7	0.9	0.0	1.0	1.3
Yellow Warbler	0.0	0.3	0.0	0.0	0.0	0.0	0.0
Yellow-throated Warbler	0.0	0.0	0.0	0.6	0.0	0.0	0.0
Pine Warbler	0.0	0.0	0.0	0.0	0.0	0.0	0.0
Prairie Warbler	0.0	0.0	0.0	0.0	0.0	0.0	0.7
Black-and-white Warbler	0.0	0.3	0.3	0.3	0.0	0.0	0.0
Worm-eating Warbler	0.0	0.0	0.0	0.0	0.0	0.0	0.0
Louisiana Waterthrush	0.0	0.0	0.0	0.0	0.0	0.0	0.7
Kentucky Warbler	0.0	2.0	1.3	1.4	0.4	1.0	7.3
Common Yellowthroat	40.0	15.3	18.0	16.6	24.0	7.0	21.3
Yellow-breasted Chat	0.0	0.7	2.3	5.1	0.8	11.0	8.7
Summer Tanager	0.0	0.0	2.0	7.7	0.4	13.0	14.0
Scarlet Tanager	0.0	0.0	0.3	0.3	0.0	0.0	2.0
Northern Cardinal	13.0	87.7	61.0	76.3	46.0	119.0	103.0
Rose-breasted Grosbeak	3.0	0.7	3.7	0.0	0.0	0.0	0.7
Blue Grosbeak	0.0	0.7	3.3	13.4	3.2	11.0	8.0
Indigo Bunting	53.0	32.0	41.3	67.1	16.0	61.0	57.3
Dickcissel	150.0	40.0	55.3	89.4	106.0	2.0	6.0
Eastern Towhee	0.0	4.0	2.3	13.1	3.2	18.0	30.0
Chipping Sparrow	0.0	0.0	4.0	21.4	2.8	10.0	20.0
Field Sparrow	0.0	6.0	25.3	82.9	45.2	27.0	44.0
Vesper Sparrow	0.0	0.0	0.0	0.0	0.0	0.0	0.0
Lark Sparrow	7.0	1.0	0.3	2.0	0.8	0.0	1.3
Grasshopper Sparrow	47.0	1.0	14.7	41.1	8.0	0.0	0.0
Song Sparrow	10.0	10.3	1.0	0.3	0.0	5.0	0.0
Bobolink	0.0	0.0	0.0	0.0	0.0	0.0	0.0
Red-winged Blackbird	217.0	561.0	328.0	191.0	210.0	42.0	78.0
Eastern Meadowlark	2.0	68.7	112.0	219.0	196.0	51.0	42.0
Western Meadowlark	0.0	0.0	0.7	0.0	0.0	0.0	0.0
Common Grackle	147.0	235.0	126.0	101.0	203.0	72.0	103.0
Brown-headed Cowbird	83.0	33.3	49.0	99.7	38.4	16.0	47.3
Orchard Oriole	7.0	3.0	0.3	4.9	2.0	6.0	8.0
Baltimore Oriole	7.0	0.3	2.0	4.6	4.0	10.0	2.7
House Finch	0.0	0.3	0.0	0.0	0.0	0.0	1.3
American Goldfinch	0.0	7.0	6.7	12.9	10.8	6.0	8.0
House Sparrow	163.0	388.0	153.0	121.0	205.0	63.0	22.7

Relative Abundance of Breeding Bird Species by Natural Division, 1986–1992

| | | | Ozark Border Route Number | | | | | |
Species	518D	561B	593C	636C	714C	753F	788D	1042B
				Average Number of Birds/100 Stops				
Great Blue Heron	0.0	0.0	0.0	0.0	0.0	3.0	0.0	0.0
Green Heron	<0.1	3.0	3.0	3.0	3.0	3.0	0.0	10.0
Canada Goose	0.0	0.0	0.0	0.0	0.0	0.0	0.0	0.0
Wood Duck	0.0	3.0	0.0	0.0	0.0	7.0	0.0	0.0
Turkey Vulture	0.0	0.0	0.0	0.0	0.0	3.0	0.0	0.0
Red-shouldered Hawk	0.0	0.0	0.0	0.0	0.0	3.0	0.0	0.0
Red-tailed Hawk	3.0	3.0	3.0	0.0	0.0	0.0	0.0	0.0
Ring-necked Pheasant	0.0	0.0	0.0	0.0	0.0	0.0	0.0	0.0
Wild Turkey	33.0	30.0	3.0	0.0	13.0	0.0	<0.1	0.0
Northern Bobwhite	333.0	23.0	73.0	113.0	43.0	23.0	27.0	63.0
Killdeer	13.0	0.0	13.0	3.0	13.0	7.0	0.0	17.0
Upland Sandpiper	0.0	0.0	0.0	0.0	0.0	0.0	0.0	0.0
Rock Dove	0.0	0.0	0.0	0.0	0.0	0.0	0.0	7.0
Mourning Dove	140.0	50.0	80.0	30.0	13.0	40.0	10.0	90.0
Yellow-billed Cuckoo	23.0	43.0	33.0	3.0	7.0	7.0	3.0	27.0
Chimney Swift	7.0	13.0	13.0	3.0	10.0	13.0	17.0	17.0
Ruby-throated Hummingbird	0.0	7.0	0.0	0.0	0.0	<0.1	0.0	0.0
Belted Kingfisher	0.0	0.0	<0.1	0.0	0.0	7.0	0.0	0.0
Red-headed Woodpecker	30.0	10.0	13.0	0.0	7.0	3.0	0.0	17.0
Red-bellied Woodpecker	40.0	57.0	30.0	10.0	23.0	13.0	13.0	30.0
Downy Woodpecker	3.0	13.0	13.0	10.0	27.0	3.0	0.0	0.0
Hairy Woodpecker	0.0	3.0	0.0	3.0	0.0	0.0	0.0	0.0
Northern Flicker	17.0	10.0	20.0	0.0	17.0	13.0	0.0	7.0
Pileated Woodpecker	0.0	20.0	0.0	7.0	3.0	7.0	0.0	3.0
Eastern Wood-Pewee	3.0	40.0	23.0	7.0	13.0	20.0	20.0	37.0
Acadian Flycatcher	0.0	0.0	<0.1	0.0	0.0	10.0	0.0	17.0
Eastern Phoebe	10.0	7.0	17.0	7.0	7.0	27.0	23.0	10.0
Great Crested Flycatcher	13.0	73.0	43.0	20.0	7.0	10.0	3.0	20.0
Eastern Kingbird	20.0	3.0	23.0	13.0	37.0	20.0	0.0	27.0
Scissor-tailed Flycatcher	0.0	0.0	0.0	0.0	0.0	0.0	0.0	0.0
Horned Lark	13.0	0.0	20.0	<0.1	0.0	0.0	0.0	0.0
Purple Martin	<0.1	3.0	10.0	3.0	10.0	20.0	0.0	0.0
No Rough-winged Swallow	0.0	0.0	3.0	0.0	<0.1	7.0	0.0	7.0
Bank Swallow	0.0	0.0	0.0	0.0	0.0	0.0	0.0	3.0
Barn Swallow	73.0	13.0	50.0	53.0	17.0	30.0	20.0	40.0
Blue Jay	30.0	67.0	103.0	43.0	30.0	33.0	20.0	70.0
American Crow	137.0	140.0	120.0	60.0	63.0	67.0	73.0	133.0
Black-capped Chickadee	13.0	13.0	3.0	0.0	0.0	<0.1	0.0	0.0
Carolina Chickadee	0.0	13.0	0.0	3.0	20.0	10.0	3.0	17.0
Tufted Titmouse	7.0	127.0	50.0	60.0	47.0	13.0	30.0	37.0
White-breasted Nuthatch	10.0	37.0	13.0	10.0	3.0	20.0	7.0	3.0
Carolina Wren	0.0	13.0	13.0	10.0	27.0	20.0	7.0	37.0
Bewick's Wren	0.0	<0.1	10.0	10.0	7.0	0.0	0.0	0.0
House Wren	0.0	0.0	0.0	0.0	3.0	0.0	17.0	0.0
Blue-gray Gnatcatcher	0.0	37.0	23.0	13.0	7.0	0.0	7.0	20.0
Eastern Bluebird	33.0	23.0	67.0	27.0	10.0	43.0	20.0	23.0
Wood Thrush	0.0	17.0	0.0	3.0	0.0	0.0	3.0	0.0
American Robin	33.0	30.0	67.0	87.0	30.0	37.0	33.0	60.0

Relative Abundance of Breeding Bird Species by Natural Division, 1986–1992

Species	518D	561B	Ozark Border Route Number 593C	636C	714C	753F	788D	1042B
			Average Number of Birds/100 Stops					
Gray Catbird	7.0	0.0	7.0	0.0	27.0	3.0	7.0	3.0
Northern Mockingbird	50.0	13.0	40.0	3.0	13.0	20.0	10.0	17.0
Brown Thrasher	37.0	7.0	23.0	13.0	7.0	23.0	3.0	13.0
Cedar Waxwing	0.0	0.0	67.0	0.0	0.0	0.0	0.0	30.0
Loggerhead Shrike	3.0	0.0	0.0	3.0	0.0	0.0	0.0	0.0
European Starling	30.0	10.0	107.0	3.0	10.0	30.0	7.0	13.0
White-eyed Vireo	0.0	0.0	0.0	3.0	0.0	3.0	10.0	13.0
Bell's Vireo	0.0	0.0	0.0	0.0	0.0	0.0	0.0	0.0
Yellow-throated Vireo	0.0	0.0	0.0	0.0	0.0	0.0	0.0	0.0
Warbling Vireo	3.0	0.0	3.0	0.0	0.0	7.0	0.0	0.0
Red-eyed Vireo	0.0	27.0	<0.1	0.0	0.0	0.0	3.0	7.0
Blue-winged Warbler	0.0	3.0	0.0	0.0	0.0	3.0	7.0	3.0
Northern Parula	7.0	0.0	3.0	0.0	3.0	10.0	3.0	7.0
Yellow Warbler	0.0	0.0	0.0	0.0	0.0	3.0	0.0	0.0
Yellow-throated Warbler	0.0	0.0	0.0	0.0	0.0	0.0	0.0	0.0
Pine Warbler	0.0	0.0	0.0	0.0	0.0	0.0	0.0	0.0
Prairie Warbler	0.0	0.0	0.0	0.0	0.0	0.0	0.0	0.0
Black-and-white Warbler	0.0	0.0	0.0	0.0	0.0	0.0	0.0	0.0
Worm-eating Warbler	0.0	0.0	0.0	0.0	0.0	0.0	0.0	0.0
Louisiana Waterthrush	0.0	0.0	0.0	0.0	3.0	0.0	0.0	0.0
Kentucky Warbler	0.0	10.0	7.0	0.0	0.0	0.0	0.0	7.0
Common Yellowthroat	60.0	10.0	13.0	13.0	10.0	10.0	17.0	17.0
Yellow-breasted Chat	3.0	17.0	3.0	3.0	3.0	13.0	10.0	33.0
Summer Tanager	0.0	37.0	13.0	7.0	3.0	0.0	7.0	3.0
Scarlet Tanager	0.0	10.0	0.0	0.0	0.0	0.0	0.0	0.0
Northern Cardinal	90.0	103.0	77.0	90.0	63.0	40.0	73.0	170.0
Rose-breasted Grosbeak	0.0	0.0	7.0	0.0	0.0	0.0	0.0	0.0
Blue Grosbeak	0.0	17.0	17.0	37.0	0.0	0.0	3.0	3.0
Indigo Bunting	23.0	57.0	93.0	33.0	53.0	97.0	30.0	130.0
Dickcissel	197.0	13.0	103.0	3.0	0.0	7.0	7.0	47.0
Eastern Towhee	7.0	27.0	13.0	20.0	30.0	3.0	53.0	7.0
Chipping Sparrow	0.0	0.0	23.0	23.0	20.0	10.0	27.0	7.0
Field Sparrow	20.0	70.0	80.0	47.0	57.0	17.0	13.0	17.0
Vesper Sparrow	0.0	0.0	0.0	0.0	0.0	0.0	0.0	0.0
Lark Sparrow	0.0	0.0	0.0	0.0	0.0	0.0	0.0	0.0
Grasshopper Sparrow	157.0	0.0	33.0	<0.1	0.0	0.0	0.0	0.0
Song Sparrow	3.0	0.0	0.0	0.0	0.0	0.0	0.0	0.0
Bobolink	0.0	0.0	0.0	0.0	0.0	0.0	0.0	0.0
Red-winged Blackbird	427.0	163.0	167.0	63.0	40.0	47.0	13.0	253.0
Eastern Meadowlark	420.0	40.0	117.0	40.0	27.0	30.0	37.0	37.0
Western Meadowlark	0.0	0.0	0.0	0.0	0.0	0.0	0.0	0.0
Common Grackle	123.0	20.0	80.0	40.0	57.0	40.0	37.0	183.0
Brown-headed Cowbird	70.0	127.0	103.0	140.0	63.0	53.0	27.0	63.0
Orchard Oriole	0.0	13.0	13.0	10.0	0.0	10.0	17.0	10.0
Baltimore Oriole	23.0	7.0	30.0	8.0	10.0	3.0	0.0	0.0
House Finch	0.0	0.0	0.0	0.0	0.0	0.0	0.0	0.0
American Goldfinch	7.0	3.0	17.0	3.0	7.0	17.0	0.0	27.0
House Sparrow	183.0	0.0	73.0	10.0	60.0	33.0	3.0	87.0

				Osage Plains Route Number					
	15	**23**	**124**	**511A**	**576C**	**580A**	**584E**	**719A**	**880D**
Species				Average Number of Birds/100 Stops					
Great Blue Heron	4.3	6.6	8.0	3.0	20.0	3.0	33.0	7.0	0.0
Green Heron	3.0	2.9	0.0	0.0	0.0	3.0	0.0	<0.1	0.0
Canada Goose	1.3	0.6	4.0	0.0	0.0	0.0	0.0	7.0	0.0
Wood Duck	3.0	1.7	0.0	0.0	0.0	0.0	0.0	0.0	0.0
Turkey Vulture	4.7	9.4	4.0	3.0	0.0	0.0	0.0	0.0	0.0
Red-shouldered Hawk	0.0	0.9	0.0	0.0	0.0	0.0	0.0	0.0	0.0
Red-tailed Hawk	4.7	7.7	2.0	3.0	7.0	7.0	0.0	17.0	0.0
Ring-necked Pheasant	0.0	0.0	0.0	0.0	0.0	0.0	0.0	0.0	0.0
Wild Turkey	2.3	2.6	0.0	7.0	0.0	0.0	<0.1	0.0	0.0
Northern Bobwhite	121.0	93.7	126.0	43.0	220.0	107.0	120.0	93.0	27.0
Killdeer	9.0	20.0	20.0	10.0	60.0	13.0	57.0	17.0	20.0
Upland Sandpiper	0.0	6.3	0.0	0.0	0.0	0.0	0.0	0.0	7.0
Rock Dove	10.7	3.4	4.0	0.0	10.0	50.0	0.0	40.0	0.0
Mourning Dove	110.0	83.7	118.0	17.0	200.0	167.0	240.0	87.0	47.0
Yellow-billed Cuckoo	37.7	31.1	42.0	7.0	27.0	40.0	17.0	37.0	17.0
Chimney Swift	7.0	7.7	22.0	0.0	33.0	27.0	27.0	17.0	0.0
Ruby-throated Hummingbird	3.7	6.6	0.0	0.0	0.0	0.0	0.0	10.0	0.0
Belted Kingfisher	1.3	1.4	2.0	0.0	3.0	0.0	0.0	0.0	0.0
Red-headed Woodpecker	9.0	12.0	12.0	7.0	10.0	10.0	0.0	17.0	0.0
Red-bellied Woodpecker	37.7	20.0	20.0	17.0	17.0	23.0	13.0	43.0	7.0
Downy Woodpecker	4.3	13.1	8.0	0.0	10.0	7.0	0.0	3.0	0.0
Hairy Woodpecker	0.3	2.6	0.0	0.0	0.0	0.0	0.0	0.0	3.0
Northern Flicker	6.0	9.1	6.0	0.0	7.0	7.0	0.0	10.0	3.0
Pileated Woodpecker	0.7	4.0	0.0	3.0	3.0	0.0	0.0	7.0	0.0
Eastern Wood-Pewee	26.3	263.0	8.0	3.0	7.0	0.0	17.0	7.0	0.0
Acadian Flycatcher	0.7	3.1	2.0	3.0	0.0	0.0	0.0	0.0	0.0
Eastern Phoebe	17.3	11.1	16.0	0.0	0.0	7.0	3.0	10.0	3.0
Great Crested Flycatcher	27.3	20.9	16.0	23.0	10.0	3.0	17.0	7.0	3.0
Eastern Kingbird	48.7	23.1	36.0	7.0	57.0	37.0	73.0	20.0	33.0
Scissor-tailed Flycatcher	4.7	6.6	6.0	0.0	0.0	0.0	13.0	<0.1	23.0
Horned Lark	7.3	18.3	0.0	0.0	17.0	13.0	3.0	3.0	<0.1
Purple Martin	58.7	7.7	2.0	7.0	0.0	3.0	3.0	10.0	0.0
No Rough-winged Swallow	6.3	13.4	0.0	<0.1	0.0	0.0	3.0	7.0	0.0
Bank Swallow	0.0	0.0	0.0	0.0	0.0	0.0	0.0	0.0	0.0
Barn Swallow	35.0	77.1	14.0	23.0	87.0	93.0	40.0	30.0	47.0
Blue Jay	68.3	36.0	42.0	20.0	30.0	37.0	30.0	10.0	3.0
American Crow	119.0	79.7	82.0	23.0	100.0	17.0	157.0	50.0	20.0
Black-capped Chickadee	0.0	1.1	40.0	17.0	27.0	17.0	3.0	0.0	0.0
Carolina Chickadee	9.7	18.6	0.0	0.0	7.0	0.0	0.0	7.0	10.0
Tufted Titmouse	54.3	38.3	50.0	50.0	30.0	27.0	13.0	20.0	7.0
White-breasted Nuthatch	8.0	11.4	0.0	20.0	7.0	0.0	3.0	10.0	0.0
Carolina Wren	3.3	13.4	4.0	0.0	7.0	3.0	0.0	0.0	7.0
Bewick's Wren	6.0	8.3	2.0	0.0	0.0	0.0	<0.1	0.0	0.0
House Wren	0.0	3.1	8.0	7.0	0.0	0.0	3.0	10.0	0.0
Blue-gray Gnatcatcher	12.0	17.4	2.0	0.0	13.0	0.0	0.0	17.0	17.0
Eastern Bluebird	57.0	34.3	54.0	13.0	150.0	73.0	37.0	20.0	7.0
Wood Thrush	0.0	0.6	0.0	17.0	0.0	3.0	0.0	0.0	0.0
American Robin	40.7	79.7	56.0	70.0	53.0	117.0	30.0	30.0	13.0

Relative Abundance of Breeding Bird Species by Natural Division, 1986–1992

Species	15	23	124	Osage Plains Route Number				719A	880D
				511A	576C	580A	584E		
				Average Number of Birds/100 Stops					
Gray Catbird	3.0	5.7	8.0	3.0	0.0	7.0	10.0	17.0	3.0
Northern Mockingbird	81.3	35.4	62.0	17.0	53.0	27.0	57.0	23.0	20.0
Brown Thrasher	24.3	15.4	28.0	3.0	10.0	63.0	33.0	37.0	17.0
Cedar Waxwing	0.0	2.0	0.0	50.0	0.0	0.0	0.0	0.0	0.0
Loggerhead Shrike	7.7	3.1	12.0	0.0	17.0	13.0	20.0	10.0	0.0
European Starling	42.7	11.1	306.0	3.0	27.0	27.0	53.0	30.0	7.0
White-eyed Vireo	0.0	1.7	0.0	0.0	0.0	0.0	0.0	0.0	0.0
Bell's Vireo	0.3	8.9	0.0	0.0	0.0	7.0	0.0	0.0	0.0
Yellow-throated Vireo	0.7	0.6	0.0	7.0	0.0	7.0	0.0	0.0	0.0
Warbling Vireo	1.7	0.9	0.0	0.0	7.0	0.0	7.0	7.0	0.0
Red-eyed Vireo	0.3	3.4	0.0	0.0	0.0	0.0	0.0	0.0	0.0
Blue-winged Warbler	0.0	0.0	0.0	0.0	<0.1	0.0	0.0	0.0	0.0
Northern Parula	4.0	6.3	2.0	0.0	0.0	7.0	<0.1	0.0	0.0
Yellow Warbler	0.0	0.0	0.0	0.0	0.0	0.0	0.0	0.0	0.0
Yellow-throated Warbler	0.0	0.0	0.0	0.0	0.0	0.0	0.0	0.0	0.0
Pine Warbler	0.0	0.0	0.0	0.0	0.0	0.0	0.0	0.0	0.0
Prairie Warbler	0.7	0.0	0.0	0.0	0.0	0.0	0.0	0.0	0.0
Black-and-white Warbler	0.7	0.6	0.0	0.0	0.0	0.0	0.0	0.0	0.0
Worm-eating Warbler	0.0	0.3	0.0	0.0	0.0	0.0	0.0	0.0	0.0
Louisiana Waterthrush	0.3	0.3	0.0	0.0	0.0	0.0	0.0	0.0	0.0
Kentucky Warbler	0.3	7.7	0.0	7.0	0.0	0.0	0.0	0.0	0.0
Common Yellowthroat	10.3	38.9	4.0	7.0	7.0	67.0	67.0	30.0	17.0
Yellow-breasted Chat	1.7	10.3	0.0	3.0	0.0	0.0	0.0	0.0	0.0
Summer Tanager	5.3	12.9	0.0	3.0	0.0	0.0	0.0	0.0	0.0
Scarlet Tanager	0.0	2.3	0.0	3.0	0.0	0.0	0.0	3.0	0.0
Northern Cardinal	83.0	73.4	106.0	47.0	100.0	83.0	77.0	50.0	50.0
Rose-breasted Grosbeak	0.0	2.6	0.0	0.0	0.0	0.0	0.0	3.0	0.0
Blue Grosbeak	19.7	27.4	10.0	10.0	10.0	13.0	20.0	30.0	10.0
Indigo Bunting	42.0	85.4	46.0	37.0	27.0	107.0	27.0	20.0	13.0
Dickcissel	180.0	256.0	110.0	10.0	290.0	110.0	233.0	77.0	43.0
Eastern Towhee	7.3	13.1	0.0	17.0	0.0	0.0	7.0	7.0	0.0
Chipping Sparrow	2.7	6.0	2.0	3.0	0.0	0.0	0.0	0.0	0.0
Field Sparrow	89.7	80.3	32.0	10.0	10.0	13.0	30.0	7.0	17.0
Vesper Sparrow	0.0	0.0	0.0	0.0	0.0	0.0	0.0	0.0	0.0
Lark Sparrow	6.0	4.0	12.0	0.0	0.0	7.0	<0.1	13.0	0.0
Grasshopper Sparrow	42.7	40.9	34.0	3.0	13.0	17.0	113.0	13.0	17.0
Song Sparrow	0.0	0.0	0.0	0.0	0.0	0.0	0.0	0.0	0.0
Bobolink	0.0	0.0	0.0	0.0	0.0	0.0	0.0	0.0	0.0
Red-winged Blackbird	119.0	158.0	228.0	67.0	393.0	287.0	193.0	137.0	53.0
Eastern Meadowlark	287.0	224.0	368.0	13.0	463.0	320.0	417.0	73.0	37.0
Western Meadowlark	0.0	0.0	0.0	0.0	0.0	0.0	0.0	0.0	0.0
Common Grackle	137.0	82.3	198.0	30.0	117.0	153.0	207.0	50.0	43.0
Brown-headed Cowbird	115.0	57.4	76.0	20.0	110.0	43.0	73.0	93.0	33.0
Orchard Oriole	5.0	9.7	4.0	3.0	0.0	3.0	3.0	7.0	0.0
Baltimore Oriole	10.3	4.6	16.0	0.0	40.0	7.0	7.0	0.0	0.0
House Finch	2.0	0.0	0.0	0.0	0.0	0.0	0.0	0.0	0.0
American Goldfinch	8.0	13.4	10.0	7.0	17.0	7.0	13.0	13.0	7.0
House Sparrow	120.0	171.0	292.0	43.0	253.0	183.0	57.0	67.0	30.0

				Ozarks Route Number				
	2	**3**	**4**	**5**	**6**	**8**	**10**	**11**	**12**
Species				Average Number of Birds/100 Stops				
Great Blue Heron	4.6	2.7	1.3	6.0	5.7	0.6	0.6	1.1	4.0
Green Heron	4.9	2.7	0.3	2.3	2.6	6.3	1.4	0.9	0.9
Canada Goose	0.3	0.0	0.0	0.0	2.3	0.0	0.0	0.0	0.0
Wood Duck	2.6	0.0	0.0	1.7	1.7	0.6	3.1	0.9	0.9
Turkey Vulture	2.9	6.7	11.0	3.1	15.7	1.4	18.6	0.6	12.0
Red-shouldered Hawk	0.6	0.0	0.7	0.0	1.4	3.4	0.0	0.3	0.9
Red-tailed Hawk	2.3	2.7	2.7	2.3	1.7	3.4	1.1	2.3	0.6
Ring-necked Pheasant	0.0	0.0	0.0	0.0	0.0	0.0	0.0	0.0	0.0
Wild Turkey	1.1	2.0	4.0	0.0	0.3	4.3	0.9	4.9	2.0
Northern Bobwhite	59.4	10.8	50.0	82.3	18.3	33.7	60.3	58.0	40.3
Killdeer	6.3	8.7	3.0	8.9	9.1	6.9	10.6	4.0	1.4
Upland Sandpiper	0.0	0.0	0.0	0.0	0.0	0.0	0.3	0.0	0.0
Rock Dove	38.0	0.0	7.3	14.9	12.9	12.3	2.0	2.0	1.1
Mourning Dove	96.9	25.7	18.0	33.1	17.4	52.9	20.0	11.4	16.0
Yellow-billed Cuckoo	18.6	23.3	9.7	29.4	23.7	34.3	18.3	36.9	15.7
Chimney Swift	47.4	10.7	3.3	5.1	26.6	7.7	7.1	8.3	6.6
Ruby-throated Hummingbird	4.0	3.3	1.7	2.3	3.7	5.7	0.3	0.6	1.4
Belted Kingfisher	1.7	0.0	0.3	1.4	2.3	2.3	0.0	0.6	0.3
Red-headed Woodpecker	5.1	1.0	1.3	6.3	0.6	0.3	6.0	0.3	0.6
Red-bellied Woodpecker	12.3	10.3	18.7	14.3	14.0	22.3	4.3	13.4	15.4
Downy Woodpecker	6.0	1.0	5.3	3.7	6.6	13.1	6.6	11.1	4.3
Hairy Woodpecker	2.6	0.3	0.3	3.1	0.9	1.7	0.3	0.0	0.9
Northern Flicker	6.6	3.3	1.0	2.9	1.4	8.3	2.6	9.4	2.0
Pileated Woodpecker	1.7	1.3	1.7	0.6	1.4	4.6	2.6	9.4	4.9
Eastern Wood-Pewee	32.9	24.3	22.0	19.1	32.6	21.7	22.6	54.0	24.9
Acadian Flycatcher	3.1	0.7	1.7	0.6	4.9	2.3	0.6	3.7	7.7
Eastern Phoebe	20.3	13.0	10.4	6.9	8.9	36.3	8.3	13.4	11.1
Great Crested Flycatcher	11.4	5.3	11.0	1.7	10.3	5.4	13.1	25.7	9.4
Eastern Kingbird	17.4	29.3	11.7	35.4	16.3	12.6	41.1	16.9	2.6
Scissor-tailed Flycatcher	0.0	0.0	0.0	5.7	2.9	0.0	0.0	0.0	0.0
Horned Lark	0.0	0.0	0.3	2.6	0.3	0.6	6.9	1.7	0.6
Purple Martin	25.1	4.3	0.3	18.9	11.4	46.3	23.4	8.0	19.7
No Rough-winged Swallow	46.6	0.0	0.3	4.0	22.9	8.0	1.7	1.4	11.1
Bank Swallow	0.0	0.0	0.0	0.0	0.0	0.0	0.0	0.0	0.0
Barn Swallow	39.7	41.3	60.3	66.6	15.7	89.1	48.6	14.0	17.7
Blue Jay	40.3	52.3	38.7	21.4	16.6	26.6	44.6	35.4	45.1
American Crow	53.7	69.0	106.0	64.0	48.0	158.0	187.0	102.0	56.6
Black-capped Chickadee	0.0	0.0	0.0	0.0	0.0	0.0	0.0	0.0	2.0
Carolina Chickadee	15.7	12.3	25.0	3.1	11.7	36.3	15.4	28.9	10.6
Tufted Titmouse	14.0	17.0	46.0	13.7	42.9	33.4	29.7	75.4	36.3
White-breasted Nuthatch	4.6	4.3	7.7	4.9	16.3	8.0	10.0	28.9	13.7
Carolina Wren	16.9	6.7	12.7	1.1	7.7	12.9	9.1	8.0	10.9
Bewick's Wren	0.0	7.7	3.7	2.3	0.6	0.0	12.9	4.9	6.0
House Wren	1.1	3.0	0.0	1.7	0.3	0.0	0.3	0.0	0.6
Blue-gray Gnatcatcher	16.0	13.3	26.0	16.6	48.9	66.3	17.7	55.4	22.9
Eastern Bluebird	22.6	24.3	38.0	48.9	16.6	33.7	46.3	11.7	19.1
Wood Thrush	5.1	1.3	2.3	0.0	2.9	16.3	13.1	6.6	16.9
American Robin	78.3	64.0	22.0	78.9	42.3	18.6	93.1	29.7	52.9

Relative Abundance of Breeding Bird Species by Natural Division, 1986–1992

Species	2	3	4	5	6	8	10	11	12

				Ozarks Route Number					
Species	2	3	4	5	6	8	10	11	12
				Average Number of Birds/100 Stops					
Gray Catbird	2.6	2.0	2.7	0.6	2.3	8.3	1.4	4.9	2.3
Northern Mockingbird	45.7	18.3	17.7	41.7	14.0	14.0	18.6	1.1	2.3
Brown Thrasher	6.3	26.3	12.7	6.9	4.0	13.7	20.0	12.9	15.1
Cedar Waxwing	0.0	0.0	0.0	0.0	0.0	2.6	0.0	1.4	0.0
Loggerhead Shrike	1.4	0.3	2.0	6.0	5.1	1.1	1.1	0.0	0.0
European Starling	43.7	27.0	9.7	57.1	28.9	14.6	67.1	7.7	11.1
White-eyed Vireo	8.6	4.0	3.0	1.1	4.3	33.1	9.4	10.9	8.9
Bell's Vireo	1.1	0.0	0.0	0.0	0.0	0.3	0.0	0.0	0.0
Yellow-throated Vireo	1.1	0.0	0.0	1.1	2.9	4.6	1.7	2.3	0.9
Warbling Vireo	2.3	0.0	0.0	0.0	0.6	7.7	0.0	0.0	0.3
Red-eyed Vireo	2.9	9.0	4.0	2.6	26.3	15.4	85.1	53.4	10.9
Blue-winged Warbler	0.6	2.3	1.3	0.0	0.0	14.6	0.0	3.4	3.7
Northern Parula	7.7	0.3	7.0	2.3	10.3	16.9	2.6	8.3	6.3
Yellow Warbler	0.0	0.0	0.0	0.0	0.3	0.0	0.0	0.0	1.1
Yellow-throated Warbler	0.0	0.0	0.0	1.1	1.7	3.1	4.3	5.7	0.9
Pine Warbler	0.0	0.0	3.3	0.0	10.0	6.9	13.7	16.3	0.0
Prairie Warbler	0.0	0.3	1.3	3.1	3.1	13.7	0.3	10.6	0.0
Black-and-white Warbler	1.1	1.7	0.7	0.0	0.6	3.7	0.3	2.0	3.4
Worm-eating Warbler	0.0	0.0	0.0	0.0	0.3	6.9	0.3	1.1	0.9
Louisiana Waterthrush	0.0	0.0	1.7	0.3	0.3	2.9	0.0	0.6	1.4
Kentucky Warbler	2.9	1.0	3.3	0.0	2.3	17.4	1.7	8.0	6.3
Common Yellowthroat	16.6	17.7	3.7	6.0	6.9	30.6	5.1	16.9	11.4
Yellow-breasted Chat	27.0	24.3	16.0	6.9	4.0	42.6	15.7	39.1	23.4
Summer Tanager	7.4	5.7	6.0	5.1	14.3	13.7	29.1	39.7	16.0
Scarlet Tanager	0.0	0.3	2.0	0.0	0.3	6.0	1.7	1.4	0.3
Northern Cardinal	100.0	57.3	79.0	44.3	43.7	54.6	46.9	26.3	45.1
Rose-breasted Grosbeak	0.0	0.0	0.0	0.0	0.0	0.0	0.0	0.0	0.0
Blue Grosbeak	6.6	5.7	10.3	6.3	8.3	6.6	26.3	16.6	10.9
Indigo Bunting	96.0	70.7	125.0	56.9	58.0	125.0	80.3	115.0	74.3
Dickcissel	28.0	13.0	11.0	55.1	10.3	7.7	45.1	2.6	0.6
Eastern Towhee	16.6	10.7	6.3	2.0	0.3	48.3	17.7	26.0	18.9
Chipping Sparrow	13.7	13.0	7.7	18.6	17.1	35.4	22.6	29.7	16.3
Field Sparrow	42.9	40.3	98.0	34.6	19.4	46.0	28.3	36.3	27.1
Vesper Sparrow	0.0	0.0	0.0	0.0	0.0	0.0	0.0	0.0	0.0
Lark Sparrow	1.1	1.0	0.7	0.3	0.6	0.0	0.6	0.0	0.0
Grasshopper Sparrow	0.3	0.3	0.0	1.4	0.3	0.3	7.7	1.1	0.3
Song Sparrow	0.0	0.0	0.0	0.0	0.0	0.0	0.0	0.0	0.0
Bobolink	0.0	0.0	0.0	0.0	0.0	0.0	0.0	0.0	0.0
Red-winged Blackbird	76.0	3.0	71.3	79.4	48.6	60.6	171.0	63.7	24.0
Eastern Meadowlark	53.7	53.3	49.0	160.0	40.9	40.0	210.0	41.4	35.4
Western Meadowlark	0.0	0.0	0.0	0.0	0.0	0.0	1.4	0.0	0.0
Common Grackle	115.0	80.0	32.3	395.0	77.1	58.0	142.0	32.0	59.7
Brown-headed Cowbird	44.9	77.3	72.0	68.3	22.9	58.3	58.6	64.0	49.7
Orchard Oriole	10.9	19.7	6.0	2.9	1.7	28.6	15.4	14.9	17.7
Baltimore Oriole	0.9	1.3	0.7	0.0	2.9	0.6	0.0	0.0	1.1
House Finch	0.0	0.0	0.0	0.0	0.0	0.0	0.0	0.0	0.0
American Goldfinch	4.6	2.3	12.0	10.3	13.7	29.1	10.6	16.3	4.9
House Sparrow	47.4	26.7	21.7	72.9	51.1	28.6	42.3	3.7	31.4

Species	Ozarks Route Number								
	13	14	16	18	20	109	692D	710D	807D
	Average Number of Birds/100 Stops								
Great Blue Heron	2.5	3.4	6.0	8.7	2.0	0.0	7.0	0.0	7.0
Green Heron	3.5	1.7	1.4	3.0	4.0	2.7	0.0	0.0	3.0
Canada Goose	0.0	1.4	0.0	0.0	0.0	0.0	0.0	0.0	13.0
Wood Duck	0.0	0.0	0.0	0.0	0.0	0.3	0.0	0.0	0.0
Turkey Vulture	7.0	1.4	0.9	2.0	4.3	7.3	0.0	0.0	0.0
Red-shouldered Hawk	0.0	0.0	0.0	0.0	1.1	1.7	0.0	0.0	0.0
Red-tailed Hawk	0.5	2.6	4.0	2.3	0.6	1.0	3.0	7.0	0.0
Ring-necked Pheasant	0.0	0.0	0.0	0.0	0.0	0.0	0.0	0.0	0.0
Wild Turkey	0.0	0.0	1.4	2.0	1.7	0.0	10.0	3.0	0.0
Northern Bobwhite	36.5	70.0	99.1	38.3	51.7	36.7	103.0	10.0	230.0
Killdeer	20.5	4.9	21.1	6.3	5.7	2.0	3.0	0.0	7.0
Upland Sandpiper	0.0	0.0	0.0	0.0	0.0	0.0	0.0	0.0	0.0
Rock Dove	11.5	0.6	11.7	0.3	6.9	0.0	0.0	0.0	0.0
Mourning Dove	31.0	34.9	100.0	38.0	30.6	14.0	77.0	13.0	160.0
Yellow-billed Cuckoo	8.5	24.6	20.9	15.7	14.6	14.7	80.0	33.0	7.0
Chimney Swift	8.5	6.0	12.9	15.7	8.3	4.7	0.0	13.0	0.0
Ruby-throated Hummingbird	1.0	0.6	0.3	0.7	0.6	4.3	0.0	7.0	0.0
Belted Kingfisher	0.0	1.7	2.0	2.7	2.0	3.0	0.0	0.0	0.0
Red-headed Woodpecker	1.5	1.4	2.0	2.3	0.6	0.7	0.0	0.0	0.0
Red-bellied Woodpecker	13.5	18.9	8.3	14.0	21.4	11.3	73.0	17.0	57.0
Downy Woodpecker	9.0	0.9	7.7	6.0	8.9	5.3	0.0	13.0	7.0
Hairy Woodpecker	1.0	0.0	0.3	1.3	0.9	4.0	0.0	3.0	0.0
Northern Flicker	5.5	0.6	2.3	5.3	4.6	2.0	0.0	23.0	0.0
Pileated Woodpecker	0.5	0.6	0.3	7.7	3.4	8.3	27.0	17.0	10.0
Eastern Wood-Pewee	12.5	9.7	4.3	17.0	13.7	15.0	20.0	17.0	23.0
Acadian Flycatcher	0.5	0.0	0.3	0.7	0.6	0.3	0.0	0.0	0.0
Eastern Phoebe	8.0	5.4	10.3	13.7	6.3	19.3	10.0	3.0	20.0
Great Crested Flycatcher	8.0	6.0	37.0	15.3	7.1	6.0	17.0	37.0	17.0
Eastern Kingbird	14.5	14.9	32.0	14.0	9.4	7.3	17.0	7.0	37.0
Scissor-tailed Flycatcher	3.0	0.6	10.6	0.0	0.9	0.0	0.0	0.0	3.0
Horned Lark	4.0	1.7	16.6	0.7	0.9	0.0	0.0	0.0	10.0
Purple Martin	52.0	5.1	2.0	9.3	11.1	17.7	0.0	0.0	0.0
No Rough-winged Swallow	9.0	1.1	5.1	8.3	5.4	1.0	0.0	0.0	0.0
Bank Swallow	0.0	0.0	0.0	0.0	0.0	0.0	0.0	0.0	0.0
Barn Swallow	42.0	39.4	89.7	20.7	20.0	33.7	27.0	17.0	27.0
Blue Jay	37.0	25.4	10.9	37.7	29.1	18.3	20.0	60.0	43.0
American Crow	41.5	50.0	66.3	94.0	94.3	40.7	290.0	130.0	220.0
Black-capped Chickadee	0.0	0.0	0.0	0.0	0.0	0.7	0.0	0.0	0.0
Carolina Chickadee	11.0	3.4	6.6	8.3	5.1	5.0	10.0	37.0	0.0
Tufted Titmouse	17.0	16.0	4.9	31.3	50.0	23.0	30.0	57.0	20.0
White-breasted Nuthatch	13.5	4.3	4.0	15.3	12.0	10.7	23.0	30.0	0.0
Carolina Wren	5.0	1.7	1.4	8.7	10.6	15.3	3.0	20.0	0.0
Bewick's Wren	9.5	0.3	0.6	5.3	4.6	1.0	0.0	0.0	0.0
House Wren	3.0	2.3	0.0	2.7	0.3	1.3	0.0	3.0	0.0
Blue-gray Gnatcatcher	8.0	1.1	0.9	14.7	8.6	13.0	7.0	30.0	0.0
Eastern Bluebird	22.0	18.3	15.7	18.0	23.4	21.3	70.0	23.0	13.0
Wood Thrush	0.0	0.0	0.0	0.7	1.4	17.0	0.0	3.0	0.0
American Robin	74.0	56.6	32.0	33.7	25.4	38.3	37.0	3.0	37.0

Species	13	14	16	18	20	109	692D	710D	807D
					Ozarks Route Number				
					Average Number of Birds/100 Stops				
Gray Catbird	4.0	0.3	1.1	3.0	7.1	11.3	3.0	3.0	3.0
Northern Mockingbird	26.5	42.3	35.1	2.3	1.7	2.3	7.0	0.0	133.0
Brown Thrasher	15.0	12.0	20.9	11.3	6.9	5.7	10.0	13.0	83.0
Cedar Waxwing	0.0	0.0	1.1	3.3	0.3	0.0	0.0	0.0	0.0
Loggerhead Shrike	8.5	2.6	7.1	0.7	0.0	0.3	0.0	0.0	3.0
European Starling	58.0	24.6	32.6	21.3	29.7	41.0	17.0	0.0	10.0
White-eyed Vireo	1.0	0.0	0.3	3.7	9.7	20.7	0.0	3.0	0.0
Bell's Vireo	1.5	0.0	0.6	0.3	1.7	0.3	0.0	0.0	7.0
Yellow-throated Vireo	1.0	0.0	0.0	0.7	0.6	1.3	0.0	0.0	0.0
Warbling Vireo	1.0	0.0	1.4	1.7	0.3	1.0	0.0	0.0	0.0
Red-eyed Vireo	3.5	0.3	0.3	7.0	6.0	14.7	0.0	40.0	0.0
Blue-winged Warbler	1.0	0.0	0.0	1.7	0.6	1.3	0.0	13.0	0.0
Northern Parula	2.5	0.3	1.4	9.3	7.1	8.0	0.0	0.0	0.0
Yellow Warbler	0.0	0.0	0.0	0.7	0.0	1.0	0.0	0.0	0.0
Yellow-throated Warbler	0.0	0.0	0.0	0.0	0.0	0.3	0.0	0.0	0.0
Pine Warbler	0.0	0.0	0.0	0.0	0.0	0.7	0.0	13.0	0.0
Prairie Warbler	0.0	0.0	0.0	4.3	0.0	5.0	13.0	0.0	0.0
Black-and-white Warbler	1.0	0.0	0.3	1.3	0.9	0.7	0.0	7.0	0.0
Worm-eating Warbler	1.0	0.0	0.0	0.0	0.0	0.7	0.0	0.0	0.0
Louisiana Waterthrush	1.0	0.0	0.0	1.0	0.0	1.0	0.0	7.0	0.0
Kentucky Warbler	3.0	0.3	0.0	5.0	6.9	11.0	7.0	0.0	0.0
Common Yellowthroat	7.5	4.9	6.6	10.3	16.0	10.7	17.0	3.0	3.0
Yellow-breasted Chat	2.0	2.6	1.1	10.3	23.4	32.7	17.0	20.0	0.0
Summer Tanager	6.0	0.6	0.6	12.3	27.4	9.3	3.0	43.0	10.0
Scarlet Tanager	0.5	0.0	0.0	1.0	0.6	3.0	0.0	0.0	0.0
Northern Cardinal	48.0	51.1	70.0	50.3	67.4	23.3	93.0	33.0	77.0
Rose-breasted Grosbeak	1.0	0.0	0.0	0.0	0.0	0.0	0.0	0.0	0.0
Blue Grosbeak	16.0	2.0	6.9	17.0	7.7	9.0	20.0	0.0	37.0
Indigo Bunting	38.0	20.6	16.6	64.3	45.1	58.7	43.0	83.0	33.0
Dickcissel	31.0	36.9	237.0	9.7	6.9	1.0	40.0	0.0	177.0
Eastern Towhee	4.5	10.0	2.0	23.3	18.3	36.3	40.0	33.0	<0.1
Chipping Sparrow	7.0	0.9	0.3	15.7	5.7	9.0	0.0	27.0	0.0
Field Sparrow	34.0	40.0	13.4	59.7	40.6	26.7	97.0	27.0	50.0
Vesper Sparrow	0.0	0.0	0.0	0.0	0.0	0.0	0.0	0.0	0.0
Lark Sparrow	2.0	0.0	2.3	0.0	0.3	0.0	10.0	0.0	3.0
Grasshopper Sparrow	8.5	3.1	23.1	7.0	6.6	1.3	0.0	0.0	30.0
Song Sparrow	0.0	0.0	0.0	0.0	0.0	0.0	0.0	3.0	0.0
Bobolink	0.0	0.0	0.0	0.0	0.0	0.0	0.0	0.0	0.0
Red-winged Blackbird	77.0	48.6	327.0	43.3	55.4	35.7	50.0	27.0	230.0
Eastern Meadowlark	92.5	113.0	304.0	43.7	34.3	20.3	113.0	10.0	237.0
Western Meadowlark	0.0	0.0	0.6	0.0	0.0	0.0	0.0	0.0	0.0
Common Grackle	82.5	73.7	249.0	29.0	46.0	74.7	13.0	0.0	40.0
Brown-headed Cowbird	43.0	26.0	38.0	56.7	26.9	18.7	47.0	23.0	23.0
Orchard Oriole	5.5	0.6	1.7	10.3	7.4	12.7	0.0	7.0	0.0
Baltimore Oriole	1.0	0.9	0.6	1.7	0.0	0.3	0.0	3.0	0.0
House Finch	0.0	0.0	0.0	0.0	0.0	0.0	0.0	0.0	0.0
American Goldfinch	4.5	11.7	5.4	14.0	7.7	21.3	7.0	3.0	<0.1
House Sparrow	92.5	43.4	69.7	33.7	35.1	16.3	20.0	0.0	13.0

Species	814A	820B	908A	Ozarks Route Number 928F	933B	940C	943F	995D
				Average Number of Birds/100 Stops				
Great Blue Heron	7.0	7.0	0.0	3.0	7.0	0.0	0.0	0.0
Green Heron	7.0	0.0	<0.1	0.0	0.0	0.0	0.0	7.0
Canada Goose	0.0	0.0	0.0	0.0	23.0	0.0	0.0	0.0
Wood Duck	7.0	0.0	0.0	23.0	0.0	7.0	0.0	0.0
Turkey Vulture	3.0	0.0	0.0	23.0	0.0	0.0	0.0	0.0
Red-shouldered Hawk	0.0	0.0	0.0	0.0	0.0	0.0	0.0	0.0
Red-tailed Hawk	3.0	7.0	0.0	3.0	0.0	0.0	3.0	0.0
Ring-necked Pheasant	0.0	0.0	0.0	0.0	0.0	0.0	0.0	0.0
Wild Turkey	20.0	83.0	3.0	0.0	0.0	3.0	7.0	0.0
Northern Bobwhite	117.0	213.0	17.0	63.0	93.0	37.0	190.0	53.0
Killdeer	10.0	20.0	7.0	10.0	7.0	0.0	3.0	3.0
Upland Sandpiper	0.0	0.0	0.0	0.0	0.0	0.0	0.0	0.0
Rock Dove	0.0	0.0	0.0	70.0	0.0	0.0	0.0	0.0
Mourning Dove	50.0	37.0	3.0	163.0	87.0	23.0	57.0	23.0
Yellow-billed Cuckoo	20.0	80.0	0.0	27.0	7.0	10.0	17.0	30.0
Chimney Swift	0.0	7.0	0.0	0.0	10.0	0.0	0.0	7.0
Ruby-throated Hummingbird	0.0	0.0	0.0	3.0	0.0	3.0	0.0	0.0
Belted Kingfisher	0.0	0.0	0.0	0.0	0.0	3.0	0.0	0.0
Red-headed Woodpecker	3.0	0.0	0.0	0.0	3.0	3.0	0.0	7.0
Red-bellied Woodpecker	47.0	30.0	3.0	37.0	3.0	27.0	13.0	7.0
Downy Woodpecker	7.0	13.0	3.0	7.0	3.0	17.0	3.0	13.0
Hairy Woodpecker	0.0	3.0	3.0	0.0	0.0	3.0	0.0	0.0
Northern Flicker	10.0	7.0	0.0	3.0	0.0	7.0	13.0	7.0
Pileated Woodpecker	3.0	0.0	10.0	3.0	3.0	7.0	7.0	3.0
Eastern Wood-Pewee	20.0	60.0	27.0	7.0	10.0	17.0	43.0	23.0
Acadian Flycatcher	0.0	0.0	13.0	0.0	0.0	0.0	0.0	3.0
Eastern Phoebe	0.0	20.0	10.0	3.0	7.0	10.0	3.0	20.0
Great Crested Flycatcher	20.0	33.0	20.0	10.0	0.0	17.0	13.0	3.0
Eastern Kingbird	20.0	70.0	17.0	50.0	40.0	17.0	70.0	13.0
Scissor-tailed Flycatcher	0.0	0.0	0.0	7.0	7.0	0.0	0.0	0.0
Horned Lark	17.0	30.0	0.0	0.0	3.0	0.0	0.0	0.0
Purple Martin	10.0	27.0	10.0	0.0	53.0	17.0	23.0	13.0
No Rough-winged Swallow	7.0	0.0	<0.1	23.0	0.0	0.0	0.0	0.0
Bank Swallow	0.0	0.0	0.0	0.0	0.0	0.0	0.0	0.0
Barn Swallow	47.0	30.0	3.0	73.0	70.0	10.0	77.0	7.0
Blue Jay	13.0	100.0	20.0	33.0	43.0	33.0	37.0	0.0
American Crow	170.0	277.0	17.0	80.0	157.0	70.0	213.0	93.0
Black-capped Chickadee	0.0	0.0	0.0	0.0	0.0	0.0	0.0	0.0
Carolina Chickadee	13.0	33.0	7.0	10.0	23.0	27.0	17.0	20.0
Tufted Titmouse	43.0	53.0	17.0	30.0	0.0	53.0	7.0	27.0
White-breasted Nuthatch	20.0	37.0	17.0	0.0	3.0	10.0	7.0	10.0
Carolina Wren	13.0	0.0	10.0	0.0	43.0	37.0	10.0	17.0
Bewick's Wren	0.0	30.0	0.0	0.0	0.0	10.0	20.0	10.0
House Wren	0.0	0.0	0.0	0.0	13.0	0.0	0.0	0.0
Blue-gray Gnatcatcher	13.0	20.0	7.0	7.0	3.0	3.0	10.0	7.0
Eastern Bluebird	47.0	77.0	10.0	40.0	60.0	10.0	47.0	17.0
Wood Thrush	0.0	0.0	23.0	0.0	0.0	0.0	0.0	17.0
American Robin	43.0	60.0	0.0	43.0	63.0	33.0	43.0	47.0

Relative Abundance of Breeding Bird Species by Natural Division, 1986–1992

Species	814A	820B	908A	Ozarks Route Number 928F	933B	940C	943F	995D
				Average Number of Birds/100 Stops				
Gray Catbird	3.0	0.0	3.0	0.0	0.0	0.0	0.0	0.0
Northern Mockingbird	37.0	67.0	0.0	43.0	127.0	23.0	63.0	10.0
Brown Thrasher	17.0	10.0	0.0	17.0	7.0	7.0	97.0	7.0
Cedar Waxwing	0.0	0.0	0.0	0.0	0.0	0.0	0.0	0.0
Loggerhead Shrike	7.0	0.0	0.0	7.0	27.0	0.0	13.0	0.0
European Starling	7.0	10.0	0.0	3.0	350.0	7.0	27.0	3.0
White-eyed Vireo	0.0	0.0	27.0	0.0	0.0	0.0	0.0	33.0
Bell's Vireo	0.0	0.0	0.0	0.0	0.0	0.0	0.0	0.0
Yellow-throated Vireo	<0.1	0.0	7.0	0.0	0.0	3.0	0.0	0.0
Warbling Vireo	0.0	0.0	3.0	0.0	0.0	0.0	0.0	10.0
Red-eyed Vireo	13.0	0.0	77.0	0.0	0.0	0.0	3.0	13.0
Blue-winged Warbler	0.0	3.0	10.0	0.0	0.0	0.0	0.0	3.0
Northern Parula	0.0	0.0	7.0	0.0	0.0	3.0	0.0	3.0
Yellow Warbler	0.0	0.0	20.0	0.0	0.0	3.0	0.0	0.0
Yellow-throated Warbler	0.0	7.0	10.0	0.0	0.0	0.0	0.0	7.0
Pine Warbler	0.0	10.0	13.0	0.0	0.0	0.0	0.0	3.0
Prairie Warbler	<0.1	0.0	3.0	0.0	0.0	0.0	0.0	0.0
Black-and-white Warbler	0.0	0.0	7.0	0.0	0.0	0.0	0.0	0.0
Worm-eating Warbler	0.0	0.0	3.0	0.0	0.0	0.0	0.0	0.0
Louisiana Waterthrush	0.0	0.0	0.0	0.0	0.0	0.0	0.0	3.0
Kentucky Warbler	0.0	3.0	10.0	0.0	0.0	3.0	0.0	20.0
Common Yellowthroat	13.0	13.0	7.0	3.0	0.0	10.0	0.0	0.0
Yellow-breasted Chat	10.0	10.0	13.0	0.0	10.0	0.0	10.0	30.0
Summer Tanager	13.0	7.0	43.0	3.0	0.0	0.0	7.0	30.0
Scarlet Tanager	0.0	0.0	13.0	0.0	0.0	7.0	0.0	3.0
Northern Cardinal	47.0	33.0	7.0	73.0	73.0	43.0	53.0	77.0
Rose-breasted Grosbeak	0.0	0.0	0.0	0.0	0.0	0.0	0.0	3.0
Blue Grosbeak	23.0	37.0	10.0	10.0	0.0	23.0	63.0	30.0
Indigo Bunting	40.0	117.0	37.0	37.0	53.0	50.0	57.0	87.0
Dickcissel	3.0	147.0	0.0	47.0	227.0	3.0	93.0	7.0
Eastern Towhee	7.0	7.0	47.0	0.0	10.0	10.0	27.0	40.0
Chipping Sparrow	<0.1	60.0	13.0	0.0	40.0	0.0	3.0	17.0
Field Sparrow	67.0	113.0	3.0	33.0	23.0	47.0	77.0	37.0
Vesper Sparrow	0.0	0.0	0.0	0.0	0.0	0.0	0.0	0.0
Lark Sparrow	0.0	0.0	0.0	20.0	0.0	0.0	0.0	0.0
Grasshopper Sparrow	10.0	37.0	0.0	57.0	60.0	0.0	47.0	3.0
Song Sparrow	0.0	0.0	0.0	0.0	0.0	0.0	0.0	0.0
Bobolink	0.0	0.0	0.0	0.0	0.0	0.0	0.0	0.0
Red-winged Blackbird	67.0	237.0	10.0	183.0	233.0	133.0	383.0	0.0
Eastern Meadowlark	70.0	270.0	7.0	203.0	437.0	73.0	447.0	20.0
Western Meadowlark	0.0	0.0	0.0	0.0	0.0	0.0	0.0	0.0
Common Grackle	53.0	23.0	17.0	210.0	273.0	40.0	127.0	7.0
Brown-headed Cowbird	47.0	93.0	27.0	93.0	87.0	17.0	77.0	13.0
Orchard Oriole	17.0	17.0	53.0	0.0	0.0	0.0	3.0	40.0
Baltimore Oriole	0.0	0.0	0.0	3.0	7.0	0.0	0.0	0.0
House Finch	0.0	0.0	0.0	0.0	3.0	0.0	0.0	0.0
American Goldfinch	10.0	13.0	7.0	0.0	<0.1	3.0	10.0	3.0
House Sparrow	30.0	27.0	0.0	30.0	33.0	50.0	43.0	17.0

Species	Ozarks Route Number						
	1034F	1053E	1057A	1060A	1069A	1083C	1147C
	Average Number of Birds/100 Stops						
Great Blue Heron	3.0	3.0	3.0	23.0	10.0	0.0	0.0
Green Heron	0.0	0.0	0.0	3.0	0.0	0.0	0.0
Canada Goose	0.0	0.0	0.0	0.0	0.0	0.0	0.0
Wood Duck	7.0	0.0	0.0	0.0	3.0	0.0	0.0
Turkey Vulture	10.0	0.0	0.0	0.0	0.0	0.0	0.0
Red-shouldered Hawk	0.0	0.0	0.0	3.0	0.0	0.0	7.0
Red-tailed Hawk	0.0	7.0	0.0	0.0	13.0	0.0	3.0
Ring-necked Pheasant	0.0	0.0	0.0	0.0	0.0	0.0	0.0
Wild Turkey	0.0	0.0	0.0	0.0	0.0	3.0	0.0
Northern Bobwhite	3.0	137.0	120.0	110.0	153.0	<0.1	13.0
Killdeer	0.0	27.0	3.0	7.0	50.0	0.0	0.0
Upland Sandpiper	0.0	0.0	0.0	0.0	0.0	0.0	0.0
Rock Dove	7.0	3.0	13.0	13.0	7.0	0.0	0.0
Mourning Dove	23.0	97.0	60.0	40.0	127.0	7.0	10.0
Yellow-billed Cuckoo	30.0	23.0	7.0	70.0	60.0	10.0	57.0
Chimney Swift	0.0	23.0	57.0	3.0	0.0	0.0	13.0
Ruby-throated Hummingbird	3.0	0.0	7.0	3.0	0.0	3.0	7.0
Belted Kingfisher	0.0	0.0	0.0	0.0	0.0	0.0	0.0
Red-headed Woodpecker	0.0	0.0	0.0	0.0	3.0	17.0	0.0
Red-bellied Woodpecker	7.0	27.0	7.0	37.0	20.0	3.0	37.0
Downy Woodpecker	7.0	13.0	0.0	7.0	17.0	3.0	3.0
Hairy Woodpecker	3.0	3.0	0.0	0.0	0.0	0.0	0.0
Northern Flicker	0.0	3.0	7.0	0.0	0.0	7.0	0.0
Pileated Woodpecker	7.0	0.0	0.0	0.0	0.0	10.0	13.0
Eastern Wood-Pewee	13.0	10.0	0.0	67.0	47.0	27.0	70.0
Acadian Flycatcher	0.0	0.0	0.0	7.0	0.0	10.0	0.0
Eastern Phoebe	10.0	13.0	0.0	17.0	7.0	<0.1	17.0
Great Crested Flycatcher	10.0	10.0	3.0	10.0	<0.1	10.0	27.0
Eastern Kingbird	7.0	40.0	13.0	30.0	93.0	3.0	20.0
Scissor-tailed Flycatcher	0.0	23.0	10.0	3.0	0.0	0.0	0.0
Horned Lark	0.0	3.0	0.0	0.0	0.0	0.0	0.0
Purple Martin	0.0	0.0	37.0	7.0	<0.1	0.0	0.0
No Rough-winged Swallow	10.0	7.0	3.0	3.0	0.0	0.0	3.0
Bank Swallow	0.0	0.0	0.0	0.0	0.0	0.0	0.0
Barn Swallow	13.0	93.0	33.0	13.0	100.0	7.0	23.0
Blue Jay	40.0	23.0	30.0	40.0	30.0	53.0	30.0
American Crow	60.0	97.0	53.0	90.0	293.0	40.0	223.0
Black-capped Chickadee	0.0	0.0	0.0	0.0	0.0	0.0	0.0
Carolina Chickadee	20.0	17.0	0.0.0	23.0	13.0	10.0	10.0
Tufted Titmouse	23.0	10.0	17.0	50.0	87.0	27.0	37.0
White-breasted Nuthatch	17.0	10.0	3.0	13.0	30.0	27.0	23.0
Carolina Wren	13.0	3.0	0.0	33.0	60.0	13.0	43.0
Bewick's Wren	0.0	10.0	3.0	7.0	23.0	0.0	3.0
House Wren	3.0	0.0	27.0	3.0	0.0	0.0	0.0
Blue-gray Gnatcatcher	33.0	0.0	0.0	83.0	43.0	10.0	73.0
Eastern Bluebird	23.0	43.0	3.0	53.0	57.0	3.0	17.0
Wood Thrush	0.0	0.0	0.0	0.0	3.0	23.0	10.0
American Robin	23.0	53.0	147.0	80.0	37.0	7.0	20.0

Species	Ozarks Route Number						
	1034F	1053E	1057A	1060A	1069A	1083C	1147C
	Average Number of Birds/100 Stops						
Gray Catbird	3.0	0.0	0.0	0.0	0.0	7.0	0.0
Northern Mockingbird	10.0	60.0	27.0	37.0	80.0	0.0	<0.1
Brown Thrasher	3.0	30.0	17.0	17.0	10.0	3.0	0.0
Cedar Waxwing	0.0	0.0	0.0	0.0	0.0	0.0	0.0
Loggerhead Shrike	0.0	0.0	0.0	0.0	7.0	0.0	0.0
European Starling	17.0	43.0	203.0	67.0	27.0	0.0	0.0
White-eyed Vireo	13.0	3.0	0.0	0.0	0.0	27.0	67.0
Bell's Vireo	0.0	0.0	0.0	0.0	0.0	0.0	0.0
Yellow-throated Vireo	0.0	0.0	0.0	0.0	0.0	3.0	0.0
Warbling Vireo	0.0	0.0	0.0	7.0	0.0	0.0	3.0
Red-eyed Vireo	27.0	<0.1	3.0	10.0	7.0	20.0	23.0
Blue-winged Warbler	10.0	0.0	0.0	0.0	0.0	10.0	3.0
Northern Parula	7.0	0.0	0.0	40.0	7.0	3.0	47.0
Yellow Warbler	3.0	3.0	0.0	0.0	0.0	0.0	0.0
Yellow-throated Warbler	0.0	0.0	0.0	7.0	0.0	3.0	0.0
Pine Warbler	0.0	0.0	0.0	0.0	0.0	33.0	0.0
Prairie Warbler	0.0	0.0	0.0	0.0	0.0	0.0	0.0
Black-and-white Warbler	10.0	0.0	0.0	0.0	3.0	7.0	0.0
Worm-eating Warbler	0.0	0.0	0.0	0.0	13.0	0.0	0.0
Louisiana Waterthrush	7.0	0.0	0.0	10.0	0.0	3.0	<0.1
Kentucky Warbler	10.0	3.0	0.0	10.0	7.0	37.0	17.0
Common Yellowthroat	17.0	3.0	0.0	7.0	7.0	10.0	27.0
Yellow-breasted Chat	40.0	0.0	0.0	0.0	27.0	43.0	80.0
Summer Tanager	17.0	0.0	3.0	13.0	0.0	43.0	17.0
Scarlet Tanager	10.0	0.0	0.0	0.0	0.0	0.0	0.0
Northern Cardinal	27.0	57.0	73.0	117.0	167.0	27.0	60.0
Rose-breasted Grosbeak	0.0	0.0	0.0	0.0	0.0	0.0	0.0
Blue Grosbeak	0.0	7.0	10.0	7.0	53.0	7.0	27.0
Indigo Bunting	50.0	27.0	27.0	77.0	123.0	67.0	153.0
Dickcissel	0.0	127.0	23.0	87.0	120.0	0.0	0.0
Eastern Towhee	3.0	0.0	3.0	10.0	20.0	27.0	0.0
Chipping Sparrow	0.0	0.0	0.0	23.0	17.0	23.0	23.0
Field Sparrow	3.0	33.0	20.0	93.0	87.0	23.0	120.0
Vesper Sparrow	0.0	0.0	0.0	0.0	0.0	0.0	0.0
Lark Sparrow	0.0	0.0	0.0	0.0	7.0	0.0	17.0
Grasshopper Sparrow	0.0	20.0	7.0	0.0	7.0	3.0	0.0
Song Sparrow	0.0	0.0	0.0	0.0	0.0	0.0	0.0
Bobolink	0.0	0.0	0.0	0.0	0.0	0.0	0.0
Red-winged Blackbird	13.0	203.0	157.0	53.0	297.0	3.0	13.0
Eastern Meadowlark	3.0	210.0	180.0	140.0	267.0	7.0	17.0
Western Meadowlark	<0.1	0.0	0.0	0.0	0.0	0.0	0.0
Common Grackle	10.0	437.0	667.0	83.0	233.0	17.0	30.0
Brown-headed Cowbird	<0.1	93.0	0.0	83.0	50.0	10.0	97.0
Orchard Oriole	0.0	0.0	0.0	7.0	3.0	3.0	0.0
Baltimore Oriole	0.0	3.0	0.0	0.0	0.0	0.0	0.0
House Finch	0.0	0.0	0.0	0.0	0.0	0.0	0.0
American Goldfinch	7.0	20.0	0.0	20.0	7.0	10.0	20.0
House Sparrow	0.0	77.0	67.0	23.0	97.0	0.0	0.0

| Species | Ozarks Route Number | | | | | |
| | 1188F | 1193E | 1198E | 1205E | 1209D | 1214F |
	Average Number of Birds/100 Stops					
Great Blue Heron	3.0	3.0	0.0	0.0	0.0	7.0
Green Heron	0.0	3.0	0.0	0.0	0.0	0.0
Canada Goose	0.0	0.0	0.0	0.0	0.0	0.0
Wood Duck	0.0	0.0	0.0	0.0	0.0	0.0
Turkey Vulture	3.0	0.0	0.0	0.0	0.0	0.0
Red-shouldered Hawk	3.0	0.0	10.0	0.0	0.0	13.0
Red-tailed Hawk	3.0	3.0	0.0	0.0	0.0	0.0
Ring-necked Pheasant	0.0	0.0	0.0	0.0	0.0	0.0
Wild Turkey	0.0	0.0	3.0	0.0	3.0	0.0
Northern Bobwhite	63.0	3.0	70.0	160.0	63.0	43.0
Killdeer	3.0	0.0	0.0	20.0	7.0	3.0
Upland Sandpiper	0.0	0.0	0.0	0.0	0.0	0.0
Rock Dove	50.0	0.0	0.0	0.0	0.0	0.0
Mourning Dove	7.0	43.0	23.0	143.0	53.0	47.0
Yellow-billed Cuckoo	67.0	43.0	37.0	100.0	63.0	67.0
Chimney Swift	10.0	13.0	0.0	0.0	0.0	0.0
Ruby-throated Hummingbird	13.0	7.0	0.0	0.0	0.0	0.0
Belted Kingfisher	0.0	3.0	0.0	0.0	7.0	0.0
Red-headed Woodpecker	0.0	0.0	0.0	0.0	7.0	0.0
Red-bellied Woodpecker	47.0	23.0	10.0	23.0	20.0	13.0
Downy Woodpecker	17.0	10.0	0.0	0.0	27.0	17.0
Hairy Woodpecker	7.0	0.0	7.0	3.0	3.0	0.0
Northern Flicker	3.0	0.0	0.0	0.0	0.0	3.0
Pileated Woodpecker	20.0	3.0	20.0	7.0	0.0	10.0
Eastern Wood-Pewee	47.0	10.0	30.0	73.0	40.0	83.0
Acadian Flycatcher	7.0	7.0	0.0	0.0	3.0	20.0
Eastern Phoebe	27.0	30.0	23.0	10.0	13.0	33.0
Great Crested Flycatcher	10.0	13.0	40.0	20.0	67.0	37.0
Eastern Kingbird	23.0	0.0	10.0	97.0	7.0	17.0
Scissor-tailed Flycatcher	0.0	0.0	0.0	0.0	0.0	0.0
Horned Lark	0.0	0.0	0.0	7.0	0.0	0.0
Purple Martin	0.0	0.0	13.0	60.0	27.0	0.0
No Rough-winged Swallow	0.0	3.0	0.0	0.0	0.0	0.0
Bank Swallow	0.0	0.0	0.0	0.0	0.0	0.0
Barn Swallow	133.0	20.0	10.0	50.0	13.0	30.0
Blue Jay	20.0	37.0	10.0	30.0	63.0	57.0
American Crow	230.0	170.0	237.0	187.0	120.0	93.0
Black-capped Chickadee	0.0	0.0	0.0	0.0	0.0	0.0
Carolina Chickadee	20.0	57.0	3.0	43.0	7.0	13.0
Tufted Titmouse	93.0	73.0	17.0	37.0	90.0	50.0
White-breasted Nuthatch	33.0	27.0	20.0	10.0	23.0	3.0
Carolina Wren	30.0	23.0	57.0	47.0	27.0	60.0
Bewick's Wren	0.0	0.0	0.0	20.0	0.0	0.0
House Wren	0.0	0.0	0.0	0.0	0.0	0.0
Blue-gray Gnatcatcher	90.0	57.0	73.0	30.0	60.0	37.0
Eastern Bluebird	60.0	50.0	53.0	70.0	13.0	13.0
Wood Thrush	0.0	13.0	0.0	0.0	3.0	3.0
American Robin	10.0	37.0	17.0	70.0	3.0	7.0

Relative Abundance of Breeding Bird Species by Natural Division, 1986–1992

Species			Ozarks Route Number			
	1188F	1193E	1198E	1205E	1209D	1214F
			Average Number of Birds/100 Stops			
Gray Catbird	3.0	0.0	0.0	3.0	0.0	0.0
Northern Mockingbird	17.0	0.0	7.0	173.0	17.0	13.0
Brown Thrasher	13.0	7.0	<0.1	40.0	10.0	7.0
Cedar Waxwing	0.0	27.0	0.0	0.0	0.0	0.0
Loggerhead Shrike	0.0	0.0	0.0	7.0	0.0	0.0
European Starling	20.0	0.0	3.0	7.0	3.0	0.0
White-eyed Vireo	27.0	17.0	17.0	0.0	3.0	13.0
Bell's Vireo	0.0	0.0	0.0	0.0	0.0	0.0
Yellow-throated Vireo	3.0	0.0	0.0	0.0	3.0	10.0
Warbling Vireo	3.0	0.0	0.0	0.0	0.0	0.0
Red-eyed Vireo	33.0	43.0	0.0	0.0	<0.1	7.0
Blue-winged Warbler	3.0	0.0	0.0	0.0	0.0	3.0
Northern Parula	63.0	30.0	7.0	0.0	10.0	7.0
Yellow Warbler	0.0	0.0	0.0	0.0	0.0	0.0
Yellow-throated Warbler	0.0	0.0	0.0	0.0	0.0	0.0
Pine Warbler	0.0	7.0	0.0	0.0	0.0	0.0
Prairie Warbler	3.0	20.0	10.0	0.0	0.0	0.0
Black-and-white Warbler	0.0	7.0	3.0	0.0	0.0	0.0
Worm-eating Warbler	0.0	<0.1	0.0	0.0	0.0	0.0
Louisiana Waterthrush	13.0	3.0	0.0	0.0	0.0	0.0
Kentucky Warbler	20.0	0.0	0.0	0.0	0.0	17.0
Common Yellowthroat	47.0	3.0	3.0	0.0	10.0	17.0
Yellow-breasted Chat	40.0	30.0	70.0	7.0	33.0	23.0
Summer Tanager	23.0	30.0	47.0	7.0	3.0	20.0
Scarlet Tanager	0.0	0.0	0.0	0.0	3.0	0.0
Northern Cardinal	113.0	120.0	93.0	97.0	113.0	170.0
Rose-breasted Grosbeak	0.0	0.0	0.0	0.0	0.0	0.0
Blue Grosbeak	3.0	7.0	7.0	87.0	10.0	3.0
Indigo Bunting	153.0	163.0	147.0	157.0	120.0	97.0
Dickcissel	3.0	0.0	40.0	177.0	7.0	33.0
Eastern Towhee	10.0	20.0	0.0	10.0	13.0	17.0
Chipping Sparrow	17.0	13.0	50.0	20.0	0.0	0.0
Field Sparrow	53.0	63.0	113.0	103.0	43.0	40.0
Vesper Sparrow	0.0	0.0	0.0	0.0	0.0	0.0
Lark Sparrow	3.0	0.0	17.0	23.0	0.0	0.0
Grasshopper Sparrow	0.0	0.0	0.0	87.0	0.0	10.0
Song Sparrow	0.0	0.0	0.0	0.0	0.0	0.0
Bobolink	0.0	0.0	0.0	0.0	0.0	0.0
Red-winged Blackbird	23.0	0.0	20.0	177.0	53.0	17.0
Eastern Meadowlark	53.0	13.0	27.0	233.0	13.0	27.0
Western Meadowlark	0.0	0.0	0.0	0.0	0.0	0.0
Common Grackle	43.0	0.0	0.0	40.0	3.0	13.0
Brown-headed Cowbird	130.0	57.0	103.0	137.0	97.0	40.0
Orchard Oriole	3.0	0.0	20.0	23.0	3.0	7.0
Baltimore Oriole	0.0	0.0	3.0	0.0	0.0	0.0
House Finch	0.0	0.0	0.0	0.0	0.0	0.0
American Goldfinch	37.0	17.0	23.0	7.0	0.0	7.0
House Sparrow	3.0	20.0	3.0	120.0	0.0	3.0

| Species | Mississippi Lowlands Route Number | | | | |
| | 1 | 1132B | 1174B | 1222C | 1249B |
	Average Number of Birds/100 Stops				
Great Blue Heron	10.3	0.0	0.0	<0.1	3.0
Green Heron	0.3	10.0	3.0	0.0	0.0
Canada Goose	0.0	0.0	0.0	0.0	0.0
Wood Duck	0.3	0.0	0.0	0.0	0.0
Turkey Vulture	0.0	0.0	0.0	0.0	0.0
Red-shouldered Hawk	0.3	0.0	0.0	0.0	0.0
Red-tailed Hawk	0.9	3.0	0.0	0.0	0.0
Ring-necked Pheasant	0.0	0.0	0.0	37.0	23.0
Wild Turkey	0.0	0.0	0.0	0.0	0.0
Northern Bobwhite	45.4	117.0	37.0	47.0	110.0
Killdeer	64.6	17.0	23.0	87.0	110.0
Upland Sandpiper	0.0	0.0	0.0	0.0	0.0
Rock Dove	33.4	0.0	0.0	57.0	0.0
Mourning Dove	175.0	90.0	50.0	213.0	47.0
Yellow-billed Cuckoo	6.3	20.0	27.0	13.0	0.0
Chimney Swift	26.6	20.0	0.0	0.0	3.0
Ruby-throated Hummingbird	0.0	0.0	0.0	0.0	0.0
Belted Kingfisher	0.3	0.0	0.0	0.0	0.0
Red-headed Woodpecker	14.0	0.0	10.0	0.0	0.0
Red-bellied Woodpecker	0.6	30.0	20.0	0.0	0.0
Downy Woodpecker	1.1	3.0	0.0	0.0	0.0
Hairy Woodpecker	0.3	3.0	0.0	0.0	0.0
Northern Flicker	3.4	3.0	0.0	0.0	0.0
Pileated Woodpecker	0.0	0.0	0.0	0.0	0.0
Eastern Wood-Pewee	0.3	13.0	3.0	0.0	0.0
Acadian Flycatcher	0.0	3.0	0.0	0.0	0.0
Eastern Phoebe	0.9	3.0	7.0	0.0	0.0
Great Crested Flycatcher	0.0	17.0	0.0	0.0	0.0
Eastern Kingbird	0.3	13.0	3.0	0.0	0.0
Scissor-tailed Flycatcher	0.0	0.0	0.0	0.0	0.0
Horned Lark	38.0	0.0	40.0	3.0	200.0
Purple Martin	20.6	<0.1	33.0	7.0	0.0
No Rough-winged Swallow	46.0	0.0	0.0	0.0	0.0
Bank Swallow	0.0	0.0	0.0	0.0	0.0
Barn Swallow	77.4	27.0	3.0	30.0	33.0
Blue Jay	3.7	70.0	27.0	3.0	0.0
American Crow	18.0	23.0	13.0	0.0	37.0
Black-capped Chickadee	0.0	0.0	0.0	0.0	0.0
Carolina Chickadee	0.3	7.0	13.0	0.0	0.0
Tufted Titmouse	2.6	10.0	7.0	0.0	0.0
White-breasted Nuthatch	0.0	0.0	0.0	0.0	0.0
Carolina Wren	0.9	83.0	20.0	3.0	10.0
Bewick's Wren	0.0	0.0	0.0	0.0	0.0
House Wren	0.6	0.0	0.0	0.0	0.0
Blue-gray Gnatcatcher	0.3	13.0	0.0	0.0	0.0
Eastern Bluebird	0.9	7.0	0.0	0.0	0.0
Wood Thrush	0.0	0.0	0.0	0.0	0.0
American Robin	41.4	13.0	10.0	7.0	23.0

Relative Abundance of Breeding Bird Species by Natural Division, 1986–1992

| Species | Mississippi Lowlands Route Number | | | | |
| | 1 | 1132B | 1174B | 1222C | 1249B |
	Average Number of Birds/100 Stops				
Gray Catbird	1.4	0.0	3.0	0.0	0.0
Northern Mockingbird	16.9.0	37.0	37.0	37.0	50.0
Brown Thrasher	2.3	10.0	3.0	0.0	3.0
Cedar Waxwing	0.0	0.0	0.0	0.0	0.0
Loggerhead Shrike	2.0	0.0	7.0	0.0	17.0
European Starling	47.1	0.0	3.0	0.0	7.0
White-eyed Vireo	0.3	17.0	0.0	0.0	0.0
Bell's Vireo	0.0	0.0	0.0	0.0	0.0
Yellow-throated Vireo	0.0	0.0	0.0	0.0	0.0
Warbling Vireo	1.7	0.0	0.0	0.0	0.0
Red-eyed Vireo	0.9	30.0	0.0	0.0	0.0
Blue-winged Warbler	0.0	0.0	0.0	0.0	0.0
Northern Parula	0.0	3.0	0.0	0.0	0.0
Yellow Warbler	0.9	0.0	0.0	0.0	0.0
Yellow-throated Warbler	0.0	0.0	0.0	0.0	0.0
Pine Warbler	0.0	0.0	0.0	0.0	0.0
Prairie Warbler	0.0	0.0	0.0	0.0	0.0
Black-and-white Warbler	0.0	0.0	0.0	0.0	0.0
Worm-eating Warbler	0.0	0.0	0.0	0.0	0.0
Louisiana Waterthrush	0.3	3.0	0.0	0.0	0.0
Kentucky Warbler	0.0	10.0	0.0	0.0	0.0
Common Yellowthroat	14.3.0	20.0	30.0	6.3	13.0
Yellow-breasted Chat	0.3	33.0	0.0	0.0	0.0
Summer Tanager	0.0	30.0	0.0	0.0	0.0
Scarlet Tanager	0.0	7.0	0.0	0.0	0.0
Northern Cardinal	24.9	90.0	80.0	27.0	20.0
Rose-breasted Grosbeak	0.0	0.0	0.0	0.0	0.0
Blue Grosbeak	0.0	0.0	0.0	0.0	0.0
Indigo Bunting	18.9	147.0	50.0	23.0	7.0
Dickcissel	152.0	30.0	57.0	170.0	320.0
Eastern Towhee	0.3	17.0	0.0	0.0	0.0
Chipping Sparrow	0.0	0.0	0.0	0.0	0.0
Field Sparrow	0.9	40.0	0.0	0.0	17.0
Vesper Sparrow	0.0	0.0	0.0	0.0	0.0
Lark Sparrow	0.0	0.0	0.0	0.0	0.0
Grasshopper Sparrow	0.0	0.0	3.0	0.0	0.0
Song Sparrow	2.9	0.0	0.0	0.0	0.0
Bobolink	0.0	0.0	0.0	0.0	0.0
Red-winged Blackbird	1279.0	180.0	103.0	470.0	1360.0
Eastern Meadowlark	54.9	77.0	57.0	53.0	20.0
Western Meadowlark	0.0	0.0	0.0	0.0	0.0
Common Grackle	132.0	27.0	23.0	247.0	23.0
Brown-headed Cowbird	17.1	107.0	3.0	137.0	0.0
Orchard Oriole	0.6	3.0	7.0	0.0	0.0
Baltimore Oriole	4.9	0.0	0.0	0.0	0.0
House Finch	0.0	0.0	0.0	0.0	0.0
American Goldfinch	2.3	7.0	0.0	0.0	0.0
House Sparrow	295.0	3.0	60.0	20.0	27.0

Appendix D
Additional Species Reported in Atlas Blocks
Historically never confirmed to breed in Missouri (Robbins and Easterla 1992)

Common Loon: This species was reported in two blocks. A single bird was seen from June 9 through August. 9, 1987 at the Trenton Municipal Watershed, Grundy County, in northwest Missouri. A late individual in winter plumage was seen on Lake Thunderhead north of Unionville, Putnam County, on June 2, 1991.

Fulvous Whistling-Duck: A pair in breeding plumage was found in Pemiscot Bayou near Denton, Dunklin County, in southeast Missouri from June 30 until July 10, 1990.

Green-winged Teal: A drake was seen at Clarence Cannon National Wildlife Refuge July 1, 1988 near Annada, in Pike County.

American Wigeon: A drake was seen on June 1, 1987 in the Camden Bottoms on the Missouri River near Wellington in Ray County. The marsh habitat had dried up by July 15.

Redhead: A drake was seen at Caruthersville in Pemiscot County on June 5, 1988.

Ring-necked Duck: A pair was observed on a pond near Trenton in southeast Harrison County on June 9-10, 1990.

Lesser Scaup: This species was reported in two blocks. One individual was seen in breeding habitat on July 1, 1988 at Clarence Cannon National Wildlife Refuge near Annada in Pike County. The other was seen on June 1, 1991, but not in breeding habitat, on Lake Thunderhead north of Unionville in Putnam County.

Common Snipe: An individual was seen near Bronaugh in southwest Missouri in breeding habitat on May 27, 1992 in Vernon County. Other non-Atlas records suggest this species may have attempted to nest in Missouri, but no confirmation exists to date.

Franklin's Gull: An individual was seen near College Mound on June 19, 1988 in Macon County, but not in breeding habitat.

Olive-sided Flycatcher: This species was reported in seven blocks.   (1) A bird was seen on May 27, 1986, but not in breeding habitat, in southwest Missouri near Sparta in Christian County.   (2) A bird was seen, but not in breeding habitat, on July 9, 1987 in a valley where a small stream flows into Lake Pomme de Terre in Hickory County. (3) A bird was seen on July 11, 1987, but not in breeding habitat, along Crane Creek in Hickory County about 5.5 miles from where this species was seen near Lake Pomme de Terre. (4) A bird was seen holding territory for a month around June 22, 1989 southeast of Higginsville in Lafayette County. (5) A bird was seen on June 1, 1990 in breeding habitat west of Smallett in Douglas County. (6) A bird was seen on June 3, 1990, but not in breeding habitat, directly north of Reger on a graded road in Sullivan County. (7) A bird was seen on June 5, 1992 in breeding habitat west of Deepwater in Henry County.

Hermit Thrush: A bird was seen in breeding habitat on July 2, 1992 near Freeburg in Osage County.

Black-throated Green Warbler: A bird was seen in breeding habitat on June 13, 1992 near Saline Creek southwest of Larrabee in Audrain County. Other non-Atlas records in Missouri plus confirmed breeding at several sites in Arkansas indicate a strong potential for this species to be a breeder in small numbers.

Bay-breasted Warbler: A pair was seen on July 9, 1987 north of Redford in Reynolds County.

White-throated Sparrow: This species was reported in two blocks. A bird was heard on June 17, 1989, but not in breeding habitat, west of Stanton in Franklin County. Three individuals were seen on May 27, 1986 in breeding habitat south of Raymondville in Texas County.

Appendix E
Rates of Brown-headed Cowbird Parasitism

Species	Parasitism Events	Parasitism/ Codes Tabulated	Parasitism Percentage	Species	Parasitism Events	Parasitism/ Codes Tabulated	Parasitism Percentage
Mourning Dove	1	1/133	<1	Kentucky Warbler	2	2/29	7
Acadian Flycatcher	1	1/10	10	Common Yellowthroat	4	4/58	7
Willow Flycatcher	1	1/6	17	Yellow-breasted Chat	2	2/37	5
Eastern Phoebe	6	6/273	2	Summer Tanager	2	2/46	4
Eastern Kingbird	1	1/155	<1	Northern Cardinal	40	40/271	15
Barn Swallow	1	1/42	2	Rose-breasted Grosbeak	1	1/26	4
Carolina Wren	1	1/74	1	Blue Grosbeak	4	4/50	8
House Wren	1	1/112	<1	Indigo Bunting	19	19/133	14
Blue-gray Gnatcatcher	1	1/93	1	Dickcissel	4	4/129	3
Wood Thrush	3	3/13	23	Eastern Towhee	3	3/25	12
American Robin	3	3/427	<1	Chipping Sparrow	7	7/105	7
Gray Catbird	2	2/79	3	Field Sparrow	16	16/162	10
Northern Mockingbird	2	2/133	2	Lark Sparrow	1	1/49	2
European Starling	1	1/246	<1	Grasshopper Sparrow	1	1/67	1
White-eyed Vireo	4	4/25	16	Song Sparrow	3	3/28	11
Bell's Vireo	3	3/9	33	Bobolink	1	1/7	14
Yellow-throated Vireo	2	2/12	17	Red-winged Blackbird	25	25/346	7
Red-eyed Vireo	3	3/27	11	Eastern Meadowlark	4	4/244	2
Blue-winged Warbler	2	2/30	7	Orchard Oriole	3	3/91	3
Northern Parula	4	4/32	13	Baltimore Oriole	1	1/108	<1
Yellow Warbler	1	1/12	8	House Sparrow	3	3/258	1
Yellow-throated Warbler	3	3/15	20				
Prothonotary Warbler	1	1/23	4				
Worm-eating Warbler	1	1/16	6				
Louisiana Waterthrush	7	7/49	14				

Literature Cited

American Ornithologists' Union. 1983. Check-list of North American birds, 6th ed. American Ornithologists' Union (AOU), Washington, DC. 877 pp.

_____, 1995. Fortieth supplement to the American Ornithologists' Union check-list of North American birds. Auk 112(3):827.

Andrle, R. F. and J. R. Carroll (eds.). 1988. The atlas of breeding birds in New York State. Cornell University Press, Ithaca, New York. 551 pp.

Auckley, J. 1996. A summary of the Missouri Department of Conservation's annual report, fiscal year 1994-1995. Missouri Conservationist 57(1):26-27.

Barksdale, T. 1987. Fall survey. The Bluebird 54(1):28-35.

Barlow, J. C. 1962. Natural history of the Bell's Vireo, *Vireo bellii* Audubon. University of Kansas Publ. 12:241-296.

Basili, G., and S. A. Temple. 1995. A perilous migration. Natural History 104(9):40-47.

Bateman, D. L. 1970. Movement-behavior in three species of colonial nesting wading birds: a radio-telemetric study. Ph.D. diss., Auburn University, Auburn, Alabama. 246 pp.

Bellrose, F. C. 1976. Ducks, geese and swans of North America. Stackpole Books, Harrisburg, Pennsylvania. 540 pp.

Bennitt, R. 1932. Check-list of the birds of Missouri. University of Missouri-Columbia. 81 pp.

Bent, A. C. 1938. Life histories of North American birds of prey. U.S. Natl. Museum Bulletin 170(2):1-482.

_____. 1939. Life histories of North American woodpeckers. U.S. Natl. Museum Bulletin 174:1-322.

_____. 1953. Life histories of North American wood warblers. U.S. Natl. Museum Bulletin 203:1-734.

Braun, M. J. and M. B. Robbins. 1986. Extensive protein similarity of the hybridizing chickadees *Parus atricapillus and P. carolinensis.* Auk 103:667-675.

Brewer, R., G. A. McPeek, and R. J. Adams. 1991. The atlas of breeding birds of Michigan. Michigan State University Press, East Lansing. 594 pp.

Briskie, J. V. 1994. Least Flycatcher (*Empidonax minimus*). *In* The birds of North America, No. 99 (A. Poole and F. Gill, eds.). Philadelphia: The Academy of Natural Sciences; Washington, D. C.: The American Ornithologists' Union.

Brown, B. T. 1993. Bell's Vireo (*Vireo bellii*). *In* The birds of North America, No. 35 (A. Poole, P. Stettenheim, and F. Gill, eds.). Philadelphia: The Academy of Natural Sciences; Washington, D. C.: The American Ornithologists' Union.

Brown, C. R., and M. B. Brown. 1989. Behavioural dynamics of intraspecific brood parasitism in colonial Cliff Swallows. Anim. Behav. 37:777-796.

Brown, C. R., and M. B. Brown. 1995. Cliff Swallow (*Hirundo pyrrhonota*). *In* The birds of North America, No. 149 (A. Poole and F. Gill, eds.). Philadelphia: The Academy of Natural Sciences; Washington, D. C.: The American Ornithologists' Union.

Brown, L. N. 1963. Status of the Roadrunner in Missouri. Condor 65:242-43.

Brown, R. E., and J. G. Dickson. 1994. Swainson's Warbler (*Limnothlypis swainsonii*). *In* The birds of North America, No. 126 (A. Poole and F. Gill, eds.). Philadelphia: The Academy of Natural Sciences; Washington, D. C.: The American Ornithologists' Union.

Bull, E. L., and J. A. Jackson. 1995. Pileated Woodpecker (*Dryocopus pileatus*). *In* The birds of North America, No. 148 (A. Poole and F. Gill, eds.). Philadelphia: The Academy of Natural Sciences; Washington, D. C.: The American Ornithologists' Union.

Butler, R. W. 1992. Great Blue Heron (*Ardea herodias*). *In* The birds of North America, No. 25 (A. Poole, P. Stettenheim, and F. Gill, eds.). Philadelphia: The Academy of Natural Sciences; Washington, D. C.: The American Ornithologists' Union.

Cabe, P. R. 1993. European Starling (*Sturnus vulgaris*). In The birds of North America, No. 48 (A. Poole and F. Gill, eds.). Philadelphia: The Academy of Natural Sciences; Washington, D. C.: The American Ornithologists' Union.

Cadman, M.D., P. F. J. Eagles, and F. M. Helleiner. 1987. Atlas of the breeding birds of Ontario. University of Waterloo Press, Waterloo, Ontario, Canada. 617 pp.

Campbell, L. W. 1968. Birds of the Toledo area. The Toledo Blade Co., Toledo, OH. 330 pp.

Campbell, R. 1991. Number of inhabitants, Missouri. University of Missouri, Columbia. 16pp.

Carey, M., D. E. Burhans, and D. A. Nelson. 1994. Field Sparrow (*Spizella pusilla*). *In* The birds of North America, No. 103 (A. Poole and F. Gill, eds.). Philadelphia: The Academy of Natural Sciences; Washington, D. C.: The American Ornithologists' Union.

Carroll, J. P. 1993. Gray Partridge (*Perdix perdix*). *In* The birds of North America, No. 58 (A. Poole and F. Gill, eds.). Philadelphia: The Academy of Natural Sciences; Washington, D. C.: The American Ornithologists' Union.

Chambers, R. J. 1994. Habitat relations of Bachman's Sparrows and other birds on Missouri glades. M.S. thesis. University of Missouri-Columbia. 60 pp.

Christisen, D. M. 1985. The Greater Prairie-Chicken and Missouri's land-use patterns, Terrestrial Series No. 15. Missouri Department of Conservation, Jefferson City. 51 pp.

Cimprich, D. A., and F. R. Moore. 1995. Gray Catbird *(Dumetella carolinensis)*. *In* The birds of North America, No. 167 (A. Poole and F. Gill, eds.). Philadelphia: The Academy of Natural Sciences; Washington, D. C.: The American Ornithologists' Union.

Colvin, B. A. 1985. Common Barn Owl population decline in Ohio and the relationship to agricultural trends. J. Field Ornithol. 56:224-235.

Cooke, M. T. 1928. The spread of the European Starling in North America (to 1928). U.S. Department of Agriculture. Circular 40. 10 pp.

Crocoll, S. T. 1994. Red-shouldered Hawk (*Buteo lineatus*). *In* The birds of North America, No. 107 (A. Poole and F. Gill, eds.). Philadelphia: The Academy of Natural Sciences; Washington, D. C.: The American Ornithologists' Union.

Curson, J., D. Quinn, and D. Beadle. 1994. Warblers of the Americas: an identification guide. Houghton Mifflin, Boston. 252 pp.

Custer, T. W. and R. G. Osborn. 1978. Feeding habitat use by colonially-breeding herons, egrets, and ibises in North Carolina. Auk 95:733-743.

Dailey, T. V. 1996. Wildlife harvest and population status report-Northern Bobwhite. Unpublished. Missouri Department of Conservation, Jefferson City. 7 pp.

Davis, W. E., Jr. 1993. Black-crowned Night-Heron (*Nycticorax nycticorax*). *In* The birds of North America, No. 74 (A. Poole and F. Gill, eds.). Philadelphia: The Academy of Natural Sciences; Washington, D. C.: The American Ornithologists' Union.

_____ and J. A. Kushlan. 1994. Green Heron (*Butorides virescens*). *In* The birds of North America, No. 129 (A. Poole and F. Gill, eds.). Philadelphia: The Academy of Natural Sciences; Washington, D. C.: The American Ornithologists' Union.

Dawson, D. K., L. J. Darr, and C. S. Robbins. 1993. Breeding forest birds in a fragmented landscape. trans. 58th N.A. Wildl. and Natur. Resour. Conf. Proc. p. 62-67.

Derrickson, K. C. and R. Breitwisch. 1992. Northern Mockingbird (*Mimus polyglottos*). *In* The birds of North America, No. 7 (A. Poole, P. Stettenheim, and F. Gill, eds.). Philadelphia: The Academy of Natural Sciences; Washington, D. C.: The American Ornithologists' Union.

Dinsmore, J. J., T. H. Kent, D. Koenig, P. C. Petersen, and D. M. Roosa. 1984. Iowa birds. The Iowa State University Press, Ames. 356 pp.

Donovan, T. M. 1994. Demography of neotropical migrant birds in habitat sources and sinks. Ph.D. dissertation. University of Missouri-Columbia. 180 pp.

Donovan, T.M., F. R. Thompson, III, J. Faaborg, and J. R. Probst. 1995. Reproductive success of migratory birds in habitat sources and sinks. Conservation Biology 9(6):1380-1395.

Duggar, B. D., K. M. Dugger, and L. H. Fredrickson. 1994. Hooded Merganser (*Lophodytes cucullatus*). *In* The birds of North America, No. 98 (A. Poole and F. Gill, eds.). Philadelphia: The Academy of Natural Sciences; Washington, D. C.: The American Ornithologists' Union.

Dunnet, G. M. 1955. The breeding of the starling *Sturnus vulgaris* in relation to its food supply. Ibis 97:619-662.

Easterla, D. A. 1964. The Bald Eagle nesting in Missouri. The Bluebird 29(2):11-13.

Eaton, S. W. 1992. Wild Turkey (*Meleagris gallopavo*). *In* The birds of North America, No. 22 (A. Poole, P. Stettenheim, and F. Gill, eds.). Philadelphia: The Academy of Natural Sciences; Washington, D. C.: The American Ornithologists' Union.

Eddleman, W. R. 1978. Selection and management of Swainson's Warbler habitat. M.S. thesis, University of Missouri-Columbia. 75 pp.

Ehrlich, P. R., D. S. Dobkin and D. Wheye. 1988. The birder's handbook: a field guide to the natural history of North American birds. Simon and Schuster, New York. 785 pp.

Elder, W. H. 1985. Survivorship in the Tufted Titmouse. Wilson Bull. 97:517-524.

Elliott, J. J., and R. S. Arbib, Jr. 1953. Origin and status of the House Finch in the eastern United States. Auk 70:31-37.

Ellison, W. G. 1991. The mechanism and ecology of range expansion by the Blue-gray Gnatcatcher. M.S. thesis, University of Connecticut, Storrs. 203 pp.

_____ . 1992. Blue-gray Gnatcatcher *(Polioptila caerulea). In* The birds of North America, No. 23 (A. Poole, P. Stettenheim, and F. Gill, eds.). Philadelphia: The Academy of Natural Sciences; Washington, D. C.: The American Ornithologists' Union.

England, A. S., M. J. Bechard, and C. S. Houston. 1997. Swainson's Hawk *(Buteo swainsoni). In* the Birds of North America, No. 265 (A. Poole and F. Gill, eds.). Philadelphia: Academy of Natural Sciences; Washington, D. C.: The American Ornithologists' Union.

Evans, K. E. 1980. A recent nesting record of the Chestnut-sided Warbler in Missouri. Bluebird 47(4):6-7.

Farrand, J., Jr. (ed.). 1983. The Audubon Society master guide to birding, loons to sandpipers. Vol 1. Alfred A. Knopf, New York. 447 pp.

_____ . 1983. The Audubon Society master guide to birding, gulls to dippers. Vol. 2. Alfred A. Knopf, New York. 398 pp.

_____ . 1983. The Audubon Society master guide to birding, old world warblers to sparrows. Vol. 3. Alfred A. Knopf, New York. 399 pp.

Fischer, R. B. 1958. The breeding biology of the Chimney Swift, *Chaetura pelagica.* New York State Mus. Sci. Serv. Bull. No. 368. Albany. 141 pp.

Fredrickson, L. H. 1971. Common gallinule breeding biology and development. Auk 88:914-919.

Friedmann, H. 1963. Host relations of the parasitic cowbirds. U. S. Natl. Museum Bulletin No. 233. Washington, D. C. 276 pp.

_____ and L. F. Kiff. 1985. The parasitic cowbirds and their hosts. Proc. Western Foundation Vert. Zool. 2:226-304.

Gehlbach, F. R. 1995. Eastern Screech-Owl *(Otus asio). In* The birds of North America, No. 165 (A. Poole and F. Gill, eds.). Philadelphia: The Academy of Natural Sciences; Washington, D. C.: The American Ornithologists' Union.

Gibbs, J. P. 1988. Forest fragmentation, mating success, and the singing behavior of the Ovenbird and Kentucky Warbler. M.S. thesis. University of Missouri-Columbia. 50 pp.

_____ , S. Melvin, and F. A. Reid. 1992. American Bittern *(Botaurus lentiginosus). In* The birds of North America, No. 18 (A. Poole, P. Stettenheim, and F. Gill, eds.). Philadelphia: The Academy of Natural Sciences; Washington, D. C.: The American Ornithologists' Union.

Graber, R. R., J. W. Graber, and E. L. Kirk. 1972. Illinois birds. Hirudinidae. Illinois Natural History Biological Notes No. 80. Ill. Nat. Hist. Surv., Urbana, Illinois. 36 pp.

_____ , _____ and _____ . 1977. Illinois birds: Picidae. Illinois Natural History Biological Notes No. 102. 73 pp.

_____ , _____ and _____ . 1978. Illinois birds: Ciconiiformes. Illinois Natural History Survey Biological Notes No. 109. 80 pp.

Greenlaw, J. S. 1996. Eastern Towhee (*Pipilo erythrophthalmus*). *In* The Birds of North America, No. 262 (A. Poole and F. Gill, eds.). Philadelphia: The Academy of Natural Sciences; Washington, D. C.: The American Ornithologists' Union.

Griffin, C. R. 1978. The ecology of Bald Eagles wintering at Swan Lake National Wildlife Refuge, with emphasis on eagle-waterfowl relationships. M.S. thesis. University of Missouri-Columbia. 185 pp.

Gross, A. O. 1942. Bank Swallow. *In* Life histories of North American flycatchers, larks, swallows, and their allies. U. S. Natl. Museum Bulletin No. 179. Washington, D. C. 555 pp.

Grubb, T. C., Jr. and V. V. Pravosudov. 1994. Tufted Titmouse (*Parus bicolor*). *In* The birds of North America, No. 86 (A. Poole and F. Gill, eds.). Philadelphia: The Academy of Natural Sciences; Washington, D. C.: The American Ornithologists' Union.

Haggerty, T. M., and E. S. Morton. 1995. Carolina Wren (*Thryothorus ludovicianus*). *In* The birds of North America, No. 188 (A. Poole and F. Gill, eds.). Philadelphia: The Academy of Natural Sciences; Washington, D. C.: The American Ornithologists' Union.

Hallett, D. L. 1990. The Ring-necked Pheasant in Missouri. Missouri Department of Conservation, Jefferson City. 9 pp.

Hamas, M. J. 1994. Belted Kingfisher (*Ceryle alcyon*). *In* The birds of North America, No. 84 (A. Poole and F. Gill, eds.). Philadelphia: The Academy of Natural Sciences; Washington, D. C.: The American Ornithologists' Union.

Hamel, P. B., R. J. Cooper, and W. P. Smith. 1996. The uncertain future for Cerulean Warblers in the Mississippi alluvial valley. Proceed. The Delta Conference, Memphis TN. August. 25 pp.

Hands, H. M., R. D. Drobney, and M. R. Ryan. 1989. Status of the Henslow's Sparrow in the north central United States. U. S. Fish and Wildlife Service, Twin Cities, Minnesota. 12 pp.

_____ , _____ , and _____. 1989. Status of the Red-shouldered Hawk in the north central United States. Missouri Coop. Fish Wildl. Res. Unit Rep. 21 pp.

Haramis, G. M. 1990. Breeding ecology of the Wood Duck: a review. Pp. 45-60 *in* Proc. 1988 N. Am. Wood Duck Symp. (L.H. Fredrickson et al., eds.). St. Louis, Missouri.

Hardin, K. I. 1977. Status and habitat preferences of the Bachman's Sparrow on southern Missouri glades. M.S. thesis. University of Missouri-Columbia. 111 pp.

Harrison, H. H. 1975. A field guide to birds' nests of 285 species found breeding in the United States east of the Mississippi River. Houghton Mifflin Company, Boston, Massachusetts. 257 pp.

Heitmeyer, M. E. 1986. Post breeding distribution and habitat use of wading birds in Oklahoma, USA. Colonial Waterbirds 9:163-170.

Hepp, G. R. and F. C. Bellrose. 1995. Wood Duck (*Aix sponsa*). *In* The birds of North America, No. 169 (A. Poole and F. Gill, eds.). Philadelphia: The Academy of Natural Sciences; Washington, D. C.: The American Ornithologists' Union.

Herkert, J. R. 1994. Status and habitat selection of the Henslow's Sparrow in Illinois. Wilson Bull. 106(1):35-45.

Hicks, L. E. 1935. Distribution of the breeding birds of Ohio. Ohio Biological Survey, Bull. No. 32, 6(3):125-190.

Hill, G. E. 1993. House Finch (*Carpodacus mexicanus*). *In* The birds of North America, No. 46 (A. Poole and F. Gill, eds.). Philadelphia: The Academy of Natural Sciences; Washington, D. C.: The American Ornithologists' Union.

Hoefler, J. E. 1979. Status and distribution of Black-crowned Night-Herons in Wisconsin. Proc. Colon. Waterbird Group 3:75-84.

Hopp, S. L., A. Kirby, and C. A. Boone. 1995. White-eyed Vireo (*Vireo griseus*). *In* The Birds of North America, No. 168 (A. Poole and F. Gill, eds.). Philadelphia: The Academy of Natural Sciences; Washington, D. C.: The American Ornithologists' Union.

Ingold, J. L. 1993. Blue Grosbeak (*Guiraca caerulea*). *In* The birds of North America, No. 79 (A. Poole and F. Gill, eds.) Philadelphia: The Academy of Natural Sciences; Washington, D. C.: The American Ornithologists' Union.

Jackson, L. S., C.A. Thompson, and J. J. Dinsmore. 1996. The Iowa breeding bird atlas, University of Iowa Press, Iowa City. 484 pp.

Jacobs, R. B. 1991. First state nesting record for Black-necked Stilts, *Himantopus mexicanus*. The Bluebird 58(1):7-11.

James, D. A., and J. C. Neal. 1986. Arkansas birds: their status and distribution. Fayetteville: University of Arkansas. 402 pp.

Johnsgard, P. A. 1979. Birds of the Great Plains breeding species and their distribution. University of Nebraska Press, Lincoln. 539 pp.

Jones, D. 1978. The House Finch winter invasion reaches Missouri. Bluebird 46(1):29-30.

Kaiser, M. S. 1982. Foraging ecology of the Green Heron on Ozark streams. M.S. thesis. University of Missouri-Columbia. 112 pp.

Kaufman, K. 1996. Lives of North American Birds. Houghton Miffin Company, New York. 675 pp.

Kilhan, L. 1983. Life history studies of woodpeckers of eastern North America. Publ. Nuttall Ornithological Club No. 20.

Kleen, V. 1987. Illinois heron colony surveys, 1987 report. Illinois Birds and Birding 3:79-82.

Kridlebaugh, A. L. 1982. An ecological study of Loggerhead Shrikes in central Missouri. M.S. thesis, University of Missouri-Columbia. 114 pp.

Kritz, K. 1989. Nesting ecology and nest site habitat of Sharp-shinned and Cooper's hawks in Missouri. M.S. thesis. University of Missouri-Columbia. 84 pp.

Lanyon, W. E. 1956. Ecological aspects of the sympatric distribution of meadowlarks in the north central states. Ecology 37:98- 108.

_____. 1957. The comparative biology of the meadowlarks in Wisconsin. Publ. Nuttall Ornithol. Club, No. 1. Cambridge, MA. 67 pp.

_____. 1994. Western Meadowlark (*Sturnella neglecta*). *In* The birds of North America, No. 104 (A. Poole and F. Gill, eds.). Philadelphia: The Academy of Natural Sciences; Washington, D. C.: The American Ornithologists' Union.

_____. 1995. Eastern Meadowlark (*Sturnella magna*). *In* The birds of North America, No. 160 (A. Poole and F. Gill, eds.). Philadelphia: The Academy of Natural Sciences; Washington, D. C.: The American Ornithologists' Union.

Lowther, P. E. 1993. Brown-headed Cowbird (*Molothrus ater*). *In* The birds of North America, No. 47 (A.Poole and F. Gill, eds.). Philadelphia: The Academy of Natural Sciences; Washington, D. C.: The American Ornithologists' Union.

Lowther, P. E., and C. L. Cink. 1992. House Sparrow (*Passer domesticus*). *In* The birds of North America, No. 12 (A.Poole, P. Stettenheim, and F. Gill, eds.). Philadelphia: The Academy of Natural Sciences; Washington, D. C.: The American Ornithologists' Union.

Marti, G.D. 1974. Feeding ecology of four sympatic owls. Condor 76:45-61.

Martin, A. C., H. S. Zim, and A. L. Nelson. 1951. American wildlife and plants: a guide to wildlife food habits. New York: Dover Publ. 500 pp.

Martin, S. G., and T. A. Gavin. 1995. Bobolink (*Dolichonyx oryzivorus*). *In* The birds of North America, No. 176 (A. Poole and F. Gill, eds.). Philadelphia: The Academy of Natural Sciences; Washington, D. C.: The American Ornithologists' Union.

McCarty, J.P. 1996. Eastern Wood-Pewee (*Contopus virens*). *In* The Birds of North America, No. 245 (A. Poole and F. Gill, eds.). Philadelphia: The Academy of Natural Sciences; Washington, D. C.: The American Ornithologists' Union.

McKenzie, P. 1996. First breeding record of Great-tailed Grackle (*Quiscalus mexicanus*) east of western Missouri. The Bluebird 63(3):33-35.

McKenzie, P., and R. B. Jacobs. 1992. First recorded nesting Least Flycatcher. The Bluebird 59(4):158-160.

McKinley, D. 1961. History of the Canada Goose as a breeding bird in Missouri. The Bluebird 28(3):6-12.

Meanley, B. 1969. Natural history of the King Rail. United States Bureau of Sport Fisheries and Wildlife. North American Fauna No. 67. 108 pp.

Meyerriecks, J. A. 1960. Comparative breeding behavior of four species of North American herons. Pub. Nuttall Ornithology Club No. 2. 158 pp.

Middleton, A. L. A. 1993. American Goldfinch (*Carduelis tristis*). *In* The birds of North America, No. 80 (A. Poole and F. Gill, eds.). Philadelphia: The Academy of Natural Sciences; Washington, D. C.: The American Ornithologists' Union.

Mirarchi, R. E. and T. S. Baskett. 1994. Mourning Dove (*Zenaida macroura*). *In* The birds of North America, No. 117 (A. Poole and F. Gill, eds.): Philadelphia: The Academy of Natural Sciences; Washington, D. C.: The American Ornithologists' Union.

Missouri Department of Conservation, MO Prairie Found., MO Nature Conservancy, MO Dept. Nat. Resources. 1991. Public prairies of Missouri. Missouri Department of Conservation, Jefferson City. 29 pp.

Missouri Department of Conservation. 1993. The Wild Turkey in Missouri, Conservation Commission, Jefferson City. 13 pp.

Mitchell, M. R. 1985. Breeding bird occurrence in mid-Missouri forest fragments. M.S. thesis. University of Missouri-Columbia. 114 pp.

Moldenhauer, R. R., and D. J. Regelski. 1996. Northern Parula (*Parula americana*). *In* The birds of North America, No. 215 (A. Poole and F. Gill, eds.). Philadelphia: The Academy of Natural Sciences; Washington, D. C.: The American Ornithologists' Union.

Moore, W. S. 1995. Northern Flicker (*Colaptes auratus*). *In* The birds of North America, No. 166 (A. Poole and F. Gill, eds.). Philadelphia: The Academy of Natural Sciences; Washington, D. C.: The American Ornithologists' Union.

Morse, D. H. 1989. American warblers: an ecological and behavioral perspective. Harvard University, Cambridge, Massachusetts. 406 pp.

Mumford, R. E. 1952. Bell's Vireo in Indiana. Wilson Bull. 64:224-233.

Musselman, T. E. 1950. European Tree Sparrow at Hannibal, Missouri. Auk 67:105

Nagy, A. C. 1972. The sparrow hawk. Hawk Mt. Sanctuary Assoc. Leaflet, No. 44.

National Geographic Society. 1987. Field guide to the birds of North America. 2nd ed. National Geographic Society, Washington, DC. 464 pp.

National Wildlife Federation. 1991. Proc. midwest raptor management symposium and workshop. Natl. Wildl. Fed., Washington, D. C.   290 pp.

Nero, R. W. 1984. Redwings. Smithsonian Institution Press, Washington, D. C. 160 pp.

Newman, D. L. 1961. House Wren and Bewick's Wren, northern Ohio. Wilson Bull. 73:84-86.

Nice, M. M. 1939. "Territorial song" and non-territorial behavior of goldfinches in Ohio. Wilson Bull. 51:123.

Norris, D. J., and W. H. Elder. 1982. Decline of the Roadrunner in Missouri. Wilson Bull. 94:354-56.

Ogden, L. J., and B. J. Stutchbury. 1994. Hooded Warbler (*Wilsonia citrina*). *In* The birds of North America, No. 110 (A. Poole and F. Gill, eds.). Philadelphia: The Academy of Natural Sciences; Washington, D. C.: The American Ornithologists' Union.

Overmire, T. G. 1962. Nesting of the Bell's Vireo in Oklahoma. Condor 64:75.

Palmer, R. S., ed. 1988. Handbook of North American birds. Vol. 5. Yale University Press, New Haven, Connecticut. 567 pp.

Parker, M. 1993. The distribution and abundance of Arkansas birds, the summer season. Arkansas Audubon Society Newsletter, 38(4): 3.

Parkes, K. C. 1964. Warbler (2). *In* A new dictionary of birds (A. L. Thompson, ed.). New York: McGraw-Hill Book Co. 928 pp.

Parnell, J. F., D. G. Ainley, H. Blokpoel, B. Cain, T. W. Custer, J. I. Dusi, S. Kress, J. A. Kushland, W. E. Southern, L. E. Stenzel, and B. C. Thompson. 1988. Colonial waterbird management in North America. Colonial Waterbirds 11(2):129-345.

Payne, R. B. 1992. Indigo Bunting (*Passerina cyanea*). *In* The birds of North America, No. 4 (A. Poole, P. Stettenheim, and F. Gill, eds.). Philadelphia: The Academy of Natural Sciences; Washington, D. C.: The American Ornithologists' Union.

Peck, G. K., and R. D. James. 1987. Breeding birds of Ontario nidiology and distribution, Vol. 2: Passerines. Royal Ontario Museum, Toronto. 387 pp.

Peterjohn, B. G. 1989. The birds of Ohio. Indiana University Press, Bloomington, Indiana. 237 pp.

_____. and D. L. Rice. 1991. The Ohio breeding bird atlas. Ohio Dept. of Natural Resources, Div. of Natural Areas and Preserves, Columbus. 416 pp.

Peterjohn, B. G, and J. R. Sauer. 1995. Population status of North American grassland birds. International Conference and Training Workshop on Conservation and Ecology of Grassland Birds, Tulsa, Oklahoma, 26-28 October. p. 26.

Porter, E. F., Jr. 1989. Tree Sparrows aren't found anywhere else in the Americas. St. Louis Post-Dispatch, 4 April D1 col. 2-6 and D5 col. 1-2.

Price, H. F. 1940. Nests and eggs of the Great Horned and Barred owls. Oologist 57:38-39.

Price, J., S. Droege, and A. Price. 1995. The summer atlas of North American birds. Academic Press, Inc., San Diego, California. 364 pp.

Renken, R. B. 1988. Habitat characteristics related to Pileated Woodpecker densities and territory size in Missouri. Ph.D. dissertation, University of Missouri-Columbia. 114 pp.

Reynolds, R. T., E. C. Meslow, and H. M. Wight. 1982. Nesting habitat of coexisting *Accipiter* in Oregon. J. Wildl. Manage. 46:124-138.

Robbins, C. S., D. Bystrak, and P. H. Geissler. 1986. The breeding bird survey: its first fifteen years 1965-1979. U.S. Fish and Wildlife Service Res. Publ. No. 157. 196 pp.

_____ , J. R. Sauer, and B. G. Peterjohn. 1992. Population trends and management opportunities for neotropical migrants. pp. 17-23 *In* Status and Management of Neotropical Migratory Birds. USDA Forest Service Gen. Tech. Report RM-229.

Robbins, M. B. 1977. Spring survey. The Bluebird 44(2):14-18.

Robbins, M. B., and D. A. Easterla. 1986. Range expansion of the Great-tailed Grackle into Missouri, with details of the first nesting colony. Bluebird 53(4):24-27.

Robbins, M. B., and D. A. Easterla. 1992. Birds of Missouri: their distribution and abundance. University of Missouri Press, Columbia. 399 pp.

Robertson, R. J., Stutchbury, B. J., and R. R. Cohen. 1992. Tree Swallow (*Tachycineta bicolor*). *In* The birds of North America, No. 11 (A. Poole, P. Stettenheim, and F. Gill, eds.). Philadelphia: The Academy of Natural Sciences; Washington, D. C.: The American Ornithologists' Union.

Robinson, S. K., J. A. Grzybowski, S. I. Rothstein, M. C. Brittingham, L. J. Petit, and F. R. Thompson. 1993. Management implications of cowbird parasitism on neotropical migrant songbirds. Pp. 93-102 *in* Status and Management of Neotropical Migratory Birds, Sept. 21-25, 1992, Estes Park, Colorado. Gen. Tech. Rep. RM-229. USDA, FS.

Robinson, S. K., F. R. Thompson, T. M. Donovan, D. R. Whitehead, and J. Faaborg. 1995. Regional forest fragmentation and the nesting success of migratory birds. Science 267:1987-1990.

Robinson, W.D. 1996. Summer Tanager (*Piranga rubra*). *In* The Birds of North America, No. 248 (A. Poole and F. Gill, eds.). Philadelphia: The Academy of Natural Sciences; Washington, D.C.: The American Ornithologists' Union.

Rohwer, S. A. 1971. Molt and annual cycle of the Chuck-Will's-Widow, *Caprimulgris carolinensis*. Auk 88(3):485-519.

Rosenfield, R. N. and J. Bielefeldt. 1993. Cooper's Hawk (*Accipiter cooperii*). *In* The birds of North America, No. 75 (A. Poole and F. Gill, eds.). Philadelphia: The Academy of Natural Sciences; Washington, D. C.: The American Ornithologists' Union.

Rothstein, S.I. 1971. Observation and experiment in the analysis of interactions between brood parasites and their hosts. Amer. Nat. 105:71-74.

Sanderson, G. C. (ed.). 1977. Management of migratory shore and upland game birds in North America. International Association of Fish and Wildlife Agencies, Washington, D. C. 358 pp.

Schroeder, M. A. and L. A. Robb. 1993. Greater Prairie-Chicken (*Tympanuchus cupido*). *In* The birds of North America, No. 36 (A. Poole, P. Stettenheim, and F. Gill, eds.). Philadelphia: The Academy of Natural Sciences; Washington, D. C.: The American Ornithologists' Union.

Schroeder, W. A. 1982. Presettlement prairie of Missouri. Natural History Series, No. 2. Missouri Department of Conservation, Jefferson City. 38 pp.

Sinnott, R. J. 1981. Ecology and management of the Eastern Bluebird in Missouri. M.S. thesis. University of Missouri-Columbia. 168 pp.

Skinner, R. M., T. S. Baskett, and M. D. Blenden. 1984. Bird habitat on Missouri prairies, Terrestrial Series #14. Missouri Department of Conservation, Jefferson City. 37 pp.

Smith, S. M. 1991. The Black-capped Chickadee: behavioral ecology and natural history. Cornell University Press, Ithaca, New York. 362 pp.

Smith, S. M. 1993. Black-capped Chickadee. *In* The birds of North America, No. 39 (A. Poole, P. Stettenheim, and F. Gill, eds.). Philadelphia: The Academy of Natural Sciences; Washington, D. C.: The American Ornithologists' Union.

Stauffer, D. F., and L. B. Best. 1986. Nest-site characteristics of open-nesting birds in riparian habitats in Iowa. Wilson Bull. 98:231-242.

Sutton, G. M. 1967. Oklahoma birds. Univ. Oklahoma Press, Norman. 674 pp.

Telfair II, R. C. 1983. The Cattle Egret: a Texas focus and world view. Kleberg Stud. Nat. Resour. Tex. Agric. Exp. Stn., Texas A & M University, College Station. 144 pp.

_____. 1994. Cattle Egret *(Bubulcus ibis)*. *In* The birds of North America, No. 113 (A. Poole and F. Gill, eds.). Philadelphia: The Academy of Natural Sciences; Washington, D. C.: The American Ornithologists' Union.

Terres, J. K. 1987. The Audubon society encyclopedia of North American birds. Alfred A. Knopf, New York. 1109 pp.

Thom, R. H. and J. H. Wilson. 1980. The natural divisions of Missouri. Transactions of the Missouri Academy of Science 14:9-23.

Thomas, B. G. 1994. Habitat selection and breeding status of Swainson's Warblers. M. S. thesis, University of Missouri-Columbia. 44 pp.

Thompson, D. H. 1978. Feeding areas of Great Blue Herons and Great Egrets nesting within the floodplain of the upper Mississippi River. Proc. Colonial Waterbirds Group. 2:202-213.

Thompson III, F. R. 1994. Temporal and spatial patterns of breeding Brown-headed Cowbirds in the midwestern United States. Auk. 111(4):979(12).

_____, D. A. Freiling, and E. K. Fritzell. 1988. Ruffed Grouse in Missouri: its ecology and management. University of Missouri Extension Division Agricultural Guide. 4 pp.

Thompson, M. C., and C. Ely. 1989. Birds in Kansas, Vol. 1. Museum of Natural History, University Press of Kansas. 404 pp.

_____ and _____. 1989. Birds in Kansas, Vol. 2. Museum of Natural History, University Press of Kansas. 424 pp.

Toland, B. 1983. The ecology and biology of the American Kestrel. M.S. thesis. University of Missouri-Columbia. 181 pp.

Voous, K. H. 1988. Owls of the Northern Hemisphere. Collins Publ. Co., London. 320 pp.

Walkinshaw, L. H. 1966. Summer observations of the Least Flycatcher in Michigan. Jack-Pine Warbler 44:150-168.

Weeks, H. P., Jr. 1994. Eastern Phoebe (Sayornis phoebe). In The birds of North America, No. 94 (A. Poole and F. Gill, eds.). Philadelphia: The Academy of Natural Sciences; Washington, D. C.: The American Ornithologists' Union.

Wheelwright, N. T., and J. D. Rising. 1993. Savannah Sparrow (Passerculus sandwichensis). In The birds of North America, No. 45 (A. Poole and F. Gill, eds.). Philadelphia: The Academy of Natural Sciences; Washington, D. C.: The American Ornithologists' Union.

Widmann, O. 1907. A preliminary catalog of the birds of Missouri. Academy of Science, St. Louis, Missouri. 288 pp.

Williams, A. B. 1950. Birds of the Cleveland region. Kirtland Soc. Bull. No. 2. The Cleveland Museum of Natural History, Cleveland, Ohio. 215 pp.

Wilson, J. D. 1984. Fall survey. The Bluebird 51(1):36.

_____. 1985. Bald Eagle nesting in Missouri. The Bluebird 52(4):22-27.

_____. 1988. Summer survey. The Bluebird 55(4):123-127.

_____. 1990. The breeding bird survey in Missouri-1967 to 1989. The Bluebird 57(1):17-27.

_____. 1991a. Spring report. The Bluebird 58(3):123.

_____. 1991b. Summer report. The Bluebird 58(4):167.

_____. 1992. Summer report. The Bluebird 59(4):191.

_____. 1995. 1995 Missouri Bald Eagle reproduction side bar. The Bluebird 62(4):191.

Yasukawa, K., and W. A. Searcy. 1995. Red-winged Blackbird (*Agelaius phoeniceus*). *In* The birds of North America, No. 184 (A. Poole and F. Gill, eds.). Philadelphia: The Academy of Natural Sciences; Washington, D. C.: The American Ornithologists' Union.

Zimmerman, J. L. 1963. A nesting study of the Catbird in southern Michigan. Jack-Pine Warbler 41:142-160.

_____. 1988. Breeding season habitat selection by the Henslow's Sparrow (*Ammodramus henslowii*) in Kansas. Wilson Bull. 100:17-24.

Index